Waterborne Pathogen
Methods and Applicat

MW00836984

Waterborne Pathogens: Detection Methods and Applications

Second Edition

Edited by

Helen Bridle

Institute of Biological Chemistry, Biophysics and Bioengineering,
Heriot-Watt University, Edinburgh, Scotland

ACADEMIC PRESS

An imprint of Elsevier

Academic Press is an imprint of Elsevier
125 London Wall, London EC2Y 5AS, United Kingdom
525 B Street, Suite 1650, San Diego, CA 92101, United States
50 Hampshire Street, 5th Floor, Cambridge, MA 02139, United States
The Boulevard, Langford Lane, Kidlington, Oxford OX5 1GB, United Kingdom

Notices
Knowledge and best practice in this field are constantly changing. As new research and
experience broaden our understanding, changes in research methods, professional practices, or
medical treatment may become necessary.

Practitioners and researchers must always rely on their own experience and knowledge in
evaluating and using any information, methods, compounds, or experiments described herein.
In using such information or methods they should be mindful of their own safety and the safety
of others, including parties for whom they have a professional responsibility.

To the fullest extent of the law, neither the Publisher nor the authors, contributors, or editors,
assume any liability for any injury and/or damage to persons or property as a matter of products
liability, negligence or otherwise, or from any use or operation of any methods, products,
instructions, or ideas contained in the material herein.

Library of Congress Cataloging-in-Publication Data
A catalog record for this book is available from the Library of Congress

British Library Cataloguing-in-Publication Data
A catalogue record for this book is available from the British Library

ISBN :978-0-444-64319-3

For information on all Academic Press publications
visit our website at https://www.elsevier.com/books-and-journals

Publisher: Nikki Levy
Acquisitions Editor: Linda Versteeg-Buschman
Editorial Project Manager: Timothy Bennett
Production Project Manager: Omer Mukthar
Cover Designer: Matthew Limbert

Typeset by SPi Global, India

Working together
to grow libraries in
developing countries

www.elsevier.com • www.bookaid.org

Dedication

**"To Paul, Ted, and Ben—thanks for your support
and love"**

Contents

Section C
Detection

Section D
Applications and evaluation

12. Conclusions

Helen Bridle

Contributors

Numbers in parenthesis indicate the pages on which the authors' contributions begin.

Helen Bridle (1,9,41,63,117,147,189,293,327,367,391), Institute of Biological Chemistry, Biophysics and Bioengineering, Heriot-Watt University, Edinburgh, Scotland

Marc Desmulliez (189), Institute of Signals, Sensors and Systems, Heriot-Watt University, Edinburgh, Scotland

Kimberley Gilbride (41,237), Department of Chemistry and Biology, Ryerson University, Toronto, ON, Canada

James Green (41), Scottish Water, Juniper House, Edinburgh, Scotland

Timothée Houssin (147), University Lille Nord de France, Villeneuve d'Ascq; Institute of Electronics, Microelectronics and Nanotechnology, Lille, France

Karin Jacobsson (63), Mikrobiologienheten, Uppsala, Sweden

Susan Lee (41), Scottish Water, Juniper House, Edinburgh, Scotland

Lauren Rowe (367), IMS Consulting, Royal London Buildings, Bristol, England

Anna Charlotte Schultz (63), DTU Food, National Food Institute, Technical University of Denmark, Søborg, Denmark

Vincent Senez (147), University Lille Nord de France, Villeneuve d'Ascq; Institute of Electronics, Microelectronics and Nanotechnology, Lille, France

Graham Sprigg (367), IMS Consulting, Royal London Buildings, Bristol, England

Waterborne pathogens preface

Waterborne Pathogens: Detection Methods and Applications were written to bring together the latest research and development in analysis methods across the broad range of potential pathogens that could be present within the drinking water. At the time of writing the first edition of the book, a lot of other works were about decade old and not focussed on detection so the book aimed to fill a gap in the market in addressing emerging detection technologies.

Since the publication of the first edition, there have been some fascinating developments in detection technologies so the book has been updated to reflect the latest research and progress, with the aim of the second edition being to offer an up-to-date summary of the state-of-the-art. The second edition also includes information relating to antimicrobial resistance as the research interest in the environmental and aquatic transmission is growing and more detailed understanding is urgently needed; there are some examples of detection technologies that are described.

The main aim of this book is to give an overview of advanced emerging technologies for the detection of a range of waterborne pathogens. However, the book will also present existing methodology, and highlight where improvements can be made, as well as have a strong focus on applications and how new technology could be applied in water management. Besides, the book will address issues of sample preparation (from sampling through to concentration and enrichment), a key stage in any detection protocol.

The focus on detection differentiates the book from a lot of other works relating to waterborne pathogens where the content discusses the types of pathogens present and treatment and control approaches. This book gives a brief overview to the different types of pathogen, describing various waterborne viruses, bacteria, and protozoa, however, the reader is referred to other works for more detailed microbiology as well as for information on water treatment techniques.

More recently, a book, 'Waterborne Pathogens: Detection and Treatment' was published which covers some aspects of detection though the division of chapters is more heavily weighted towards treatment techniques. This book covers a wider range of detection technologies with five chapters dedicated to different techniques along with one covering the essential area of sample processing.

Some of the main differences and benefits of this book in comparison to others include:

1. Focus on a whole range of waterborne pathogens (including viruses, bacteria, and protozoa along with antimicrobial resistance in this new edition) with an interdisciplinary approach (including epidemiology, public health, sampling, detection—biosensors, molecular methods, impedance, other emerging technologies) therefore ensuring interest for students and water utilities with a general interest in the topic.
2. Focus on emerging advanced technologies for detection and prediction of future developments and trends: up-to-date, cutting edge research, and technology presented.
3. Application focus: with a section covering risks, costs, trade-off, and implications for water management as well as a discussion of remote and on-site monitoring and source tracking.

The book hopes to provide a concise, conceptual approach to the detection of waterborne pathogens in water for human consumption, with a particular focus on new emerging technologies.

Little knowledge is assumed and a background is given to waterborne pathogens and existing approaches to ensure those without a professional knowledge of the topic can understand the rest of the text. Technical introductions are given to each detection approach explaining the basic principles and mathematical knowledge is not necessary to understand the methods described. Basic levels of science/engineering are assumed though the biology of the pathogens are provided and the physics/chemistry/engineering of the sample processing/detection methods are given—for further details the reader is always referred to the relevant research publications. The goal is that those wishing to understand the status of the field at present can obtain a basic understanding of the principles of different techniques and an overview of the research results obtained.

Chapter 1

Introduction

Helen Bridle

Institute of Biological Chemistry, Biophysics and Bioengineering, Heriot-Watt University, Edinburgh, Scotland

Access to potable water is essential to life, a right enshrined in the Universal Declaration of Human Rights, and is critical to meeting all of the Millennium Development Goals [1]. This is also recognised by the United Nations (UN) General Assembly who have recently stated that safe and clean drinking water and sanitation is a human right essential to the full enjoyment of life and all other human rights, and designated the period from 2005 to 2015 as the International Decade for Action, 'Water for Life'.

Due to population growth, increased industrialisation and climate change, water scarcity, especially of safe, pollutant-free drinking water sources, is a major problem. Numerous countries across the world already face severe water shortages, and many more are considered water-stressed (see Fig. 1.1). It is estimated that two-thirds of the population of the world will suffer water scarcity by 2025. In terms of waterborne contaminants, which include inorganic and organic chemicals as well as pathogens/microbes, the World Health Organisation (WHO) considers microbial hazards as the primary concern in both developing and developed countries [2].

Safe drinking water is that which poses no significant risk during a lifetime of consumption. Globally, in 2017, WHO statistics indicate that 2.2 billion lack access to safely managed water services. This is despite the announcement on 6 March 2012 that the Millennium Development Goal, of halving the proportion of people without access to an improved water source, had been achieved. In addition to acknowledging the fact that numerous people are still without access to safe drinking water, the WHO press release recognised that an improved source of drinking water was not necessarily an assurance of water quality and that work remains to ensure that these sources are, and remain, safe.

Contaminated drinking water is one of the most important environmental contributors to the human disease burden being responsible for an estimated 1.9 million deaths each year, predominantly in children under 5 years. Although most of this disease impact occur in low-income countries, where it is estimated that more than 3000 children die daily from diarrhoeal diseases, waterborne

Waterborne Pathogens: Detection Methods and Applications. https://doi.org/10.1016/B978-0-444-64319-3.00001-0

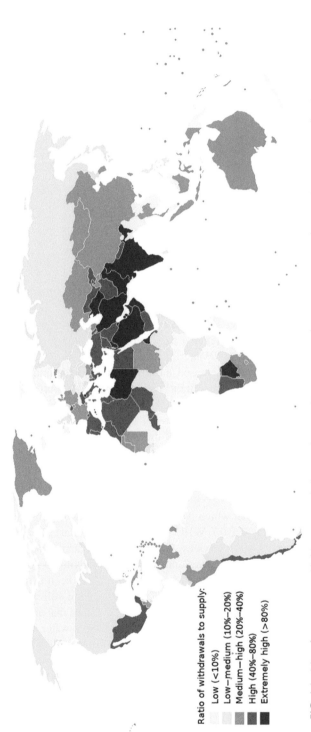

Ratio of withdrawals to supply:

Low (<10%)
Low–medium (10%–20%)
Medium–high (20%–40%)
High (40%–80%)
Extremely high (>80%)

FIG. 1.1 Map of water stress worldwide (grey areas indicate no data available). (Reproduced from https://en.wikipedia.org/wiki/Water_scarcity#/media/File:Water_stress_2019_WRI.png. Compiled by in 2019 using aqueduct 3.0 data from World Resources Institute. See table "National Water Stress Rankings" at https://www.wri.org/blog/2019/08/17-countries-home-one-quarter-world-population-face-extremely-high-water-stress and https://www.wri.org/applications/aqueduct/country-rankings/.

disease is still a threat to citizens in the developed world. For example, one of the largest recent outbreaks was the cryptosporidiosis outbreak in Milwaukee in 1993 in which approximately 400,000 people were infected [2]. Furthermore, it has been estimated that 10% of total hospital patients in US contract diseases due to poor water, significantly increasing morbidity, mortality, and financial burden. Overall, in the US lost productivity due to waterborne diseases is estimated at $20 million per year [3] (Fig. 1.2).

Waterborne pathogens also have a significant economic impact. Although very necessary the building, upgrading, and maintaining water treatment plants to remove pathogens from the supply is one such cost. Therefore, it has been suggested that catchment management strategy, which attempts to reduce the pathogen load entering the water source, is a more cost-effective strategy than adopting more and more advanced and expensive treatment technologies. It has been shown though that, in some regions, investment in the provision of water treatment returns a net economic benefit through alleviation of health-related effects. The health impact of waterborne pathogens is considerable in terms of both healthcare costs and lost productivity. This is true for both outbreaks and endemic diseases. For example, the outbreak example mentioned previously in Milwaukee was estimated to cost US$96 million [4]. An *Escherichia coli O157:H7* outbreak in Canada in 2000, killed seven and infected 2300 residents and cost $155million [5].

An overview of the most important types of microbial contamination that pose risks to health is given in Chapter 2. Based on a List of Relevant Pathogens published by the WHO, this chapter describes in detail the most problematic waterborne pathogens focussing upon those where transmission occurs via ingestion of water, for example via drinking water. Antimicrobial resistance in the aquatic environment is also an emerging issue and improved monitoring technologies are required to understand the fate and transmission of antimicrobial resistance in the water. Some of these approaches will overlap with traditional and emerging drinking water monitoring techniques.

To alleviate the problems associated with waterborne pathogens, the WHO recommends the adoption of a water safety framework (WSF) approach, tailoring the design of water safety plans (WSPs), which nations or regions design to suit the relevant environmental, social, economic, and cultural conditions. Further information on WSF and WSPs is given in Chapter 3, where the role of monitoring and detection of pathogens in the delivery of a WSF is also considered.

Monitoring for pathogens is important for many reasons:

● Investigative monitoring allows analysis of the source water enabling selection of appropriate barriers (e.g. catchment management or particular water treatment technologies) to remove the identified type of pathogen from the water. Another example of investigative monitoring includes the identification of sources of an outbreak, which contributes to halting and preventing

Deaths from unsafe water, sanitation and hygiene

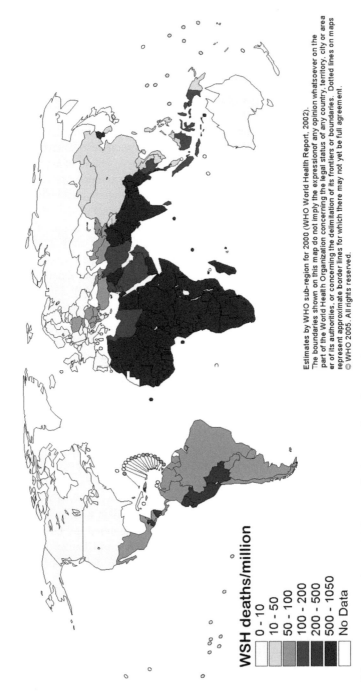

WSH deaths/million

- 0 - 10
- 10 - 50
- 50 - 100
- 100 - 200
- 200 - 500
- 500 - 1050
- No Data

FIG. 1.2   Illustration of the worldwide mortality impacts of waterborne pathogens.

the reoccurrence of an outbreak. Furthermore, the analysis of patient samples contributes to the understanding of the disease burden and setting appropriate health targets. The key important factor in this type of monitoring is the delivery of sufficient information to quantify and characterise the pathogen load.

- In contrast, operational monitoring is undertaken to provide timely indications of the performance of any implemented drinking water treatment process, enabling the possibility of taking appropriate action to remediate any potential problems. The key factor here is rapid measurements, delivering information in time for action to be taken. According to the WHO, monitoring for pathogens is of limited use for operational purposes as existing methods of detection take too long [2].
- Surveillance or verification monitoring provides information to assess the functioning of adopted WSPs and contributes to the effective management and rational allocation of resources to improve water supplies.

Chapter 3 provides a brief overview of existing, widely-adopted methods for pathogen detection, discussing their advantages and disadvantages. Generally, the problem with existing techniques is that the use of faecal indicator organisms is not always correlated with pathogen presence and therefore, while a test may be negative for the faecal indicator, pathogens may still be present. Although cheap and easy to perform, the use of culture-based methods is time-consuming.

The main focus of this book is on alternative, developing, and emerging technologies, which are presented in Chapters 5–10. This main detection section of the book gives background along with an explanation of how the various methods and techniques work followed by a comprehensive literature review detailing how each method has been applied to the detection of waterborne pathogens. Since the first edition of the book was published there have been many exciting and interesting developments in research focussed on the sample processing and detection of waterborne pathogens, which are now included in this second edition. Performance evaluation according to a range of criteria (e.g. detection limit, ability to determine species or viability, etc.) is provided where the information was reported in the literature and at the end of each chapter, a summary and comparison is given. The introduction to this detection section explains the rationale behind the chapter division and discusses the essential considerations for any detection technology in this application. Chapter 11 builds on this by considering what is required for a technology to translate from successful laboratory results to wide-spread adoption by the water sector. A final summary of the state-of-the-art and a future outlook is provided in Chapter 12.

Finally, it is important to remember that sample processing is an important part of any monitoring strategy and as more direct detection methods are developed specific sample enrichment techniques suitable for returning particular pathogens with a high recovery rate are essential. An explanation of existing procedures and a review of the latest literature are given in Chapter 4.

Consideration is also given to how different sample processing approaches may be required depending upon the chosen detection technology.

Several commentators have presented the idea of a distributed network of sensors, continuously reporting on water quality within a catchment area, enabling appropriate operational decisions. Back in 2001 when Rose and Grimes summarised the view of a colloquium panel of water experts they reported that 'water quality monitoring is mired in the past' and envisioned a future in which pathogen detection is performed in real-time feeding into operational decision making. This book will address the question of what progress has been made towards delivering novel monitoring technologies that meet the needs of the water sector and comment upon directions for future work.

References

[1] A. Rahman, Towards an Arsenic Safe Environment in Bangladesh, BCAS, 2010.

[2] World Health Organisation, Guidelines for Drinking-Water Quality, 2011.

[3] T.M. Straub, D.P. Chandler, Towards a unified system for detecting waterborne pathogens, J. Microbiol. Methods 53 (2) (2003) 185–197.

[4] P.S. Corso, et al., Cost of illness in the 1993 waterborne cryptosporidium outbreak, Milwaukee, Wisconsin, Emerg. Infect. Dis. 9 (4) (2003) 426–431.

[5] P.L. Meinhardt, Recognizing waterborne disease and the health effects of water contamination: a review of the challenges facing the medical community in the United States, J. Water Health 4 (2006) 27–34.

Section A

Overview and background

Chapter 2

Overview of waterborne pathogens

Helen Bridle

Institute of Biological Chemistry, Biophysics and Bioengineering, Heriot-Watt University, Edinburgh, Scotland

Waterborne pathogens can be divided into three main categories: viruses, bacteria, and parasites, the latter of which comprises protozoan and helminths. Such pathogens reach water sources when infected people or animals shed microbes in faeces. For example, untreated, undertreated, or accidental release, of sewage allows pathogens to enter water sources. An alternative mechanism is through run-off to source water or permeation into groundwater from animal faeces or sewage utilised as fertiliser. Many waterborne pathogens are zoonotic, capable of infecting both humans and animals (Fig. 2.1).

The persistence of pathogens in the environment depends upon many factors. For example, pathogens may be removed from the water supply due to settling in lakes, possibly augmented by interactions of pathogens with sediment (Fig. 2.2). Furthermore, inactivation by light, temperature, or chemical conditions, such as salinity or ammonia, can occur [1]. Understanding the fate and transport of pathogens in the environment is essential for risk management. However, the survival and transport of pathogens are beyond the scope of this book and the reader is referred to other articles [1] and books for more information on this topic. It is sufficient to note that some of the pathogens discussed in this book remain infective in the environment for long periods (months-years).

As waterborne pathogens are transported through the environment, they are often diluted to low concentrations. However, while these low concentrations may prove challenging for detection purposes, they may still present a considerable public health risk, as several of the pathogens discussed here have extremely low infectious doses.

One final factor, in addition to environmental persistence and infectious dose, which renders some waterborne pathogens particularly problematic is that of resistance to disinfection methods. Most viruses, the spores, or a vegetative phase of a bacterium and the cysts, oocysts, or ova of parasites are capable of some degree of resistance to chlorination [2].

Waterborne Pathogens: Detection Methods and Applications. https://doi.org/10.1016/B978-0-444-64319-3.00002-2

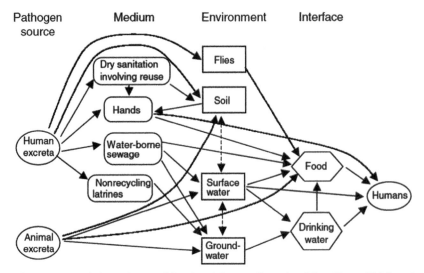

FIG. 2.1 Transmission pathways of faecal-oral disease. *(Reproduced from Figure 16.1 from A. Pruss-Ustun, et al., Unsafe water, sanitation and hygiene, in: Comparative Quantification of Health Risks, WHO.)*

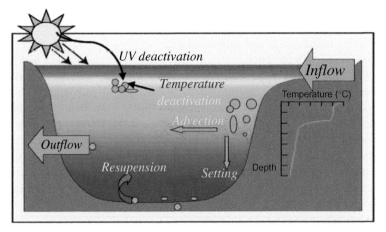

FIG. 2.2 Schematic indicating the fate and transport of pathogens in a reservoir. *(Reproduced from A. Pruss-Ustun, et al., Unsafe water, sanitation and hygiene, in: Comparative Quantification of Health Risks, WHO.)*

The World Health Organisation (WHO) has compiled a list of relevant waterborne pathogens, considering the risk represented by the three factors discussed earlier [3]. An adaption of this list forms the basis of this chapter, where each category of the waterborne pathogen is discussed, in turn, with examples of the most problematic pathogens.

In general, the focus is on those pathogens where the route of disease transmission occurs through the ingestion of water. This chapter is divided into sections, according to main categories of pathogens, and incorporates a final section focussing on antimicrobial resistance in the aquatic environment.

Environmental persistence

Environmental persistence is the length of time a pathogen can survive in the environment, and retain infectivity.

Host-dependent pathogens gradually lose viability and their ability to infect after they are shed from a host. Pathogens with low persistence are unlikely to be spread by drinking water since they would be nonviable or noninfectious by the time they reach a new host. Some waterborne pathogens are capable of growth in water. For example, *Legionella*, *Vibrio cholerae*, and *Naegleria fowleri* will grow in conditions of warm water, containing relatively high amounts of biodegradable organic carbon, which can occur in some surface waters or water distribution systems. Other pathogens, e.g. Norovirus or *Cryptosporidium* are unable to multiply in water but are robust enough to survive for a considerable length of time.

The persistence of pathogens in water is influenced by many factors including temperature, exposure to sunlight (UV), or certain chemical conditions (e.g. salinity or ammonia) all of which could inactivate pathogens as well as for settling or interaction with sediment in surface waters.

In this chapter, and the WHO list, the persistence in water is defined as short for survival periods of less than 1 week, moderate for times between a week and a month and high persistence being assigned to those pathogens capable of survival in the environment for periods greater than 1 month.

Microbiology definitions

Virus: Viruses are the smallest of the microorganisms, typically 20–300 nm, and are the only living organisms not to have a cell membrane; they consist of a small amount of nucleic acid (DNA or RNA) coated with and protected by a layer of protein.

Bacteria: Bacteria are prokaryotic microorganisms with a typical size of a few microns; prokaryotes are cells with little intracellular organisation which reproduce asexually via cell division producing two daughter cells.

Protozoa: Protozoa are a diverse group of unicellular eukaryotic organisms, including sporozoa (intracellular parasites), flagellates (which possess tail-like structures for movement), amoeba (which move using temporary cell body projections called pseudopods), and ciliates (which move by beating multiple hair-like structures called cilia).

Helminth: Helminths are worms, with the name coming from the Greek word for worm.

Genera: A taxonomic category ranking used in a biological classification that is below a family and above a species level, and includes the group(s) of species that are structurally similar or phylogenetically related. In binomial nomenclature, the genus is used as the first word of a scientific name. The genus name is always capitalised and italicised, e.g. *Cryptosporidium*. For a genus to be descriptively useful, it must have monophyly, reasonable compactness, and distinctness.

Species: It is a taxonomic category subordinate to a genus (or subgenus) and superior to a subspecies or variety, e.g. *Cryptosporidium parvum*. An exact definition is difficult but could be thought of as populations of organisms with a high degree of genetic similarity. In terms of sexually reproducing organisms, a species could be thought of as a group of organisms that could potentially interbreed and produce fertile offspring.

Strains: A simple definition is that a strain is a genetic variant or subtype of a microorganism. According to the first edition of Bergey's Manual of Systemic Bacteriology, 'A strain is made up of the descendants of single isolation in pure culture and is usually made up of a succession of cultures ultimately derived from a single colony'. For a more detailed discussion see Ref. [4].

Spores: A dormant, reproductive cell formed by certain organisms. It is thick-walled and highly resistant to survive under unfavourable conditions so that when conditions revert to being suitable it gives rise to a new individual.

Cysts/oocysts: A resting or dormant stage of a microorganism/the encysted zygotic stage in the life cycle of some sporozoans. Encystment helps the microbe to disperse easily, from one host to another, or to a more favourable environment. When the encysted microbe reaches an environment favourable to its growth and survival, the cyst wall breaks down by a process known as *excystation*.

Prokaryotic: This cell type is a simple structure and is confined to the bacteria. They are typically less than 5 μm long and 1 μm wide and have little structural organisation within the cell. The single DNA molecule is in direct contact with the cytoplasm. Prokaryotes reproduce asexually by division.

Eukaryotic: This cell type is more complex than the prokaryotes, containing several distinct intracellular compartments. These structures are known as organelles, e.g. a nucleus, and are surrounded by membranes, enabling the maintenance of chemical conditions within organelles, different from the cytoplasm. There is no such thing as a typical eukaryotic cell as they exhibit considerable diversity. Eukaryotes can reproduce sexually or asexually.

Culture: Concerning microorganisms this term refers to the in vitro cultivation of cells. Culturing of microorganisms requires specific culture media containing the nutrients and growth factors necessary for microorganism growth; this is highly specific to different microorganisms, thus allowing selective growth of certain microorganisms from a mixed sample. However, many microorganisms are nonculturable, meaning it is not possible to grow them in the laboratory.

Self-limiting: This term is used to describe some of the disease resulting from infection with microorganisms. In self-limiting cases, the immune response of the human body will eventually kick-in thus preventing further multiplication of the pathogen and a return to health.

For more detail, the reader is referred to standard microbiology textbooks.

Infectious dose

The infectious dose does not refer to a minimum, threshold value above which infection occurs. Every single pathogen ingested can initiate infection. However, between different pathogens, the probability that a single organism will initiate infection can vary widely. The infectious dose attempts to characterise this by providing a probability of infection. The infectious dose can, therefore, be thought of as the dose above which the probability of infection exceeds a certain value. The probability of illness to develop the following infection depends upon the degree of host damage and whether this is sufficient to result in clinical symptoms.

Sometimes, clinical or epidemiological data is available from previous outbreaks. Alternatively, to determine the infective dose, dose–response experiments can be undertaken with healthy adult volunteers to determine the probability of infection at different dose levels. This data is not available for all pathogens and does not indicate the susceptibility of vulnerable subpopulations to any of the studied pathogens. In addition, experiments are often performed with laboratory strains of pathogens, which may differ from the wild-type. However, these studies provide strong evidence that the infective dose of many waterborne pathogens is very low.

In this chapter and the WHO list, the relative infectivity is categorised as low if the infective dose is greater than 10^4 pathogens, moderate for doses between 10^2 and 10^4 and high for doses between 1 and 10^2. There is no indication in the WHO list of what probability of infection this infective dose represents.

Resistance to disinfection

The resistance to disinfection describes the likelihood that a pathogen will retain infectivity following exposure to a disinfectant. The standard method of disinfection in the water industry has relied upon chemicals and in particular chlorine. While highly effective for bacteria, viruses and protozoa display different levels of resistance. Due to the robustness of some of these pathogens alternative mechanisms such as exposure to UV or ozone have been explored, and adopted, within the water industry. The resistance to disinfection can be characterised in terms of the percentage of organisms inactivated at a particular time, which can be studied experimentally.

In this chapter, and the WHO list, the resistance to disinfection is considered low if 99% of organisms are inactivated in 1 min, moderate if 99% inactivation takes 1–30 min and high for times longer than 30 min (for freely suspended organisms, at conventional doses and pH 7–8).

Prevalence

Prevalence can either refer to the abundance of a certain pathogen in source waters or to the prevalence of infection within a certain population.

The waterborne pathogens discussed are ubiquitous in source waters across the globe and concentrations may vary considerably over time, and short-term events can cause large spikes in concentration. Determination of the specific pathogen load in any source water involves sampling and monitoring using a method of direct detection.

Prevalence in a population indicates the level of disease and can be utilised in epidemiological studies over time to determine whether public health interventions are reducing disease incidence. Prevalence of infection can be measured by testing either faecal samples for excreted pathogens or serum samples for the presence of antibodies. Analysis of faecal samples is useful in an outbreak whereas serological studies provide an overview of the proportion of a population who have been exposed to a particular pathogen, assuming that pathogen confers long-lived immunity, which is not the case for all waterborne pathogens.

2.1. Viruses

Viruses are the smallest of the waterborne pathogens, with typical sizes ranging between 20 and 300 nm. A virus consists of a small amount of nucleic acid (DNA or RNA) coated with and protected by a layer of protein (Fig. 2.3). Viruses replicate only within host cells, where the viral nucleic acid directs the infected cell to produce progeny viruses. Viral infections are initiated when the virus protein outer layer specifically interacts with cell surface receptors.

Gastroenteritic viruses replicate in the upper third of the intestine, destroying the mature enterocyte covering the villi and therefore, disrupting the reabsorption of water from the gut, resulting in diarrhoea. Eventually, the villi are repopulated with immature, undifferentiated cells that are resistant to the virus infection. However, it takes time for these undifferentiated cells to mature sufficiently to develop the necessary ion uptake capabilities.

Viruses are in general quite species-specific and also tissue-specific (known as tropism) in which cells they can infect. Several of the waterborne viruses have secondary tropisms beyond the intestine, e.g. enteroviruses which can cause poliomyelitis or meningitis, or hepatitis which can infect the liver. Besides, there is speculation that some waterborne viruses may be zoonotic, with potential reservoirs being cattle, pigs, poultry, or rodents. However, this has not been confirmed, possibly due to limited diagnostics, in identifying viruses in human or animal faeces and water. Many viruses are long-lived in aqueous environments and have been shown to contaminate deep aquifers, a water source often considered as free from microbial contamination [5].

A selection of waterborne viruses, as identified by the WHO, are summarised in Table 2.1. We will concentrate our more detailed descriptions of each pathogen upon those transmitted by the ingestion of water as opposed to inhalation or skin contact. Where available, information is presented regarding the disease caused by each of these pathogens, in terms of symptoms, prevalence, and treatment. Besides, information is given on the number of different species of each pathogen, their size and appearance with the reader being referred to original scientific articles for more in-depth information on aspects such as the

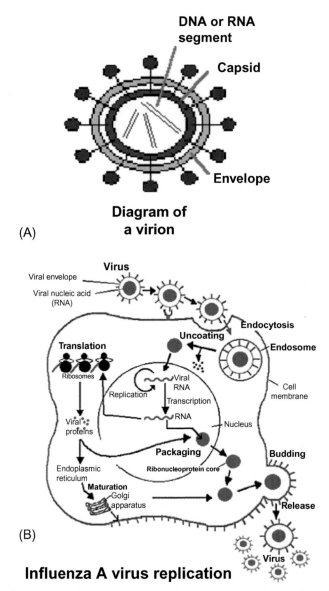

DNA or RNA segment

Capsid

Envelope

Diagram of a virion

(A)

Virus

Viral envelope

Viral nucleic acid (RNA)

Endocytosis

Uncoating

Translation

Endosome

Ribosomes

Viral RNA

Replication

Transcription

Cell membrane

Viral proteins

RNA

Nucleus

Packaging

Budding

Endoplasmic reticulum

Ribonucleoprotein core

Maturation

Golgi apparatus

Release

(B)

Virus

Influenza A virus replication

Virus

FIG. 2.3 Schematic of a virus and a typical reproduction process. (A) A virus schematic showing the simple structure of a virus. The molecular material (DNA or RNA) is surrounded by a protein coating. (B) Schematic of the reproduction process of the influenza virus as an illustration of the way that viruses are taken up by cells and how they take over the cellular processes to direct replication of viral molecular material generating numerous virus copies which are expelled from the cell.

TABLE 2.1 Overview of waterborne viruses.

Pathogen	Disease	Relative infectivity	Persistence in water	Resistance to disinfection	Route of transmission
Adenoviruses	Gastroenteritis, respiratory infection	High	Long	Moderate	Ingestion, inhalation
Astroviruses	Gastroenteritis	High	Long	Moderate	Ingestion
Enteroviruses	Gastroenteritis	High	Long	Moderate	Ingestion, inhalation
Hepatitis viruses A and E	Hepatitis	High	Long	Moderate	Ingestion
Noroviruses	Gastroenteritis	High	Long	Moderate	Ingestion
Rotavirus	Gastroenteritis	High	Long	Moderate	Ingestion
Sapoviruses	Gastroenteritis	High	Long	Moderate	Ingestion

Reproduced with permission from the WHO list.

cell biology, pathogen life cycle, etc. Brief details of recent outbreaks of disease attributed to waterborne viruses are summarised in Ref. [6].

2.1.1 Adenovirus

Adenovirus consists of double-stranded DNA in a nonenveloped icosahedral capsid with a diameter of 80 nm. It is thought to have a low infectious dose and cause gastroenteritis, mainly in infants and children. There are 51 antigenic types of human adenovirus, of which 30% are pathogenic to humans [7]. The majority of these infect the upper respiratory tract and spread primarily by droplets. Subtypes 40 and 41 are the major cause of gastroenteritis. Adenovirus accounts for 5%–20% of US hospital admissions for diarrhoea and affects primarily young children [7]. The incubation period typically lasts between 3 and 10 days with illness persisting for up to a week.

Adenovirus has been detected in sewage, raw water sources, and treated drinking water in large numbers. It is exceptionally resistant to disinfection procedures, even UV treatment [8]. Further information regarding adenoviruses can be found in a review article by Jiang [9].

2.1.2 Astrovirus

Astrovirus is a single-stranded RNA in a nonenveloped icosahedral capsid, with a diameter of 28 nm. It is typically described as rounded with a smooth margin and may have a five- or six-point star motif in the centre. However, the appearance of astrovirus is highly variable, making visual identification of this virus particularly challenging.

There are eight different serotypes of human astrovirus, of which the most common is serotype number one. Over 80% of children between the ages of 5 and 10 have been found to have antibodies against serotype number one. The virus causes mainly mild, self-limiting gastroenteritis, lasting 2–3 days, with a similar length of incubation time. Although the illness is normally mild, astrovirus is the second most common viral cause of diarrhoea, responsible for 5% of US hospital admissions for diarrhoea. Astrovirus has been detected in water sources and supplies.

2.1.3 Enteroviruses

The enteroviruses are among the smallest of all viruses, with a diameter of 20–30 nm, and consisting of single-stranded RNA in a nonenveloped icosahedral capsid. They are one of the most common causes of human infection, reported to be responsible for 30 million cases annually in the US alone. There are many viruses under this group, including poliovirus and enterovirus, and therefore an equally broad range of different diseases and symptoms. Enteroviruses have been found in both raw water sources and treated water.

2.1.4 Hepatitis viruses A and E

Hepatitis A and E belong to different groups. Hepatitis A is similar in size and structure to enteroviruses, though it has its genus (Hepatovirus), of which it is the sole member. The infectious dose for hepatitis A virus is estimated to be about 10–100 infectious particles [10] and it can persist for a very long period in the water environment. The infection proceeds via the epithelial cells, lining the intestine into the bloodstream and eventually infecting the liver. The incubation period is normally around 48 days [7]. In around 90% of cases, there is no liver damage, especially in very young children, and lifelong immunity is incurred. However, the severity of the illness increases with age at which it is contracted [7].

The size and structure of hepatitis E are not so different from hepatitis A. It also consists of single-stranded RNA in a nonenveloped icosahedral capsid, with a diameter of 27–34 nm. The symptoms of infection are also fairly similar to hepatitis A, though often symptoms are more severe and those infected are slightly older. It also results in a 25% mortality rate among infected pregnant women. The incubation period is 60 days. Several large waterborne outbreaks, infecting tens of thousands, were reported last century, in India and China [7].

2.1.5 Norovirus

Norovirus is one of the four genera of caliciviruses. Two of these caliciviruses are infective to humans, with sapoviruses being the other. Norovirus (NoV) consists of single-stranded RNA in a nonenveloped icosahedral capsid. The size of this virus is around 35–40 nm in diameter. Unlike the other caliciviruses, NoV does not usually display a cup-like structure in the surface morphology.

This virus causes acute gastroenteritis in all age groups and can be particularly serious for the elderly or immunocompromised. The incubation period is normally around 48 h with the self-limiting illness lasting 24–48 h. In around 40% of cases, the main symptom is diarrhoea, whereas other patients exhibit no diarrhoea but suffer instead of vomiting. The infection is sometimes referred to as the winter vomiting sickness. Infection with NoV confers a short-lived immunity.

The infectious dose for NoV is estimated to be about 10–100 infectious particles [11], and therefore, although the viral load in water samples may be low it may still be a source of infection and illness. Besides, NoV is very persistent in the water environment [12] and has been shown to resist chlorination [7].

Of the 28 outbreaks, and associated 12,000 cases, of waterborne virus infections in the United States in the late 19th century, 75% were attributed to NoV [7]. In 2008; in Lilla Edet, Sweden, 2400 people contracted gastroenteritis, with the majority of cases ascribed to NoV [13]. Other recent outbreaks include at a Dutch scout camp [14] and in Turkey [15].

2.1.6 Rotavirus

Rotavirus is the most important single cause of infant mortality and accounts for 50%–60% of all hospitalisations of children with acute gastroenteritis. The disease burden varies significantly between high- and low-income countries which are estimated at 14 and 480 disability-adjusted life years (DALYs), respectively, per 1000 cases.

Rotavirus is a segmented double-stranded RNA in a nonenveloped icosahedral capsid. This structure has a diameter of 50–65 nm and is then further enclosed in a layered shell, taking the final virus size to 80 nm diameter. The outer shell looks wheel-like thus inspiring the naming of this virus. A schematic of the virus structure can be seen in Fig. 2.4.

There are two genera in the Reoviridae virus family typically associated with human infection, of which rotavirus is the most problematic from a public health viewpoint. There are seven groups, incorporating further subgroups. For information on progress towards a rotavirus vaccine see the 2007 review in Nature Reviews Microbiology by Angel et al. [16].

Patients typically shed 10^{11} virus particles per gram of faeces for 8 days. Therefore, it is unsurprising that rotavirus has been found in large numbers in sewage and both source and treated waters.

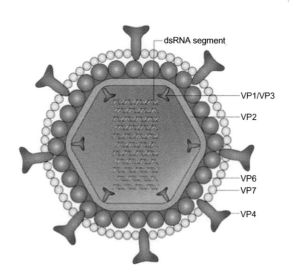

Nature Reviews | Microbiology

FIG. 2.4   A schematic of a rotavirus virion. The 11 segmented dsRNA is surrounded by a 3-layered protein shell, made up of 6 different proteins [16].

2.1.7 Sapovirus

Like NoV, this is a member of the caliciviruses and causes acute gastroenteritis in all age groups. Sapovirus usually displays the typical calicivirus morphology. Little is known about prevalence worldwide.

2.2. Bacteria

Bacteria are prokaryotic microorganisms of typically a few microns in size. They are bigger than most viruses but smaller than most protozoa and helminths. There is extraordinary diversity in the bacterial kingdom, with a wide range of morphologies exhibited by different bacterial species (Fig. 2.5). There are typically a million bacteria in every millilitre of water, of which the majority are not pathogenic. Bacteria can be categorised into gram-negative and gram-positive.

A selection of waterborne bacterial pathogens, as identified by the WHO, are summarised in Table 2.2. We will concentrate on more detailed descriptions of each pathogen upon those transmitted by the ingestion of water as opposed to inhalation or skin contact. Where available, information is presented regarding

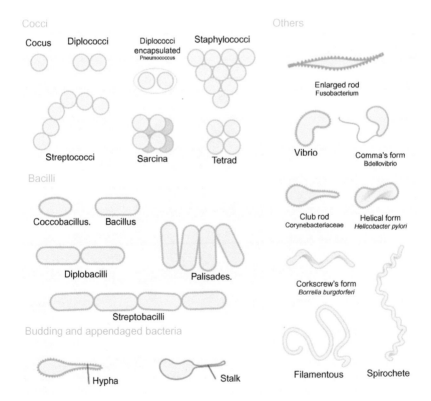

FIG. 2.5 Illustration of a variety of bacteria morphologies.

TABLE 2.2 Overview of waterborne bacteria.

Pathogen	Disease	Relative infectivity	Persistence in water	Resistance to disinfection	Route of transmission
Campylobacter jejuni, C. coli	Gastroenteritis	Moderate	Moderate	Low	Ingestion
Escherichia coli—pathogenic	Gastroenteritis	Low	Moderate	Low	Ingestion
E. coli O157:H7 (enterohaemorrhagic)	Gastroenteritis, haemolytic-uraemia	High	Moderate	Low	Ingestion
Legionella spp.	Legionnaires' disease	Moderate	May multiply	Moderate	Inhalation
Pseudomonas aeruginosa	Pulmonary disease, skin infection	Low	May multiply	Low	Skin contact
Salmonella typhi	Typhoid fever	Low	Moderate	Low	Ingestion
Salmonella enterica	Salmonellosis	Low	May multiply	Low	Ingestion
Shigella	Shigellosis	High	Short	Low	Ingestion
Vibrio cholerae	Cholera	Low	Bioaccumulates	Low	Ingestion

Reproduced with permission from the WHO list.

the disease caused by each of these pathogens, in terms of symptoms, prevalence, and treatment. In addition, information is given on the number of different species of each pathogen, their size and appearance with the reader being referred to original scientific articles for more in-depth information on aspects such as the cell biology, pathogen life cycle, etc.

2.2.1   Campylobacter

There are several species of this type of bacteria. *Campylobacter jejuni* is the one most commonly isolated from infected patients, although *Campylobacter coli*, *Campylobacter laridis*, and *Campylobacter fetus* have also all been found. These bacteria are microaerophilic (require reduced oxygen) and capnophilic (require increased carbon dioxide). They are gram-negative curved spiral rods with a single unsheathed polar flagellum.

The infectious dose for *Campylobacter* is low for a bacterium at around 1000 cells. The resulting illness is generally self-limiting, typically lasting 3–7 days with common symptoms of abdominal pain, diarrhoea, vomiting, chills, and fever. The occurrence of this pathogen in water is associated with high rainfall, water temperature, and the presence of waterfowl. The animal reservoirs of this pathogen are thought to be poultry, wild birds, and cattle. Outbreaks have resulted from inadequate or unchlorinated supplies or when reservoirs are faecally contaminated by birds.

There are millions of cases annually, of what is generally mild, self-limiting diarrhoea though this still results in significant productivity and days of work, losses particularly in developing countries [17]. In the United States, it is estimated that there are around 250,000 cases annually and several deaths [18] from both food and water sources.

2.2.2   Escherichia coli

Large numbers of *Escherichia coli* bacteria exist naturally in the human intestine. However, a limited number of enteropathogenic strains can cause acute diarrhoea (Fig. 2.6). These strains have been classified according to virulence, and include enterohaemorrhagic *E. coli* (EHEC), enterotoxigenic *E. coli* (ETEC), enteropathogenic *E. coli* (EPEC), enteroinvasive *E. coli* (EIEC), enteroaggregative *E. coli* (EAEC), and diffusely adherent *E. coli* (DAEC).

E. coli O157:H7 (an EHEC) is a well known example of one of these strains, ingestion of which can lead to symptoms ranging from mild to highly bloody diarrhoea and in around 2%–7% of cases haemolytic uraemic syndrome. This latter disease, more likely in children under five is potentially fatal. EHEC is the main zoonotic source of pathogenic *E. coli*, known to infect cattle and sheep and to lesser extent goats, chickens and pigs. Although infection with the enteropathogenic *E. coli* strains is far less common than, e.g. *Campylobacter* infection, the symptoms can be much more severe. Besides, the infective dose is low at less than 100 organisms.

In 2000, an outbreak in Walkerton, Canada caused 7 deaths and 2300 illnesses. More recently, the largest ever verocytotoxigenic *E. coli* outbreak

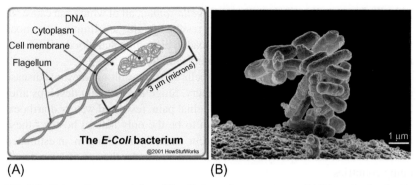

FIG. 2.6 Schematic representation of *E. coli* plus a microscope image. (A) Schematic showing the major structures of *E. coli*. (B) Low-temperature electron micrograph of a cluster of *E. coli* bacteria, magnified 10,000 times. Each bacterium is oblong shaped. *(Authors: Photo by Eric Erbe, digital colourisation by Christopher Pooley, both of USDA, ARS, EMU. Credit: http://commons.wikimedia. org/wiki/File:E_coli_at_10000x_original.jpg.)*

occurred in Europe in 2011, with 3929 illnesses and 47 deaths. This outbreak was linked to the consumption of sprouts from contaminated seeds [19], which may have become contaminated through the use of poor quality water for irrigation purposes.

2.2.3 Salmonella

Salmonella is a motile, gram-negative bacteria, which does not ferment lactose. Most produce hydrogen sulphide or gas from carbohydrate fermentation. It is now thought that there are 2 main species, which are *enterica* and *bongori*, with other 2000 subspecies. Some species show host-specificity, particularly *Salmonella typhi* which infects humans. Animal reservoirs of other species include poultry, cows, sheep, pigs, and birds.

Various types of diseases result from different species, including gastroenteritis, septicaemia and typhoid fever. *Salmonella enterica* generally results in self-limiting diarrhoea with an onset of 6–72 h following consumption of contaminated food or water (or contact with infected animals/people) and typically lasting 3–5 days. The nontyphodial subspecies rarely cause waterborne outbreaks. Infection with typhoid species is the most common waterborne disease and following a 1–14 day incubation period, can result in a more serious typhoid fever which can be fatal. There are millions of cases every year, though typhoid occurrence is generally limited where sanitation systems are good. Besides, *Salmonella* is known to be sensitive to chlorine disinfection.

2.2.4 Shigella

These gram-negative, nonmotile and nonspore-forming rod-like bacteria can grow with or without oxygen. There are four species, *Shigella dysenteriae*,

Shigella flexneri, *Shigella boydii*, and *Shigella sonnei*, all of which can cause severe disease. However, *S. sonnei* usually results in mild, self-limiting diarrhoea whereas infection by *S. dysenteriae* can lead to ulceration and bloody diarrhoea, in which the production of the Shiga toxin is implicated.

The infectious dose is low at 10–100 cells and the resulting intestinal disease can be severe, including bacillary dysentery. Shigellosis usually develops after 24–72 h, with symptoms including abdominal pain, fever and watery diarrhoea. Humans and other higher primates seem to be the only natural hosts of these bacteria. Around 2 million infections occur annually, resulting in an estimated 600,000 deaths, the majority occurring in children under the age of 10 in developing countries.

2.2.5 Cholerae

The *Vibrio* bacteria incorporate several species of which *V. cholerae* is the only pathogenic species associated with freshwater. These bacteria are small, curved (comma-shaped), and gram-negative, possessing a single polar flagellum.

Many of the serotypes can cause diarrhoea though it is certain strains of the O1 and O39 serotypes which cause the classical cholera symptoms. These strains produce the cholera toxin that acts to change ionic fluxes across the intestinal mucosa, resulting in severe fulminating and watery diarrhoea. These toxigenic strains are far less common than the others, which are widely distributed in water sources. The infectious dose is high at around 10^8 organisms, and many infections are asymptomatic. Those who suffer symptomatic cholera vary in the extent and severity of their symptoms, with some losing up to 10–15 L of water a day. *V. cholerae* is highly sensitive to chlorine disinfection.

2.3. Protozoa

Protozoa are a diverse group of unicellular eukaryotic organisms. This group includes sporozoa (intracellular parasites), flagellates (which possess tail-like structures for movement), amoeba (which move using temporary cell body projections called pseudopods) and ciliates (which move by beating multiple hair-like structures called cilia).

A selection of waterborne protozoan pathogens, as identified by the WHO, are summarised in Table 2.3. We will concentrate on more detailed descriptions of each pathogen upon those transmitted by the ingestion of water as opposed to inhalation or skin contact. Of these pathogens, *Cryptosporidium* is the most widely studied. Where available, information is presented regarding the disease caused by each of these pathogens, in terms of symptoms, prevalence, and treatment. In addition, information is given on the number of different species of each pathogen, their size and appearance with the reader being referred to original scientific articles for more in-depth information on aspects such as the cell biology, pathogen life cycle, etc. For an overview of recorded waterborne

TABLE 2.3 Overview of waterborne protozoa.

Pathogen	Disease	Relative infectivity	Persistence in water	Resistance to disinfection	Route of transmission
Acanthamoeba spp.	keratitis, encephalitis	High	May multiply	Low	Inhalation, skin or eye contact
Cryptosporidium spp.	Cryptosporidiosis	High	Long	High	Ingestion
Cyclospora cayetanensis	Gastroenteritis	High	Long	High	Ingestion
Entamoeba histolytica	Amoebic dysentery	High	Moderate	High	Ingestion
Giardia lamblia	Giardiasis	High	Moderate	High	Ingestion
Naegleria fowleri	primary amoebic meningoencephalitis	Moderate	May multiply	Low	Inhalation
Toxoplasma gondii	Toxoplasmosis	High	Long	High	Ingestion

Reproduced with permission from the WHO list.

protozoan outbreaks we recommend [20–22], which covers 2011–16 reporting 381 outbreaks of which *Cryptosporidium* accounted for the majority.

2.3.1 *Cryptosporidium*

Cryptosporidium was recently placed on the WHO's Neglected Diseases Initiative and is responsible for a significant proportion of childhood mortality in developing countries [1] and several outbreaks of disease, associated with water treatment failure, in developed countries, including the United States [23, 24], United Kingdom [25], Australia [26], and Sweden [27]. If ingested, this pathogen can cause acute self-limiting gastroenteritis, cryptosporidiosis, in immunocompetent hosts and potentially fatal protracted disease in immunocompromised ones. Research into this pathogen intensified in the 1980s after its association as a major opportunistic pathogen in patients with AIDS [28]. There is no recognised safe and effective treatment for human cryptosporidiosis [29].

In the developing world, persistent diarrhoea, caused by agents such as *Cryptosporidium*, accounts for 30%–50% of mortality for children under the age of 5, and it is estimated that 250–500 million cases of cryptosporidiosis occur each year [23]. In the developed world, cryptosporidiosis presents a high risk mainly to the very young, the elderly and immunocompromised individuals and accounts for most gastrointestinal disease outbreaks, where water supplies are chlorinated. The unreported rate of disease from *Cryptosporidium* in England alone has been estimated at >60,000 cases per year [30], with tap water being the most common risk factor in recorded cases. *Cryptosporidium* is particularly problematic to the water industry since it is resistant to both environmental stress and standard chlorination disinfection procedures and can survive for up to 16 months in water [31]. Besides, these organisms are ubiquitous in the environment and have an extremely low infectious dose. For some *C. parvum* isolates, one of the human pathogenic species, less than 10 oocysts can be required to cause infection [32, 33]. This number should be compared against the billions of oocysts that an infected host could shed during an episode of infection [29]. During a clinical infection, a calf may shed around 10,000 millions of oocysts, which would provide enough parasites to infect the whole human population of Europe.

In addition to the risk of disease, *Cryptosporidium* has a significant economic impact. For example, a *Cryptosporidium* contamination in the water supply for Sydney, Australia costs US$45 million in direct emergency measures [34], despite no recorded increase in the cryptosporidiosis case rate. Medical expenses and the cost of lost productivity for the Milwaukee outbreak, the largest documented outbreak with over 400,000 people infected, were estimated at US$96 million [35]. There are also substantial economic costs involved in upgrading water treatment plants to deal with the issue of *Cryptosporidium*.

Not all of the >20 *Cryptosporidium* species, and more than 44 genotypes, are pathogenic to humans. Given this, it is clear that species identification is an essential characteristic to assess the public health risk arising from the detection of any oocysts in a sample. There are no antibodies currently available that can distinguish species differences on the oocyst wall surface [33] and thus genetic comparisons using molecular techniques become important. *Cryptosporidium hominis* and *C. parvum* are the most commonly detected in human clinical cases [29], though several others have been shown to infect humans [36]. *C. hominis* and *C. parvum* have dimensions of 4.5 × 5.5 μm and contain four sporozoites (Fig. 2.7, which also shows the typical life cycle of this pathogen). *C. parvum* is the major zoonotic species, which causes acute neonatal diarrhoea in livestock and is a major contributor to environmental contamination with oocysts [29]. The characteristics of different *Cryptosporidium* species, including oocyst size, host preference, and infection sites have been reviewed by Smith and Nichols [29] and we recommend both this review and an earlier one by the same author [37] for further information on this pathogen (Fig. 2.7).

2.3.2 Cyclospora

Cyclospora cayetanensis infect the intestine of humans, causing diarrhoea in both immunocompetent and immunocompromised hosts. Most cases of the disease occur in young children. No other hosts are known, and the life cycle is not well- characterised. It is known that excreted oocysts, 8 × 10 μm in size, undergo a process of differentiation, called sporulation, in 7–10 days, depending upon environmental conditions. The sporulated oocysts are round, containing two sporozoites. Little information is available on the prevalence of *Cyclospora* in water, though it is known to be spread worldwide. Drinking water-associated disease outbreaks have been reported in a hospital in the United States, in Nepal, and infected an estimated 200 people in Turkey in 2005 [20].

2.3.3 Entamoeba

Entamoeba histolytica is responsible for amoebic dysentery and colitis. The pathogen may also spread to other organs in the body, e.g. the liver or the brain, with possibly fatal results. Worldwide, an estimated 12% of the population has been infected, though up to 90% of these will not show clinical symptoms. Some studies have claimed infections in dogs, cats, pigs, rats, and cattle whereas a 2008 report from the WHO reports humans as the sole reservoir of this pathogen. This pathogen has a feeding, replicative trophozoite of size 10–60 mm that can develop into a dormant cyst, of diameter 10–20 mm, under unfavourable conditions. Amoebic dysentery infects over 500 million people each year, resulting in approximately 100,000 deaths, with a higher incidence in tropical and subtropical regions. In contrast, amoebic encephalitides is far rarer with only 500 cases reported worldwide between 1960 and 2000 (Fig. 2.8) [38].

(A)   *Cryptosporidium* oocyst

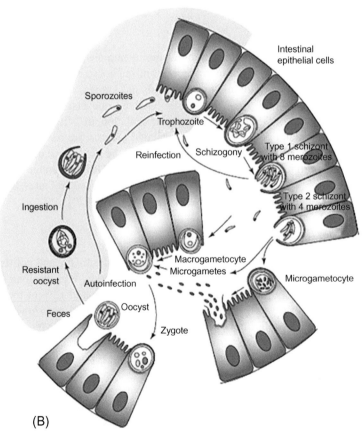

(B)

FIG. 2.7   Image of *Cryptosporidium* and the reproductive life cycle. (A) SEM micrograph of *Cryptosporidium*. (B) Schematic of the reproductive lifecycle of *Cryptosporidium*. *(Figure 1 from H.V. Smith, R.A.B. Nichols, A.M. Grimason, Cryptosporidium excystation and invasion: getting to the guts of the matter, Trends Parasitol. 21(3) (2005) 133–142.)*

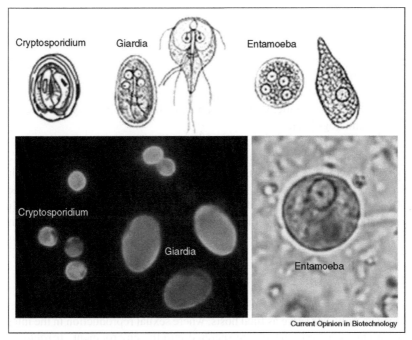

FIG. 2.8 Image of *Entamoeba*, compared to *Cryptosporidium* and *Giardia*. *(Reproduced from M. Bouzid, D. Steverding, K.M. Tyler, Detection and surveillance of waterborne protozoan parasites, Curr. Opin. Biotechnol. 19(3) (2008) 302–306.)*

2.3.4 Giardia

Giardia duodenalis contaminates water supplies across the globe and inges-tion of its cysts can cause giardiasis [39], acute self-limiting gastroenteritis. This species is the only one in the genus associated with human infection. Like *Cryptosporidium*, *G. duodenalis* has a low-infectious dose (1–10 cysts) [40] and those most at risk are the young or immunocompromised [41]. Treatment of giardiasis varies depending on the patient, as does the effectiveness of different drugs, whose side effects are very common [42]. This pathogen causes a major problem for the water industry as it is resistant to disinfection by chlorine treat-ment [43] and can also pass with up to 30% efficiency through more advanced membrane filters [44]. A 2019 review gives an overview of the current status of *G. duodenalis* knowledge and implications for waterborne disease, focussing on the United Kingdom and recommending that further epidemiological data is gathered [45].

The prevalence of *G. duodenalis* is around 20%–30% in the developing world [46], with up to 100% of children acquiring the infection before the age of 3 [42]. In the developed world, where water treatment is better and more wide-spread, the prevalence is lower but outbreaks do occur. In 1985, there were particularly serious cases in both the United Kingdom [47] and the United States [48]. In the United States, *G. duodenalis* was the most common intestinal protozoan infection in the early 2000s [41] and it has been estimated that there are 2 million cases annually, although many individuals are asymptomatic [40]. More recently, in late 2004, over 1000 cases were reported in Norway, resulting from leaking sewage and ineffective water treatment [49]. *G. duodenalis* is zoonotic, with common animal reservoirs being beavers, cattle, cats, and dogs [41].

G. duodenalis cysts are around $8 \times 12\,\mu m$ in size, containing two trophozoites, and the cyst structure lacks the presence of a noticeable suture line or operculum, which demarcates the site of excystation in other intestinal parasite ova stages [40]. The cyst wall is robust and protects the internal trophozoites [50]. A review of the details of the cell biology of *Giardia* is given by Gillin and Reiner [51] and studies of the makeup of the cyst wall are still ongoing (Fig. 2.9) [40].

2.3.5 *Toxoplasma*

Toxoplasma gondii is capable of infecting over 350 vertebrate species. However, only cats serve as final hosts, where sexual reproduction in the intestine produces oocysts, which are secreted into the environment. *Toxoplasma* oocysts are approximately $10 \times 12\,\mu m$ in size. The initially noninfectious oocysts undergo sporulation, upon exposure to air, moisture, and adequate temperatures, thus becoming infectious. Ingestion of sporulated oocysts can lead to infection in a wide variety of hosts. The lifecycle of this parasite is complex, involving many stages and we refer the reader to a recent review (Fig. 2.10) [52]. While many people are asymptomatic or experience mild flu-like symptoms, toxoplasmosis is a serious disease that can cause blindness, encephalitis, and death in immunocompromised individuals and the developing foetus.

Major outbreaks include that linked to the municipal water supply in Vancouver in 1995, when greater than 7000 people became infected with 100 cases of acute toxoplasmosis recorded [53]. More recently, in Brazil, 176 people contracted toxoplasmosis after a drinking water reservoir was contaminated with cat faeces. Four outbreaks were reported between 2001 and 2010 with the largest infecting around 250 people in India in 2005.

Toxoplasma has been found worldwide, with prevalence, based on serological studies, among adults in the United States and the United Kingdom estimated as being around 30%, and in continental Europe, about 50%–80%. Although drinking water is recognised as a transmission route for this pathogen, there are limited studies on the prevalence in water supplies, or the impact of the water treatment process, possibly due to challenges in pathogen detection. For more information, we recommend two recent reviews [52, 54].

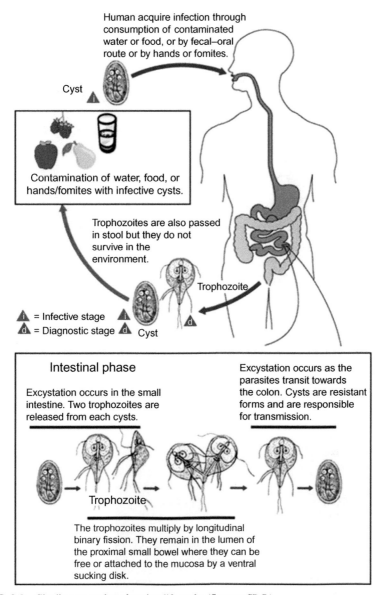

FIG. 2.9 Giardia cysts and trophozoites life cycle. *(Source: CDC.)*

2.4. Helminths

Helminth comes from the Greek word for worm and while worms, in general, infect large numbers of people and animals worldwide, the only two worms which have drinking water as a major mode of transmission are *Dracunculus medinensis* and *Fasciola* (Table 2.4).

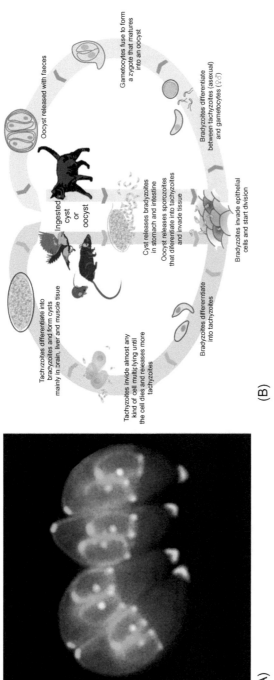

(A)

(B)

FIG. 2.10 Image of *Toxoplasma* and schematic of life cycle. (A) *Toxoplasma gondii*, an obligate intracellular human parasite, has a unique cytoskeletal apparatus that is probably used for invading host cells and for parasite replication. Shown here are images of *T. gondii* constructing daughter scaffolds within the mother cell. *Cyan:* YFP-α-Tubulin; *yellow:* mRFP-TgMORN1. (B) Schematic of life cycle in definitive host (the cat) and otherhosts (not shown here, but including humans). ((A) From Wikicommons: *http://commons.wikimedia.org/wiki/File:Toxoplasma_gondii.jpg;* (B) From Wikicommons: *http://commons.wikimedia.org/wiki/File:Toxoplasmosis_life_cycle_en.svg.)*

TABLE 2.4 Overview of waterborne helminths.

Pathogen	Disease	Relative infectivity	Persistence in water	Resistance to disinfection	Route of transmission
Dracunculus medinensis	Dracuncliasis	High	Moderate	Moderate	Ingestion
Fasciola spp.	Fascioliasis	Unknown	Moderate	High	Ingestion
Schistosoma spp.	Schistosomiasis	High	Short	Moderate	Skin contact

Reproduced with permission from the WHO list.

2.4.1 *Dracunculus medinensis*

This worm, a member of the phylum Nematoda, is commonly known as guinea worm. Major effects to eradicate this disease which is known to have a huge economic and health impact have reduced the incidence from 3.3 million cases in 1986 to only 3190 in 2009. The disease is geographically restricted to sub-Saharan Africa and mainly occurs in rural areas, lacking a piped water supply. It is easily prevented by filtration to remove the *Cyclops*, which are the water-borne hosts of the larvae. When *Cyclops* is swallowed the larvae are released in the stomach penetrating the intestinal wall and inhabiting the cutaneous and subcutaneous tissue. Here, the larvae grow into worms, where the males reach 25 mm in length and the females can reach 700 mm. When ready the female emerges from a blister or ulcer to release a large number of larvae into the water. Mortality is rare but complications can lead to disability.

2.4.2 *Fasciola*

There are two species, *Fasciola hepatica*, and *Fasciola gigantica*, implicated in waterborne human disease. These trematode worms, also known as liver flukes, infect an estimated 2.4–17 million people globally. There is also a water host, the snail. The snail releases infectious metacercariae who, when ingested, ex-cyst in the human stomach and migrate to the liver and bile ducts. The acute phase, associated with the maturation of these excysted flukes, results in symptoms such as dyspepsia, nausea, and vomiting, abdominal pain, and high fever. Subsequently, the mature adult flukes release eggs that are excreted back into the water environment and can live inside the host for 9–14 years. The metacercariae exhibit high resistance to disinfection by chlorine but can be removed by filtration. In most regions, fascioliasis is spread through eating wet grass or aquatic plants, such as watercress, though drinking-water may be a significant transmission route for fascioliasis in certain locations, e.g. the Andean Altiplano region in South America (Fig. 2.11).

2.5. Antimicrobial resistance

Water is recognised as an important reservoir of antimicrobials and antimicro-bial resistance genes (ARGs), contributing to the maintenance of resistance in the environment and transfer of ARGs between pathogenic and nonpathogenic bacteria [55]. The topic has been recently reviewed [56], as well as the subject of an EU JRC technical report. The JRC report surveyed the presence of 45 anti-biotics in 13 countries, finding sulfamethoxazole, trimethoprim, and ciprofloxa-cin to be the most prevalent antibiotics in water, with most of the data coming from Europe [55], though the issue has been reported worldwide [57–59], with India being referred to as the 'ARM capital of the world' [60]. While increasing prevalence of ARGs in water is not debated, there remains the question of to what extent this has implications for human health [61].

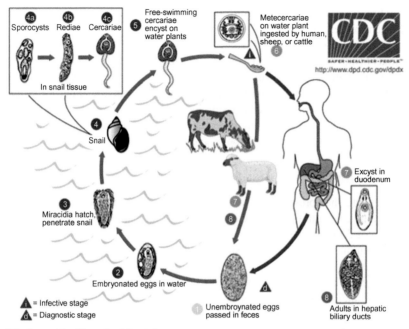

FIG. 2.11   The life cycle of *Fasciola*.

ARGs have been detected in all water types, being most prevalent in wastewater but also showing up in groundwater, drinking water, and even bottled water [56]. Drinking water treatment processes are not optimised to remove ARGs and may even promote their presence, though further research is needed to understand in more depth these effects [56]. Further research is also required to develop methods for ARG removal at wastewater treatment plants [62].

Bacteria are the main focus of antimicrobial resistance (AMR) and ARG studies in water. Drug resistance is also observed for protozoa, though there is no evidence of environmental spread [63, 64]. Antiviral drug resistance is also observed [65, 66], and antivirals have been detected in aquatic environments, which is difficult to remove in rivers and sewage treatment plants [67], with one study finding implications for influenza circulating in ducks [68]. There was a recent global review on the occurrence of antivirals in water, published in 2020, to which the reader is referred for more in-depth information [69]. To summarise, the review found that in many countries there is a significant lack of data on antiviral occurrence in water, with the majority of studies coming after an influenza pandemic, and that more monitoring is required to understand the hazards posed by antiviral in water, particularly focussing on the gaps in knowledge on environmental transmission and risks as well as screening a wider variety of antivirals [69].

Disease outbreaks have been attributed to the occurrence of antibiotic-resistant bacteria in the water in the past few years, e.g. typhoid fever, caused by *Salmonella*, in Pakistan [70], and *Shigella* in China [71]; the general risk to human health of antibiotic resistance infections via drinking water is thought to be low though further information is needed [45, 55]. Current data is considered insufficient to enact water regulations and there is an urgent need for more monitoring technologies to detect antibiotic resistance in drinking water systems, particularly low-cost methods to enable frequent monitoring and facilitate improved data gathering in developing countries [56].

2.6. Summary

This chapter has presented an overview of the different types of waterborne pathogens concentrating mainly on those who cause infection through the ingestion of drinking water.

In summary, viruses are the smallest of these pathogens, typically around 20–300 nm in diameter, making them difficult both to remove and detect. Many viruses exhibit considerable resistance to chlorination as well as longevity and therefore their presence in water does not correlate well with that of indicator bacteria, one of the monitoring methods described in the next chapter. Often highly infectious, viruses are in general more species-specific than the other pathogens with less-known examples of zoonoses.

In general, bacteria are the next smallest, with sizes on the order of a few micrometres, followed by the slightly larger protozoa and finally the helminths. Bacteria are commonly less infectious than the other types of the pathogen, with a few notable exceptions (e.g. *E. coli* O157:H7 and *Shigella*) and more susceptible to chlorine disinfection. However, some of them can enter a spore-like state. Some waterborne bacteria are also capable of multiplication in the water environment, enhancing numbers.

Protozoa have a low infectious dose and many are zoonotic. Like viruses, these organisms are unable to reproduce outside a host. Besides, protozoa have the similarity with viruses of poor indicator bacteria correlation, due to chlorine resistance and robustness. Finally, the incidence of disease caused by the guinea worm helminth has been significantly reduced over the last decades, whereas liver fluke cases are not well-characterised. Both helminths rely upon a life cycle involving a water host and the causative agents are easily removed by filtration.

All of these pathogens contribute to the large disease burden, caused by waterborne pathogens, impacting both health and economic productivity worldwide. The effects of waterborne pathogens are greater in developing countries where poor water treatment and sanitation coverage prevents the delivery of safe drinking water. However, even in developed countries, these pathogens have a significant impact, particularly in small supplies [72, 73]. The following chapter will describe the guidelines and regulations developed to ensure drinking water safety as well as the existing methods of monitoring to detect these pathogens. There is also growing awareness of the issue of AMR and the role of the aquatic environment in the spread of AMR, and some of the sample processing methods and detection approaches are also applicable to the challenge of monitoring for AMR in water.

Acknowledgement

We would like to acknowledge the WHO Guidelines and Microbial Fact Sheets which have provided an extremely useful source of information for this chapter: Ref. [3].

References

[1] J.D. Brookes, et al., Fate and transport of pathogens in lakes and reservoirs, Environ. Int. 30 (2004) 741–759.

[2] J.T. Connelly, A.J. Baeumner, Biosensors for the detection of waterborne pathogens, Anal. Bioanal. Chem. 402 (2012) 117–127.

[3] World Health Organisation, Guidelines for drinking-water quality, 2011.

[4] L. Dijkshoorn, B.M. Ursing, J.B. Ursing, Strain, clone and species: comments on three basic concepts of bacteriology, J. Med. Microbiol. 49 (2000) 397–401.

[5] M.A. Borchardt, et al., Human enteric viruses in groundwater from a confined bedrock aquifer, Environ. Sci. Technol. 41 (18) (2007) 6606–6612.

[6] I.A. Hamza, et al., Methods to detect infectious human enteric viruses in environmental water samples, Int. J. Hyg. Environ. Health 214 (2012) 426–436.

[7] M.J. Carter, Enterically infecting viruses: pathogenicity, transmission and significance for food and waterborne infection, J. Appl. Microbiol. 98 (6) (2005) 1354–1380.

[8] World Health Organisation, Guidelines for drinking-water quality, 2008.

[9] S.C. Jiang, Human adenoviruses in water: occurrence and health implications: a critical review, Environ. Sci. Technol. 40 (2006) 7132–7140.

[10] P.F. Teunis, et al., Norwalk virus: how infectious is it? J. Med. Virol. 80 (2008) 1468–1476.

[11] S. Yezli, J.A. Otter, Minimum infective dose of the major human respiratory and enteric viruses transmitted through food and the environment, Food Environ. Virol. 3 (2011) 1–30.

[12] S.R. Seitz, et al., Norovirus infectivity in humans and persistence in water, Appl. Environ. Microbiol. 77 (2011) 6884–6888.

[13] N.P. Nenonen, et al., Marked genomic diversity of norovirus genogroup I strains in a waterborne outbreak, Appl. Environ. Microbiol. 78 (6) (2012) 1846–1852.

[14] H.t.L.G. ter Waarbeek, et al., Waterborne gastroenteritis outbreak at a scouting camp caused by two norovirus genogroups: GI and GII, J. Clin. Virol. 47 (3) (2010) 268–272.

[15] N. Albayrak, D.Y. Caglayik, G. Korukluoglu, PVII-13 Waterborne outbreaks of acute gastroenteritis in January-May 2009, Turkey, J. Clin. Virol. 46 (Suppl. 1(0)) (2009) S40.

[16] J. Angel, M.A. Franco, H.B. Greenberg, Rotavirus vaccines: recent developments and future considerations, Nat. Rev. Microbiol. 5 (7) (2007) 529–539.

[17] S. Vandeplas, et al., Contamination of poultry flocks by the human pathogen Campylobacter spp. and strategies to reduce its prevalence at the farm level, Biotechnol. Agron. Soc. Environ. 12 (2008) 317–344.

[18] R.E. Levin, *Campylobacter jejuni*: a review of its characteristics, pathogenicity, ecology, distribution, subspecies characterization and molecular methods of detection, Food Biotechnol. 21 (2007) 271–374.

[19] U. Buchholz, H. Bernard, D. Werber, German outbreak of *Escherichia coli* O104:H4 associated with sprouts, N. Engl. J. Med. 365 (2011) 1763–1770.

[20] S. Baldersson, P. Karanis, Waterborne transmission of protozoan parasites: review of worldwide outbreaks—an update 2004–2010, Water Res. 45 (2011) 6603–6614.

[21] F.Y. Ramírez-Castillo, et al., Waterborne pathogens: detection methods and challenges, Pathogens (Basel, Switzerland) 4 (2) (2015) 307–334.

[22] A. Efstratiou, J.E. Ongerth, P. Karanis, Waterborne transmission of protozoan parasites: review of worldwide outbreaks—an update 2011–2016, Water Res. 114 (2017) 14–22.

[23] W.J. Snelling, et al., Cryptosporidiosis in developing countries, J. Infect. Dev. Ctries. 1 (3) (2007) 242–256.

[24] W.R. MacKenzie, et al., Cryptosporidium infection from Milwaukee's public water supply, N. Engl. J. Med. 331 (3) (1994) 161–168.

[25] A.P. Davies, R.M. Chalmers, Cryptosporidiosis, BMJ 339 (2009) b4168. https://doi.org/10.1136/bmj.b4168.

[26] J.S.Y. Ng, et al., Molecular characterisation of Cryptosporidium outbreaks in Western and South Australia, Exp. Parasitol. 125 (4) (2010) 325–328.

[27] SmittskyddsInstitutet, 2010. Available from: http://www.smittskyddsinstitutet.se/sjukdomar/cryptosporidium-infektion/.

[28] S. Tzipori, G. Widmer, A hundred-year retrospective on cryptosporidiosis, Trends Parasitol. 4 (2008) 184–189.

[29] H.V. Smith, R.A.B. Nichols, Cryptosporidium: detection in water and food, Exp. Parasitol. 124 (2010) 61–79.

[30] C.C. Tam, et al., Longitudinal study of infectious intestinal disease in the UK (IID2 study): incidence in the community and presenting to general practice, Gut (2011). https://doi.org/10.1136/gut.2011.238386. Published on line.

[31] F. Chen, et al., Comparison of viability and infectivity of *Cryptosporidium parvum* oocysts stored in potassium dichromate solution and chlorinated tap water, Vet. Parasitol. 150 (1–2) (2007) 13–17.

[32] B.J. King, P.T. Monis, Critical processes affecting Cryptosporidium oocyst survival in the environment, Parasitology 134 (2007) 309–323.

[33] P.C. Okhuysen, et al., Virulence of three distinct *Cryptosporidium parvum* isolates for healthy adults, J. Infect. Dis. 180 (4) (1999) 1275–1281.

[34] J.W. Bridge, et al., Engaging with the water sector for public health benefits: waterborne pathogens and diseases in developed countries, Bull. World Health Organ. 88 (2010) 873–875.

[35] P.S. Corso, et al., Cost of illness in the 1993 waterborne Cryptosporidium outbreak, Milwaukee, Wisconsin, Emerg. Infect. Dis. 9 (4) (2003) 426–431.

[36] G. Robinson, K. Elwin, R.M. Chalmers, Unusual Cryptosporidium genotypes in human cases of diarrhea, Emerg. Infect. Dis. 11 (2008) 1800–1802.

[37] M. Smith, K.C. Thompson, Cryptosporidium: The analytical challenge, Royal Society of Chemistry, 2001.

[38] M. Bouzid, D. Steverding, K.M. Tyler, Detection and surveillance of waterborne protozoan parasites, Curr. Opin. Biotechnol. 19 (3) (2008) 302–306.

[39] S.M. Caccio, et al., Unravelling Cryptosporidium and Giardia epidemiology, Trends Parasitol. 21 (9) (2005) 430–437.

[40] C.D. Karr, Identification of a novel ß1,3-NAcetylgalactosaminyltransferase activity and its unique ß1,3-GalNAc Homopolymer that forms the Giardia cyst wall, Department of Biology, Northeastern University, 2009.

[41] B. Lebwohl, R.J. Deckelbaum, P.H.R. Green, Giardiasis, Gastrointest. Endosc. 57 (7) (2003) 906–913.

[42] G.P. Heresi, J.R. Murphy, T.G. Cleary, Giardiasis, Semin. Pediatr. Infect. Dis. 11 (3) (2000) 189–195.

[43] R.L. Owen, The ultrastructural basis of Giardia function, Trans. R. Soc. Trop. Med. Hyg. 74 (4) (1980) 429–433.

[44] C.C. Falk, et al., Bench scale experiments for the evaluation of a membrane filtration method for the recovery efficiency of Giardia and Cryptosporidium from water, Water Res. 32 (3) (1998) 565–568.

[45] B. Horton, et al., *Giardia duodenalis* in the UK: current knowledge of risk factors and public health implications, Parasitology 146 (4) (2018) 413–424.

[46] M.J.G. Farthing, Giardiasis as a disease, in: R.C.A. Thompson, J.A. Reynoldson, A.J. Lymbery (Eds.), Giardia: From molecules to disease, CABI Publishing, Wallingford, 1994, pp. 15–37.

[47] A.E. Jephcott, N.T. Begg, I.A. Baker, Outbreak of giardiasis associated with mains water in the United Kingdom, Lancet 327 (8483) (1986) 730–732.

[48] G.P. Kent, et al., Epidemic giardiasis caused by a contaminated public water supply, Am. J. Public Health 78 (2) (1988) 139–143.

[49] K. Nygard, et al., A large community outbreak of waterborne giardiasis-delayed detection in a non-endemic urban area, BMC Public Health 6 (2006) 141.

[50] G.J. Gerwig, et al., The *Giardia intestinalis* filamentous cyst wall contains a novel beta(1-3)-N-acetyl-D-galactosamine polymer: a structural and conformational study, Glycobiology 12 (8) (2002) 499–505.

[51] F.D. Gillin, D.S. Reiner, Cell biology of the primitive eukaryote *Giardia lamblia*, Annu. Rev. Microbiol. 50 (1996) 679–705.

[52] J.L. Jones, J.P. Dubry, Waterborne toxoplasmosis—recent developments, Exp. Parasitol. 124 (2010) 10–25.

[53] W.R. Bowie, et al., Outbreak of toxoplasmosis associated with municipal drinking water, Lancet 350 (1977) 173–177.

[54] E.A. Innes, A brief history and overview of *Toxoplasma gondii*, Zoonoses Public Health 57 (2010) 1–7.

[55] I. Sanseverino, et al., State of the art on the contribution of water to antimicrobial resistance, JRC Technical Reports, P.O.o.t.E. Union, 2018.

[56] E. Sanganyado, W. Gwenzi, Antibiotic resistance in drinking water systems: occurrence, removal, and human health risks, Sci. Total Environ. 669 (2019) 785–797.

[57] C. Ng, K.Y.-H. Gin, Monitoring antimicrobial resistance dissemination in aquatic systems, Water 11 (1) (2019) 71.

[58] B. Ram, M. Kumar, Correlation appraisal of antibiotic resistance with fecal, metal and microplastic contamination in a tropical Indian river, lakes and sewage, NPJ Clean Water 3 (1) (2020) 3.

[59] A.I. Moreno-Switt, et al., Antimicrobial resistance in water in Latin America and the Caribbean: a scoping review protocol, JBI Evid. Synth. 17 (10) (2019) 2174–2186.

[60] N. Taneja, M. Sharma, Antimicrobial resistance in the environment: the Indian scenario, Indian J. Med. Res. 149 (2) (2019) 119–128.

[61] A.C. Singer, et al., Review of antimicrobial resistance in the environment and its relevance to environmental regulators, Front. Microbiol. (2016) 7(1728).

[62] F. Barancheshme, M. Munir, Strategies to combat antibiotic resistance in the wastewater treatment plants, Front. Microbiol. (2018) 8(2603).

[63] R. Capela, R. Moreira, F. Lopes, An overview of drug resistance in protozoal diseases, Int. J. Mol. Sci. 20 (22) (2019) 5748.

[64] C.G. García, et al., Drug resistance mechanisms in *Entamoeba histolytica, Giardia lamblia, Trichomonas vaginalis*, and opportunistic anaerobic protozoa, in: D.L. Mayers, et al. (Eds.), Antimicrobial Drug Resistance: Mechanisms of Drug Resistance, vol. 1, Springer International Publishing, Cham, 2017, pp. 613–628.

[65] W.-J. Shin, B.L. Seong, Novel antiviral drug discovery strategies to tackle drug-resistant mutants of influenza virus strains, Expert Opin. Drug Discovery 14 (2) (2019) 153–168.

[66] K.K. Irwin, et al., Antiviral drug resistance as an adaptive process, Virus Evol. 2 (1) (2016) vew014. https://doi.org/10.1093/ve/vew014.

[67] T. Azuma, et al., Fate of new three anti-influenza drugs and one prodrug in the water environment, Chemosphere 169 (2017) 550–557.

[68] J.D. Järhult, et al., Environmental levels of the antiviral oseltamivir induce development of resistance mutation H274Y in influenza A/H1N1 virus in mallards, PLoS One 6 (9) (2011) e24742.

[69] C. Nannou, et al., Antiviral drugs in aquatic environment and wastewater treatment plants: a review on occurrence, fate, removal and ecotoxicity, Sci. Total Environ. 699 (2020) 134322.

[70] F.N. Qamar, et al., Outbreak investigation of ceftriaxone-resistant *Salmonella enterica* serotype Typhi and its risk factors among the general population in Hyderabad, Pakistan: a matched case-control study, Lancet Infect. Dis. 18 (12) (2018) 1368–1376.

[71] Q. Ma, et al., A waterborne outbreak of *Shigella sonnei* with resistance to azithromycin and third-generation cephalosporins in China in 2015, Antimicrob. Agents Chemother. 61 (6) (2017) e00308–e00317.

[72] A. Munene, D.C. Hall, Factors influencing perceptions of private water quality in North America: a systematic review, Syst. Rev. 8 (1) (2019) 111.

[73] M.J. Gunnarsdottir, et al., Status of small water supplies in the Nordic countries: characteristics, water quality and challenges, Int. J. Hyg. Environ. Health 220 (8) (2017) 1309–1317.

Chapter 3

Existing methods of detection

Helen Bridle[a], Kimberley Gilbride[b], James Green[c], and Susan Lee[c]
[a]Institute of Biological Chemistry, Biophysics and Bioengineering, Heriot-Watt University, Edinburgh, Scotland, [b]Department of Chemistry and Biology, Ryerson University, Toronto, ON, Canada, [c]Scottish Water, Juniper House, Edinburgh, Scotland

Chapter 2 gave an overview of the different microbial pathogens, implicated in waterborne outbreaks of diseases, with specific emphasis given to those where drinking water is an important mode of transmission. This chapter considers the guidelines and regulations to reduce or prevent adverse health effects associated with unsafe drinking water.

Safe drinking water is considered to be that which poses no significant risk to health over a lifetime of consumption, considering the increased sensitivity that some elements of the population (e.g. infants and young children, the elderly, or the debilitated) may display. There is no single, universally applicable approach towards managing the delivery of safe drinking water [1].

The chapter focuses specifically on the role of monitoring the delivery of safe drinking water and gives an overview of the existing methods of detection, which are routinely adopted.

3.1. WHO guidelines

The World Health Organisation (WHO) has published Guidelines for Drinking Water Quality, previously the International Standards for Drinking Water, since 1958. There have been several editions and various addendums over time, reflecting developments in understanding.

The more recent Guidelines for Drinking Water Quality (with the 4th edition published in 2011) have not set international standards, recognising 'the advantage of risk-benefit approach in the establishment of national standards and regulations' [1]. The guidelines are intended to support the development and implementation of risk management strategies at local, national, or regional levels with this approach designed to allow countries to adopt standards and monitoring according to local environmental, social, economic, and cultural conditions to determine legislative and regulatory frameworks. For example, it is unlikely to be an efficient use of resources for developing countries to spend scarce resources on determining standards and monitoring for substances with minor implications for public health.

Waterborne Pathogens: Detection Methods and Applications. https://doi.org/10.1016/B978-0-444-64319-3.00003-4

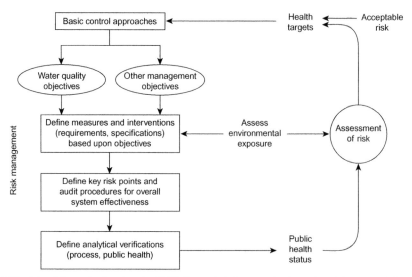

FIG. 3.1 Schematic showing the WSF: Expanded framework. *Reproduced from Fig. 1.2 in A. Davidson, G. Howard, M. Stevens, et al. Water Safety Plans: Managing Drinking-Water Quality from Catchment to Consumer: World Health Organisation, 2005, originally from (Bartram et al., 2001).*

Microbial contamination is considered to pose the greatest threat to health in both developing and developed countries, with faecally derived pathogens being the principal concern [1]. The WHO guidelines comment that the presence of pathogens in water can vary rapidly and by the time pathogen presence is detected many people may have been exposed. The WHO, therefore, advocate that monitoring alone is not sufficient and should be incorporated into a wider water safety framework (WSF) and adoption of water safety plans (WSP), implementing multiple barriers from catchment to the consumer to deliver sufficient protection (see Fig. 3.1). So far there have been few studies that have assessed the impact of WSPs on water safety, though those that have, have been supportive [2]. A full discussion of WSF and WSPs is beyond the scope of this book and the reader is referred to the WHO guidelines [1] for more detail as well as books published by the WHO (Fig. 3.2).

3.2. Types of monitoring

There are many different reasons why water monitoring might be undertaken. The key to the selection of the most appropriate monitoring strategy is to consider what questions are being asked; there is no 'one-size-fits-all' solution to monitoring and certain tools and techniques will be more suited for certain types of monitoring. Different authors describe various types of monitoring approaches. We have chosen to classify monitoring strategies according to the below classification, adapted from studies [1, 3]:

(1) *surveillance* monitoring aimed at assessing long-term water quality changes and providing baseline data on raw water sources allowing the

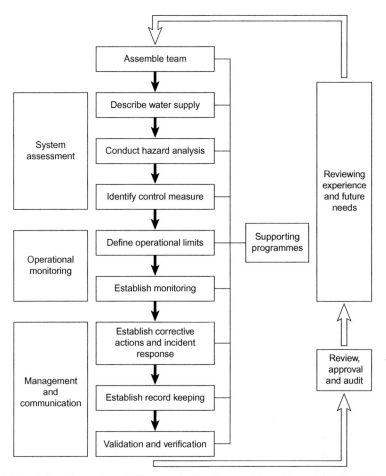

FIG. 3.2 Schematic showing a WSP approach: Steps in assembling a Water Safety Plan. *Adapted from A. Davidson, G. Howard, M. Stevens, et al. Water Safety Plans: Managing Drinking-Water Quality from Catchment to Consumer: World Health Organisation, 2005.*

design and implementation of other types of monitoring; or the monitoring undertaken by water providers, or the regulatory bodies, to ensure water treatment processes are providing the required level of public health protection; or the verification of treatment processes, validating them for future use.

(2) *operational* monitoring aimed at providing additional and essential data on water bodies at risk or failing environmental objectives of a WSF/WSP; this operational monitoring should provide information for instant action.

(3) *investigative* monitoring aimed at assessing causes of such failure; or at understanding the pathogen load in source waters; or tracking the source of an outbreak.

Surveillance and verification of treatment systems are performed by water providers who need to maintain constant vigilance over the security of their water supply systems. Once treated water enters a distribution network it may be vulnerable to ingress of contamination in several ways. Besides, even though treated to the required standard, it may still undergo deterioration in quality during transit to the consumers' taps. Where a disinfectant residual is applied, it may decline in concentration until it is no longer sufficient to control the growth of organisms. With aging assets, water providers must be able to maintain confidence in the integrity of their networks and have the means for early detection of a vulnerability in their infrastructure and its adverse impacts on water quality. Traditional monitoring is not appropriate since it can take, at best, nearly a day before the outcome of a routine test is known for a bacteriological indicator organism. The ability to monitor more frequently and for specific waterborne pathogens of concern would greatly improve public health protection. Water companies would be able to respond much more rapidly to potential incidents before drinking water reaches the consumer or is used for food production.

Operational monitoring is a key part of the water safety plan approach, enabling water providers to take up-to-date decisions about the quality of the delivered water and adjust processes accordingly. The WHO considers that this type of monitoring should comprise of simple observations and tests, to rapidly confirm that control measures are continuing to work. As above traditional microbial monitoring suffers from the speed at which results can be provided since finished waters will have reached consumer taps before results can be acted upon.

Investigative monitoring covers a range of scenarios. For example, at present, the assessment of source water quality relies upon estimates of the numbers of waterborne pathogens present, either by extrapolation from literature data or applying models to predict the transport of pathogens into the water source from surrounding land. Both methods involve many assumptions and therefore, lead to uncertainty in the risk assessment. Investigative monitoring to accurately identify the numbers and types of pathogens in particular sources of drinking water would enable water providers to properly assess risks and reliably estimate the disease burden associated with a particular water supply. This type of testing will require pathogen speciation or even a more detailed investigation. For example, many waterborne pathogens, such as all Vibrio species, have to be identified beyond the species level to determine their pathogenic potentials [4, 5]. This can also be known as microbial or faecal source tracking (MST) [6, 7]. When MST is applied to the identification of the source of an outbreak, rapid monitoring methods are also required.

3.3. Faecal indicator monitoring

During the twentieth century, the measurement of indicator bacteria in the finished water has contributed to a significant improvement in drinking water safety. This strategy assumes that faecal indicators, showing that waters have been faecally contaminated, will, therefore, also indicate the presence of any pathogens. Commonly used indicator bacteria studies include thermotolerant coliforms, faecal coliforms, and more recently *Escherichia coli* [8]. The advantages of this approach are that these tests are relatively cheap and easy to perform. However, there are concerns about the degree of correlation between the detection of indicators and the presence of microbial pathogens as well as the fact that this approach does not allow a valid identification of the pathogen [9].

Indicator monitoring takes time for the growth by culturing and therefore, it is retrospective and some contaminated drinking water may have been consumed before a problem has been detected. This has led to the view that this type of monitoring is more compatible with a quality assurance scheme, whereby the absence of indicator provides reassurance that the hygienic quality of the water supply system has not been compromised.

Routine monitoring of drinking and recreational waters for bacterial pathogens is based on detecting indicators organisms to assess water quality [10]. The bacteria that probably satisfies these criteria best is *E. coli*. It is a member of the coliform group of bacteria. The coliform group comprises all of the aerobic and facultatively anaerobic gram-negative, nonsporing, rod-shaped bacteria that ferment lactose with gas formation within 48 h at 35°C. Even though all the coliforms will not come from human faeces, the total coliform count is used as an index of faecal pollution. The representative species of the faecal group is *E. coli* which may be differentiated from members of the nonfaecal group utilising four biochemical tests—indole, methyl red, Voges-Proskauer, and Citrate Coliforms from the intestine are also able to ferment lactose at an elevated temperature of 44.5°C with the production of acid and gas. It is reasoned that if many coliforms are present in a given water sample, there is a good likelihood that enteric pathogens may also be present.

Streptococcus fecalis, Lancefield's group D enterococci, are also normal inhabitants of the intestine and may be used as an indicator of faecal contamination of water. The ratio of *E. coli* to *Streptococci* has been used to indicate whether the contamination comes from human or animal sources.

One method for measuring coliform numbers (quantity) in water is the membrane filter technique. This technique involves filtering a known volume of water through a special sterile membrane filter, with a pore size of 0.45, thus small enough to retain bacteria. When the water sample is filtered, bacteria are trapped on the surface. The filter is then placed on media in a petri dish to allow the bacteria to grow. The use of selective media allows for easy detection of the representative indicator. A wide range of selective media has been developed and validated for use in membrane filtration techniques for Coliform and Enterococci/ faecal streptococci analyses. On m-ENDO plates, coliform colonies will appear pink to dark red with a golden green metallic sheen or lustre. On m-FC plates, faecal coliforms will appear blue. And if necessary, m-enterococcus plates can be used to detect faecal streptococci colonies which will appear light pink and flat or dark red on this media. These colonies can be directly counted and are typically reported as CFU (colony forming units)/100 mL.

There are several official published methods based on the above approach, notably a series of ISO methods, such as ISO 9308-1 for coliforms and *E. coli* and ISO 7899-2 for enterococci. The US Environmental Protection Agency (EPA) has also published official microbiological methods for water testing.

The heterotrophic plate count (HPC) is also another method for monitoring the overall bacteriological quality of drinking water. Although they do not directly relate to the safety of water, concentration changes in these counts can indicate overall changes in the quality of the water. This analysis is performed by mixing a volume, typically 1 mL, of the sample water to a volume of molten, cooled, nonselective agar (for example Yeast Extract Agar) in a petri dish. The agar is allowed to set then incubated at the desired temperature and duration. Following incubation, bacterial and fungal colonies will grow within the agar and can be counted and reported as (colony forming units)/millilitre ($cfu\,mL^{-1}$) [11] (Fig. 3.3).

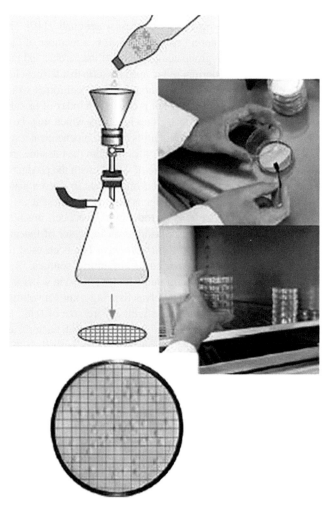

FIG. 3.3 Flowchart illustrating the stages of faecal indicator monitoring.

The main focus of this book is upon the description, review, and evaluation of emerging methods of detection, as covered in Chapters 5–10. However, the field of faecal indicator monitoring is also developing, with the identification of new indicator organisms, offering improved correlation with groups of pathogens [12], or the design of new technology. Many different approaches are still published today to provide new methods or to suggest new indicators for faecal source detection [7].

Alternative indicators include faecal anaerobes (genera *Bacteroides* and *Bifidobacterium*, spore-forming *Clostridium perfringens*), viruses (*B. fragilis* phage, coliphages (FRNA phage)), and faecal organic compounds (coprostanol) as described in more detail in a 2006 review [12] and summarised in a table adapted from that publication (Table 3.1.). The WHO Guidelines present how some of these different strategies can be applied to different types of monitoring (Table 3.2.) [1].

Alternative technologies include adenosine triphosphate, ATP, sensor system. ATP is an energy-carrying molecule found in all living cells—and, as such, it can be used as an indicator for living bacteria in drinking water [13]. It is assumed that a high level of ATP correlates well with a high level of bacteria. No limit of detection (LOD) was reported. The detection can be performed rapidly, within a few seconds and with high sensitivity [14]. The analysis is well established and is used for monitoring water quality [15]. This early warning system was developed as part of the EU Sensowaq project, which also designed a novel chemiluminescent method to detect *Enterococci* [16].

3.3.1   Correlation of indicators with pathogen presence

As described above the traditional method of water monitoring, aimed at ensuring water safety, was microbiological monitoring, especially for *E. coli* (syn. Faecal coliform counts or thermotolerant coliform counts). However, in recent years it has been recognised that there is not always a strong correlation between faecal indicator positive results and pathogen presence. This was highlighted by the emergence of *Cryptosporidium* as a major cause of waterborne outbreaks, where often the faecal indicator testing had shown the absence of *E. coli* [17]. This is likely to be because *E. coli* is sensitive to chlorination, whereas *Cryptosporidium* demonstrates considerable resistance. Many viruses also demonstrate chlorine resistance [18] and norovirus outbreaks have been reported where *E. coli* was shown to be absent [17]. The lack of correlation between indicator *E. coli* and *Cryptosporidium* as well as *Salmonella* was again demonstrated in a Canadian study in 2007 and 2008 (Fig. 3.4). Another reason for a lack of correlation may be different degrees of persistence in water. The short persistence time of indicator organisms has also been quoted as one further disadvantage of this approach [12].

Most of the above examples deal with a lack of correlation between faecal indicator bacteria and viral and protozoan pathogens. Recently, however, it has

TABLE 3.1 Characteristics of conventional and alternative faecal indicators in terms of prediction of faecal pollution and associated pathogens.

Criteria for faecal indicators	Conventional faecal indicators		Faecal streptococci and enterococci
	Total and faecal coliforms	E. coli	
Presence in the faeces and environmental waters	Present in faeces of warm-blooded animals; positive correlation with faecal material; however, a nonfecal source in wastewaters and tropical areas	Present in faeces of warm-blooded animals; a more specific indicator of faecal contamination; however, a nonfecal source in tropical areas	Present in faeces of warm-blooded animals; not exclusively of faecal origin
Ability to multiply and survive in a water environment	Regrowth in the aquatic environment; limited survival in water; weakness to the water treatment process	Variable survival in nonhost environments; possibility for regrowth in tropical and temperate environments	Rapid inactivation in the first contact with the environment; regrowth caused by tides and sediments
Correlation with pathogens or waterborne disease	Poor correlations with enteric pathogens	Present at higher concentrations than pathogens	Similar persistence to those of some waterborne pathogens; high correlation with gastrointestinal symptoms
Applied methodology	Time-consuming culture-dependent detection methods; inability to identify the source of faecal contamination	Normally not pathogenic to humans; ability to identify the origin of faecal contamination by phenotypic and molecular methods	Inability to provide a selective medium for all groups with taxonomical and ecological heterogeneity and different levels of sanitary significance; ability to identify the origin of faecal contamination by phenotypic methods (ARA)

Alternative faecal indicators

	Bacteroides spp.	Bifidobacterium spp.	Clostridium perfringens
Presence in the faeces and environmental waters	A significant portion of faecal bacteria in warm-blooded animals; host-specific distribution of species; wide geographical distribution of human-specific genetic marker	A significant portion of faecal bacteria in warm-blooded animals; host-specific distribution of species; similar frequency and distribution of species of human faecal isolates and isolates from raw sewage	Entirely of faecal origin from warm-blooded animals; consistently present in sewage and aquatic sediments; species vary among different animals; may indicate remote or old faecal pollution
Ability to multiply and survive in a water environment	Short survival time in nonhost environments; unable to replicate after releasing into a water body; seasonable changes in persistence; effect of temperature and predation on survival rates; long persistence of DNA marker in the water	Short survival time in nonhost environments; unable to multiply in oxygenated nonhost environments; low frequency of detection and recovery from warm waters; inhibition of the recovery due to background bacteria	Extreme stability to environmental factors (more stable in environmental waters than pathogens); no effect of temperature and predation
Correlation with pathogens or waterborne disease	No sufficient data on survival and correlation with pathogens	No sufficient data on survival and correlation with pathogens	Good correlation with some pathogens; might indicate the efficiency of inactivation and removal of viruses, cysts, and oocysts; high correlation with diarrhoea disease levels
Applied methodology	Increasing usage of molecular methods to identify the origin of recent or extensive faecal contamination using host-specific genetic markers	Identification of the source of recent faecal pollution by molecular methods using host-specific genetic markers	Commonly culture-based methods

Continued

TABLE 3.1 Characteristics of conventional and alternative faecal indicators in terms of prediction of faecal pollution and associated pathogens.—cont'd

	Alternative faecal indicators		
	F-specific RNA coliphage	B. fragilis phage	Faecal sterols
Presence in the faeces and environmental waters	Present in human and animal faeces as well as in sewage; well correlated with the source of faecal contamination	Main y found in human faecal samples, therefore, could serve as a specific index for human faecal pollution; isolated from sewage and fecally polluted waters	Considerable amount in animal faeces; indicate relatively fresh faecal pollution in the water column; presence in sediments may indicate old or remote faecal pollution; suitable for tropical and temperate regions
Ability to multiply and survive in a water environment	Possibility to replicate in sewage; relatively sensitive to high temperatures, sunlight at high salinity; similar persistence with enteric viruses especially in freshwaters	Do not replicate in the environment; high survival in the environment; always outnumber human enteric viruses	Aerobically degraded in the water column; under anoxic condition readily incorporated into sediments
Correlation with pathogens or waterborne disease	Valuable model for viral contamination and inactivation due to their physical resemblance	Relate to the level of human enteric viral pollution	No available data on correlations with the presence of pathogens and public health risk
Applied methodology	Expensive and labor-intensive recovering methods complicated by low number in faeces and environments; molecular- and culture-based analytical methods	Complex methodology and large sample volume for recovery with a low number in environmental samples at low levels of faecal pollution; molecular- and culture-based analytical methods	A sensitive and simple analytical method using a gas chromatography–mass spectrometry (GC–MS); lack of studies on host specificity

Adapted from Table 1 from reference O. Savichtcheva, S. Okabe, Alternative indicators of fecal pollution: relations with pathogens and conventional indicators, current methodologies for direct pathogen monitoring and future application perspectives, Water Res. 40 (13) (2006) 2463–2476.

TABLE 3.2 Application of different types of indicator organisms for different types of monitoring approaches.

Microorganism(s)	Type of monitoring		
	Validation of process	Operational	Verification and surveillance
E. coli (or thermotolerant coliforms)	Not applicable	Not applicable	Faecal indicator
Total coliforms	Not applicable	Indicator for cleanliness and integrity of distribution systems	Not applicable
Heterotrophic plate counts	Indicator for the effectiveness of the disinfection of bacteria	Indicator for the effectiveness of disinfection processes and cleanliness and integrity of distribution systems	Not applicable
Clostridium perfringens[a]	Indicator for the effectiveness of disinfection and physical removal processes for viruses and protozoa	Not applicable	Not applicable[b]
Coliphages; *Bacteroides fragilis* phages; enteric viruses	Indicator for the effectiveness of disinfection and physical removal processes for viruses	Not applicable	Not applicable[b]

[a] The use of Clostridium perfringens for validation will depend on the treatment process being assessed.
[b] It could be used for verification where source waters are known to be contaminated with enteric viruses and protozoa or where such contamination is suspected as a result of impacts of human faecal waste.
Adapted from Table 7.9 from reference World Health Organisation, Guidelines for Drinking-Water Quality, World Health Organisation, 2011.

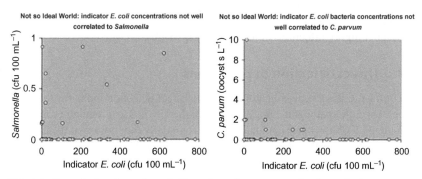

FIG. 3.4   Bacteria detection: schematic illustrating culture versus molecular methods.

been recognised that this approach is not necessarily suitable for the assessment of emerging bacterial pathogens. Furthermore, the absence of coliform does not necessarily indicate the absence of these pathogens. Several reports have indicated the presence of *Salmonella* and *Shigella* in waters where *E. coli* was not recovered [10]. Well maintained water purification and distribution systems can reduce the levels of *Salmonella* and *Shigella* to levels not associated with human illness. Likewise, *Yersina enterocolitica* and *Campylobacter jejuni* have also been found to be poorly correlated to coliform levels and HPC numbers but are reduced to innocuous levels during water treatment.

Legionella spp. are respiratory pathogens that are naturally found in water environments including the cooling tower, hot tubs, and showerheads. Although *E. coli* presence can be used to indicate their potential presence, the ubiquitous nature of *Legionella* spp. makes them hard to correlate to indicator levels at low quantities. Furthermore, although filtration and disinfection methods are successful at reducing their numbers in water treatment, *Legionella*'s ability to become establish biofilms in water systems makes them hard to detect and control. *Mycobacterium avium* presents a similar story as they are organisms with an ability to survive and grow under varied conditions and can persist in biofilms. They are typically more resistant to disinfection normally used in water treatment and for that reason, their numbers are not adequately represented by indicator numbers.

Aeromonas species have been found in a wide variety of water environments including, rivers, lakes, storages reservoirs, sewage effluents, and drinking water sources. Their numbers have been found not to show a relationship to coliform numbers which can lead to false-negative results. *Helicobacter pylori* can be isolated from environmental sources including water. It has been associated with gastritis, duodenal ulcers, and around 50% of gastric cancer. Currently, there are no regulations for the presence of the bacteria in drinking water. Since the coccoid form of this bacteria is non culturable, drinking water is not directly tested for the organisms instead *E. coli* is used as an indicator of their presence [10].

Overall, faecal indicator bacteria (FIB) are still used as the standard for faecal source tracking and managing water quality. More and more studies each year, however, show that faecal bacteria presence and numbers often do not correlate well with pathogens numbers which may be low and patchy in distribution but highly infectious at low doses (Fig. 3.5).

3.4.   Direct detection of pathogens

In contrast to the faecal indicator approach, direct detection looks for the presence of specific pathogens. For example, in the UK it was a regulatory requirement until 2008 for water companies to perform daily monitoring of the finished water to check for the presence of *Cryptosporidium*. Methods for direct detection have become more popular with the growing awareness that faecal indicator organism monitoring alone cannot guarantee water safety.

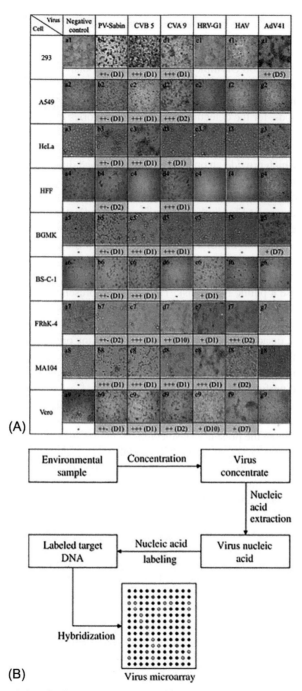

FIG. 3.5   Correlation of indicator organisms with pathogen presence.

However, direct microbiological testing presents a unique challenge, due to low numbers of pathogens. The requirement to detect a single organism in a 100 mL water sample has been compared to the problem of finding a single coffee bean in 40,000 Olympic-sized swimming pools.

This section will cover some of the existing microbiological approaches to direct detection in addition to mentioning other established approaches, such as molecular methods, for the detection of different categories of pathogens. However, advances in molecular methods will be covered in Chapter 8 and the main focus of this book in Chapters 5–10 is upon these advances and other new technologies/techniques which provide an opportunity for improved direct detection.

As described above the traditional method of detection and enumeration of pathogens has relied on selective culture and standard biochemical methods. This has been applied for both the monitoring of indicator organisms as well as the direct detection of specific pathogens. The one potentially important advantage is that, since the monitoring method relies on multiplication and growth of the pathogen, the methods detect viable pathogens. This is very useful if performing validation monitoring of finished water since it is only viable pathogens that pose a public health risk. However, the presence of nonviable pathogens in finished water may also be useful information since it could indicate problems with either the source water or some part of the treatment process. Furthermore, disadvantages are that culturing of many organisms is challenging or at present not yet possible, and each pathogen requires specific conditions, thus rendering a full-scale screen of all potential pathogens a large and expensive assay to perform.

An alternative approach is the application of molecular methods. The principal tool of most molecular-based studies is the use of polymerase chain reaction (PCR) to evaluate the presence of selected pathogens by detection of its specific pathogenic genes or small-subunit ribosomal RNA (SSU rRNA). This enables the detection of unculturable organisms, though one drawback is viability determination. It is important to note that the molecular-based technique could be also used to detect conventional faecal indicators including *E. coli*.

3.4.1 Viruses

Enteric viruses are not routinely monitored in water samples due to practical reasons. Limitations include the need to concentrate large volumes due to low numbers in the samples, the need for specialised equipment for cell cultures, the need for highly trained personal, and the overall cost of the analysis. The prediction of viral contamination relies on the presence of various indicators such as *E. coli*, total coliforms, *Enterococcus*, *Clostridium perfringens* spores, and bacteriophage. However, if viral infections are suspected due to water contamination, direct detection of the enteric viruses can be carried out. Direct testing provides invaluable information to public health authorities and researchers. Standard enteric viral recovery and detection methods are available [19–21].

Monitoring for viruses requires a two-step process, first, the concentration of the sample and second, detection of the virus. Since viral numbers are usually very low, volumes over 100 L for surface water and 1000 L for drinking water have been used to ensure the ability to detect the virus if present [22]. Concentration is usually achieved by filtration through positively charged filters [19] or ultrafiltration although the latter method usually cannot handle larger volumes than 20 L [23]. Often a second filtration method may be necessary to reduce the volume of the sample down to 1–2 mL via organic flocculation, polyethylene glycol precipitation, or ultracentrifugation [23, 24]. Sampling and analytical procedures for the virological analysis of water are well documented [25].

It is difficult to culture viruses since they do not grow outside their host cell. Methods documenting virus infection, and multiplication, in established cell lines, have been developed. This approach indicates infectivity though the methods are time-consuming and require highly-trained technicians.

Historically, cell culture follows the concentration step and is considered the gold standard for determining the infectivity of the isolated virus. A variety of cell lines have been used such as buffalo green monkey kidney (BGMK) cells and human lung carcinoma cells (A549), however, since no single line can propagate all viruses even in the same viral group, many different cell lines have been employed [24]. Furthermore, although some enteric viruses are easy to propagate (Enteroviruses), some are difficult (rotavirus and hepatitis A virus) and for some no cell line is available (Noroviruses) [24].

Cell culture propagation of wild-type strains of hepatitis A (HAV) is a complex and tedious task, which requires virus adaptation before its effective growth. Even then the virus usually establishes persistent infections, which results in low virus yields. In 2006, the use of a cell line that allows the growth of a wild-type HAV isolates from stool was reported by Konduru and Kaplan [26], although its validity for broad-spectrum isolation of HAV is not yet demonstrated. For norovirus (NoV) the prospects for cell culture are even worse and despite numerous attempts, NoV propagation has only been shown in an intestinal epithelium which provided unconfirmed expectations for infectious NoV detection [27].

Typically the cell cultures are monitored daily by microscopy for cytopathogenic or cytopathic effect (CPE) for up to 1–3 weeks [24]. CPE can be manifested in a variety of ways such as swelling, shrinking, rounding of cells, or any clustering or destruction of the monolayer. If the suspension of the virus is serially diluted and inoculated unto the cell cultures, then quantification of the virus can be carried out with a most probable number (MPN), or plaque assay method [24]. Standard methods for enteric viral detection are available and used by laboratories that are equipped to handle such analysis [19–21]. A major limitation of using standard cell culture assays for routine pathogen detection in water is that they are laborious, expensive, and the conditions for propagation of cell lines not well established for some pathogenic strains.

Nucleic acid amplification techniques are currently the most widely used methods for detection of viruses in water and food. The advent of molecular methods enabled the development of diagnostic tools with exquisite sensitivity for the detection of human pathogenic viruses that do now grow in cell cultures. In this way, the significance of the enteric viruses, for which cell culture methods were troublesome, such as hepatitis A and E viruses, rotaviruses, and, particularly, noroviruses as water and foodborne agents could be ascertained [28]. Immunological detection systems for NoV detection in faecal specimens do exist [29]. However, even when kits for antigen detection are available, their sensitivity is not high enough to be employed in scenarios of low virus concentration, such as water samples. A 2012 review of both molecular methods and other approaches is given in [24] covering both established and developing techniques.

3.4.2 Bacteria

As described above there are many problems with indicator bacteria testing. First, there is the possibility to underestimate bacterial density especially when bacteria are physically injured or stressed. For instance, slow lactose-fermenting or lactose-negative *Enterobacteriaceae*, including pathogenic *Salmonella* spp. and *Shigella* spp. could be underestimated by standard coliform test, which leads to false-negative conclusions. Furthermore, some microorganisms in the environment are viable but either difficult to culture or nonculturable.

This section briefly introduces alternative methods to detect bacteria, especially those looking at the direct detection of different bacterial species. Many waterborne pathogens, such as all Vibrio species, have to be identified beyond the species level to determine their pathogenic potential [5, 30]. Molecular methods provide a good approach to determine species, though viability is often challenging to obtain.

Application of the PCR-based methods detected the presence of *Salmonella* spp., *Shigella* spp., *Campylobacter* spp., and different pathogenic strains of *E. coli*. More details on molecular methods are given in Chapter 8.

3.4.3 Protozoan detection

The major protozoan pathogens found in water contamination are *Cryptosporidium*, *Giardia*, microsporidia, *Entamoeba histolytica*, and *Cyclospora*. Routine monitoring of water samples for protozoan parasites relies on the detection of faecal indicator bacteria (FIBs) although direct detection of the protozoan can be monitored using immunofluorescent antibodies and microscopy [31].

Since many outbreaks of disease have been attributed to *Cryptosporidium* and *Giardia*, especially in the UK and the US (most likely not indicative of worse water treatment regimes here, but perhaps detailed public health surveillance), there are standard, validated monitoring methods which are widely adopted, especially at all UK water suppliers.

The standard recovery and detection protocol includes a concentration step, usually through filtration, followed by a purification step, involving density gradients, and finally an immunofluorescent staining step which increases the chances that the (oo)cysts can be detected by microscopy. This protocol, 'information collection rule' (ICR) for protozoa, was developed as a standardised method for the collection and quantification of *Giardia* cysts and *Cryptosporidium* oocysts [19]. It is still being used by some labs and has been adopted as a standard method for the detection of other parasites as well from effluents, surface and groundwater sources. However, due to low recovery rates associated with the ICR method, a newer method called Method 1623 was developed. The main difference is the use of immunomagnetic separation (IMS) which allowed for more efficient separation of the organisms from other debris resulting in a cleaner slide for microscopy.

Method USEPA 1623 describes the filtration, concentration, and subsequent detection of these pathogens, and illustrated in Fig. 3.6. Like with faecal indicator organisms, filtration is used to capture the organisms. However, since infectious doses are so low and the organisms will not multiply outside a host, large water volumes are sampled to increase detection likelihood. Attempts to develop culture methods have not yet met with great success [32]. The first stage thus comprises filtration of 1000 L, over 24 h. This cartridge filter is then eluted to release the (oo)cysts and a second filtration step is performed, this time with a membrane filter. Next, the (oo)cysts are subjected to centrifugation, followed by specific separation using immunomagnetic beads. Finally, the (oo)cysts are removed from the beads, placed on a microscope slide and fluorescently stained before they are examined under the microscope by a highly trained technician.

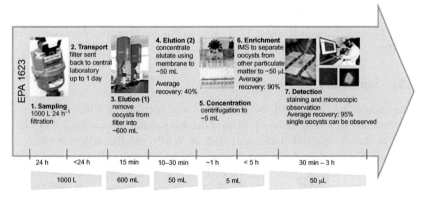

FIG. 3.6 Overview of the US EPA 1623 Method for the detection of *Cryptosporidium*. To clarify in Step 3 a minimum of 2 washes of 600 mL each should be performed. Thanks to Scottish Water for the provision of the images. *Adapted from H. Bridle, M. Kersaudy-Kerhoas, B. Miller, et al. Detection of Cryptosporidium in miniaturised fluidic devices, Water Res. 46 (6) (2012) 1641–1661.*

Each stage of the monitoring process reduces the initial 1000 L to smaller and smaller volumes eventually reaching the 50–100 µL for the microscope slide. Throughout the process, there are significant losses of (oo)cysts with acceptable recovery rates for this method at just over 10%. Water companies in the UK report around 30% recovery as an average. It is clear from the above discussion that sample processing is key to the effective monitoring of waterborne protozoan. It is extremely unlikely that any detection technology could reliably detect the single (oo)cyst level in 1000 L. This topic is discussed further in Chapter 4.

Drawbacks of this approach are the time required for detection, the potentially low recovery rates, the need of expensive fluorescent reagents as well as the requirement for highly trained technicians and lab equipment (especially the microscope). The protocol can take up to 3 days and no indication of species or viability is given. Molecular methods have been developed for these pathogens though there is not yet agreement over what are the most appropriate nucleic acid targets, and whether it is possible to determine viability simultaneously. Furthermore, reliable single (oo)cyst detection is not yet available. Developments in molecular methods are covered in Chapter 8.

Cyclospora can be recovered with the methods used for *Cryptosporidium*, however, no vital dye assay is currently available so samples require a 2 week incubation period to check for sporulation to determine viability [31]. Acid-fast staining techniques for the diagnosis of the organism are also problematic and not recommended for the detection of *Cyclospora* in environmental samples.

Microsporidium is harder to detect with the current methods due to the small size of their spore although [33] was able to detect *Microsporidium* from drinking water treatment and wastewater treatment plants using Weber's stain. Identification of the species, however, was not possible and currently relies on PCR methodology with primers specific for *Enterocytozoon bieneusi*, *Encephalitozoon intestinalis*, *Encephalitozoon hellum*, or *Encephalitozoon cuniculi* for specific identification [33]. This technique, however, does not assess viability.

3.5. Summary

This chapter has presented the guidelines and regulations which aim to ensure resource-effective delivery of safe drinking water. Monitoring plays key roles in the design and implementation of WSPs and the different types of monitoring have been described. It is clear that different techniques are more appropriate to certain types of monitoring and that existing technologies need improvement to enable wider application in all areas, though especially in operational monitoring. One question this book will address is what new detection methods are available or developing which could meet this need and will assess the current state-of-the-art, forming a judgement upon how feasible the vision of networked sensors reporting on microbial aspects of water supply for operational decision making is. Is this something we might see in the next 10 years or the next 50?

This chapter has presented the existing detection technology, based on indicator monitoring and highlighted many of the drawbacks and shortcomings of this approach. However, this is a well established, widely adopted method that is relatively cheap and easy to use. Therefore, new methods will have to offer significant advantages and be available at a reasonable cost for water sector adoption. This topic is covered further in Chapter 11.

This chapter has also described the direct detection of the various classes of pathogens. Both the existing culture and microscopy-based detection approaches as well as emerging molecular methods, which have been explored for a wide range of pathogens, have been mentioned.

At several points throughout this chapter, the sampling and sample processing part of monitoring has been highlighted as key to the recovery rate and detection limit achieved. Therefore, Chapter 4 will describe the state-of-the-art in sample processing for waterborne pathogen monitoring before the detection section, which focuses on new detection methods (Chapters 5–10).

References

[1] World Health Organisation, Guidelines for Drinking-Water Quality, World Health Organisation, 2011.

[2] M.J. Gunnarsdóttir, L.R. Gissurarson, HACCP and water safety plans in Icelandic water supply: preliminary evaluation of experience, J. Water Health 6 (2008) 377–382.

[3] I.J. Allan, B. Vrana, R. Greenwood, G.A. Mills, B. Roig, C. Gonzalez, A toolbox for biological and chemical monitoring requirements for the European Union's Water Framework Directive, Talanta 69 (2) (2006) 302–322.

[4] E. Harth-Chu, R.T. Espejo, R. Christen, C.A. Guzman, M.G. Höfle, Multiple-locus variable number of tandem-repeats analysis for clonal identification of *Vibrio parahaemolyticus* isolates using capillary electrophoresis, Appl. Environ. Microbiol. 75 (2009) 4079–4088.

[5] A.D. Robinson, D.E. Falush, J. Feil, Bacterial Population Genetics in Infectious Diseases, Wiley-Blackwell/John Wiley & Sons, Hoboken, NJ, 2010.

[6] K.G. Field, M. Samadpour, Fecal source tracking, the indicator paradigm, and managing water quality, Water Res. 41 (16) (2007) 3517–3538.

[7] C. Hagedorn, A.R. Blanch, V.J. Harwood, Microbial Source Tracking: Methods, Applications, and Case Studies, Springer Ed., New York, 2011.

[8] C. Gleeson, N.F. Gray, The Coliform Index and Waterborne Disease: Problems of Microbial Drinking Water Assessment, Taylor & Francis, London, 1996.

[9] I. Brettar, M.G. Höfle, Molecular assessment of bacterial pathogens—a contribution to drinking water safety, Curr. Opin. Biotechnol. 19 (2008) 274–280.

[10] Health Canada, Guidelines for Canadian Drinking Water Quality – Summary Table, Water, Air and Climate Change Bureau, Healthy Environments and Consumer Safety Branch, Ottawa, Ontario, 2006.

[11] Part 7—Methods for the Enumeration of Heterotrophic Bacteria by Pour and Spread Plate Techniques. The Microbiology of Drinking Water, 2002.

[12] O. Savichtcheva, S. Okabe, Alternative indicators of fecal pollution: relations with pathogens and conventional indicators, current methodologies for direct pathogen monitoring and future application perspectives, Water Res. 40 (13) (2006) 2463–2476.

[13] F. Hammes, F. Goldschmidt, M. Vital, Y. Wang, T. Egli, Measurement and interpretation of microbial adenosine tri-phosphate (ATP) in aquatic environments, Water Res. 44 (2010) 3915–3923.

[14] A. Lundin, Use of firefly luciferase in ATP-related assays of biomass, enzymes, and metabolites, Methods Enzymol. 305 (2000) 346–370.

[15] P.W. van der Wielen, D. van der Kooij, Effect of water composition, distance and season on the adenosine triphosphate concentration in unchlorinated drinking water in the Netherlands, Water Res. 44 (2010) 4860–4867.

[16] A.S. Bukh, P. Roslev, Characterization and validation of a chemiluminescent assay based on a 1,2-dioxetane for rapid detection of enterococci in contaminated water and comparison with standard methods and qPCR, J. Appl. Microbiol. 111 (2) (2011) 407–416.

[17] P.R. Hunter, Waterborne Disease: Epidemiology and Ecology, Wiley, Chichester, 1997.

[18] S. Percival, M. Embry, P. Hunter, R. Chalmers, J. Sellwood, Microbiology of Waterborne Diseases: Microbiological Aspects and Risks, Academic Press, London, 2004.

[19] USEPA, ICR Microbial Laboratory Manual, EPA-600-R-95-178, United States Environmental Protection Agency, Washington, DC, 1996.

[20] APHA, Standards Methods for the Examination of Water and Wastewater, American Public Health Association, Washington, DC, 1998.

[21] ASTM, ASTM D244-92: Standard Practice for Recovery of Enteroviruses from Waters, American Society for Testing and Materials, Philadelphia, PA, 2004.

[22] T.M. Straub, D.P. Chandler, Towards a unified system for detecting waterborne pathogens, J. Microbiol. Methods 53 (2) (2003) 185–197.

[23] Health Canada, Canadian Drinking Water Guidelines, Government of Canada, 2012.

[24] I.A. Hamza, L. Jurzik, K. Überla, M. Wilhelm, Methods to detect infectious human enteric viruses in environmental water samples, Int. J. Hyg. Environ. Health 214 (2012) 426–436.

[25] C.J. Hurst, K.A. Reynolds, Detection of viruses in environmental waters, sewage and sewage sludges, in: C.J. Hurst, R.L. Crawford, G.R. Knudsen, M.J. McInerney, L.D. Stetzenbach (Eds.), Manual of Environmental Microbiology, ASM Press, Washington D.C, 2002, pp. 244–253.

[26] K. Konduru, G. Kaplan, Stable growth of wild-type hepatitis a virus in cell culture, J. Virol. 80 (2006) 1352–1360.

[27] T.M. Straub, K. Höner zu Bentrup, P.O. Orosz-Coghlan, et al., In vitro cell culture infectivity assay for human noroviruses, Emerg. Infect. Dis. 13 (2007) 396–403.

[28] B.A. Lopman, M.H. Reacher, Y. van Duijnhoven, F.X. Hanon, D. Brown, M. Koopmans, Viral gastroenteritis outbreaks in Europe, 1995–2000, Emerg. Infect. Dis. 9 (2003) 90–96.

[29] R.L. Atmar, M.K. Estes, Diagnosis of noncultivatable gastroenteritis viruses, the human caliciviruses, Clin. Microbiol. Rev. 14 (2001) 15–37.

[30] E. Harth-Chu, R.T. Especio, R. Christen, C.A. Guzman, M.G. Höfle, Multiple-locus variablenumber of tandem-repeats analysis for clonal identification of *Vibrio parahaemolyticus* isolates using capillary electrophoresis, Appl. Environ. Microbiol. 75 (2009) 4079–4088.

[31] W. Quintero-Betancourt, E.R. Peele, J.B. Rose, *Cryptosporidium parvum* and *Cyclospora cayetanensis*: a review of laboratory methods for detection of these waterborne parasites, J. Microbiol. Methods 49 (2002) 209–224.

[32] P. Karanis, H.M. Aldeyarbi, Evolution of *Cryptosporidium* in vitro culture, Int. J. Parasitol. 41 (12) (2011) 1231–1242.

[33] F. Izquierdo, J.C. Hermida, S. Fenoy, M. Mezo, M. González-Warleta, C. del Aguila, Detection of microsporidia in drinking water, wastewater and recreational rivers, Water Res. 45 (2011) 4837–4843.

Section B

Sample processing

Chapter 4

Sample processing

Helen Bridle[a], Karin Jacobsson[b], and Anna Charlotte Schultz[c]

[a]*Institute of Biological Chemistry, Biophysics and Bioengineering, Heriot-Watt University, Edinburgh, Scotland,* [b]*Mikrobiologienheten, Uppsala, Sweden,* [c]*DTU Food, National Food Institute, Technical University of Denmark, Søborg, Denmark*

While this book focuses on emerging methods for the detection of waterborne pathogens, this chapter is devoted to sample processing which plays a key role in the ultimate success of any detection technology. Many of the techniques discussed in the following chapters process µL to a few mL, whereas it may be necessary to sample hundreds of mL to thousands of litres. Large sample volumes are required to gain a more representative sample of a large water volume as well as to increase the likelihood of detecting pathogens present at very low concentrations. Sampling of a large volume over a period of time will to some extent compensate for some spatial and temporal variations in pathogen distribution. Additionally, the concentration aspect of sample processing is necessary to bring the pathogen concentration into the detection limit of monitoring methods. Furthermore, while some techniques would detect single pathogens, the time required for a single pathogen in a large volume of water to reach the detection area/surface could be prohibitively long. Sample processing is also important for isolation of the pathogen of interest and for the removal of interferents which could disturb the detection process. The nature of problematic interferents varies with the final monitoring technique employed and therefore different sample processing methods may be required depending on the full scheme of detection.

This chapter will introduce different methods of sampling and give an introduction to the basic processes behind different sample processing techniques. The chapter is organised according to process order. First, initial methods of concentration are presented followed by secondary concentration steps, which further reduce the volume and in some cases isolate the pathogen of interest. There is some overlap between techniques used in these stages. Next, a section is devoted to the extraction of nucleic acids for detection by molecular methods. Analytical controls are important to determine how well a process is working, in terms of recovery rate, and different approaches to this are discussed. The chapter concludes with a summary identifying what has so far been achieved in the area of sample processing, what has so far been neglected and what needs to

Waterborne Pathogens: Detection Methods and Applications. https://doi.org/10.1016/B978-0-444-64319-3.00004-6

be done. This summary will discuss the challenges and limitations, timescales for the adoption of new technology and in addition to suggesting what improvements are most needed will consider how likely they are.

4.1. Background

In this section, we will first present different methods of sampling and give an overview of sample processing as applied to the different categories of pathogen. More detailed information on the techniques, including emerging research-stage methods, as they are applied to waterborne pathogens is given in subsequent sections. The basics of the physical and chemical processes underlying the techniques are also described in the various sections, with suggestions for more detailed reading.

To enable the detection of low amounts of microbial pathogens in water, large volumes (40–100 L) need to be analysed. This can be achieved by various filtration techniques that reduce the water volume while retaining the microorganisms. Available filtration methods to concentrate waterborne pathogens have many disadvantages; they are either too costly for studies requiring large numbers of samples, limited to small sample volumes, or not very portable for routine field applications.

Glass wool filtration is a cost-effective and easy-to-use method mainly used to retain viruses, though one recent study reports application to bacteria and protozoa [1], however, its efficiency and reliability are not adequately understood [2–4]. NanoCeram are cost-effective newly developed electropositive pleated microporous filters composed of microglass filaments coated with nanoalumina fibres. Most bacteria are large enough to be mechanically filtered or retained by, e.g. adsorption of *Bacillus* spores [5] while smaller particles such as viruses are retained principally by electroadhesion, which has proved useful for the isolation of viruses from water [6, 7]. Hollow fibre ultrafilters normally have a cutoff of ~30 kDa and will retain parasites, bacteria, and viruses. They are disposable thus avoiding the risk of cross-contamination and filtration can either be achieved by coupling the filter directly to a tap or pumping an environmental water sample through the filter. They have mainly been used for filtration of tap water [8, 9] but have also been used for surface waters [10, 11].

For all three techniques, the microorganisms need to be eluted from the filters. Filtration, as well as elution, needs to be optimised to give a sufficient recovery of a control panel of model organisms representing bacteria, viruses, and parasites, from treated as well as untreated ground and surface waters.

Secondary concentration procedures will differ between the kingdoms. Thus, it may not be possible to deliver a universal protocol from this step, for both this reason and, as discussed earlier, the fact that different detection methods place different requirements on sample processing. For example, bacteria and parasites can be concentrated by, e.g. pelleting by centrifugation and viruses by polyethylene glycol (PEG)-precipitated, flocculated or (ultra)filtrated [10, 11] by various techniques from the supernatant.

Significant amounts of organic and inorganic inhibitors are known to be enriched during the concentration of water samples. However, molecular detection and quantification require purified nucleic acids with the absence of reverse transcription (RT) polymerase chain reaction (PCR) inhibitors. Various kits for the preparation of ribonucleic acid (RNA)/deoxyribonucleic acid (DNA), post-lysis additives, and PCR facilitators have been investigated [12] but no universal method has been presented.

Different types of water (quality, source, natural environment, etc.) can be expected to contain different types, and concentrations, of inhibitors. Further, some organisms, such as bacterial endospores and parasite oocysts/cysts, are difficult to lyse and this needs to be considered in the development of extraction techniques.

Important factors in the analysis of any sample processing technique are the degree of concentration achieved, the pathogen recovery rate, the amount of operator skill required, the number of steps, what types of water the method is appropriate for, and the specificity of isolation. Ideally, a good sample processing procedure for water samples should meet the following criteria (adapted from Ref. [13]):

1. Consistent recovery rate of all types of organisms analysed;
2. Removal of inhibitory or cross-reacting substances that are widely present in natural water to avoid false negative and positive results when molecular detection is used;
3. Generation of an end-product that can be utilised directly for bacterial cultivation, microscopic detection of parasites, in viral culture or for molecular detection;
4. Equivalent recovery for DNA and RNA viruses;
5. Applicable to a wide range of environmental water matrices, e.g. drinking water and wastewater
6. Simple and inexpensive procedure that can be carried out in most environmental or water quality laboratories (Box 4.1).

The Water Framework Directive (WFD) is based on a risk assessment exercise, where precision and confidence are described respectively as how close the measured value(s) or indicator is to the true value and the likelihood that this measured value is within the defined precision. Adequate precision and confidence in values obtained through monitoring are essential to allow an acceptable level of risk in the decision-making process [14].

4.2. Sampling

Sampling is a critical step in the monitoring procedure because it could represent the main contribution to an error in the whole analytical process [15]. As a consequence, appropriate sampling tools and methods are required, which are easy to use and reliable, taking into account flow variations and allowing

BOX 4.1

Definitions of:

Recovery Rate: this refers to the efficiency of the process in capturing and concentrating all of the pathogens of interest; i.e. a highly successful process would have a recovery rate of 100% indicating that all of the pathogens in the original sample are in the concentrated sample; recovery rates are in general significantly lower than this and in some cases rates as low as 10%–20% are considered acceptable. Recovery rate is particularly important for pathogens to present in low numbers as if the rate is low any pathogen in the sample might be lost during the sample processing stage.

Degree of Concentration: this indicates to what extent the sample has been concentrated, for example, a concentration factor of 10^2 means that the volume of the original sample has been reduced 100 times, so if a 100 mL sample was used it would be concentrated to 1 mL.

Specificity of Isolation: this term refers to how selective the method is for the pathogen of interest and applies to techniques aiming to enrich the presence of a certain pathogen from mixed samples, e.g. immunomagnetic separation that depends upon the cross-reactivity of the antibody to confer specificity; antibody specificity can vary between different pathogens.

Water/sample matrix: this describes the type of water the sample is taken from, e.g. raw source water from rivers, lakes, reservoirs or groundwater; finished drinking water; wastewater. Evidently, these different water types will contain different amounts of interferents (inorganic, organic, microbial matter) and have different physicochemical characteristics (T, pH, conductivity, etc.), which could influence the performance of different sample processing techniques.

False positives: results indicating pathogen presence when in fact the sample is pathogen-free

False negatives: resulting indicating that the sample is pathogen-free when in fact it contains pathogens

Precision: how close the measured value(s) or indicator is to the true value

Confidence: the likelihood that this measured value is within the defined precision

an adequate number of samples to be taken. A review of this topic by Roig and colleagues in 2011 provides an excellent further introduction to the topic [15].

When sampling procedures for microorganisms in drinking, irrigation, and food processing water are considered, the questions which arise are (i) will a specific volume of water be representative, and (ii) will the water quality influence the sampling procedure? Pathogen presence in water often displays a high degree of spatiotemporal variability. For example, heavy rainfall and associated floods, if occurring, can flush microorganisms into waterways and aquifers. Greater water flow contributes to increasing the pathogen load and the penetration speed of pathogens into resource waters and the drinking-water supply [15].

Furthermore, the techniques able to process relatively clean finished, drinking water differ from those suitable for potentially 'dirty' source waters, which might contain high amounts of particulate and other matter. There is also a high degree of variability between different source waters and techniques which work well in one catchment, or even section of a catchment might perform less well in other locations.

The selection of an appropriate sampling protocol depends primarily on a clear identification of monitoring objectives (Fig. 4.1). This will influence what to monitor, where, when, and how often. Additionally, the monitoring aims will determine the type of detection technology that is appropriate, thus having knock-on effects upon the sample size and processing techniques. The rest of the chapter will describe sample processing methodologies whereas this section will focus on the sample collection aspect. An overview of the types of field-based sampling/monitoring is given in Box 4.2.

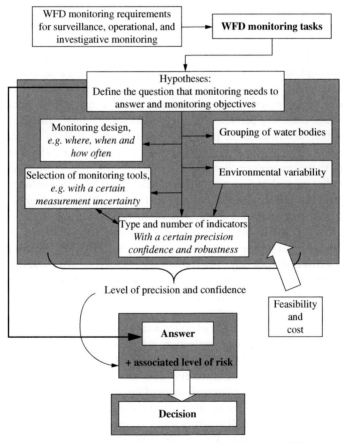

FIG. 4.1   Framework and parameters that need to be considered for successful implementation of biological and chemical monitoring under the Water Framework Directive (WFD) of the European Union (EU). *(Overview for Figure 1 from I.J. Allan, et al., Strategic monitoring for the European water framework directive, Trends Anal. Chem. 25(7) (2006).)*

BOX 4.2

Three main field methods are commonly differentiated: in situ; on-line; and, off-line [15]:

(1) In situ techniques require the use of sensors placed in the water body with no sampling step.
(2) On-line methods involve a sampling step, followed generally by pretreatment and measurement; all these steps could be automated.
(3) Off-line methods generally require a conventional spot-sampling step, followed by an analytical method that is faster and/or less expensive than commonly-used laboratory methods.

(Adapted from B. Roig, et al., Analytical issues in monitoring drinking-water contamination related to short-term, heavy rainfall events, Trends Anal. Chem. 30(8) (2011) 1243–1252.)

Water samples can be obtained through grab or spot, sampling of a certain volume at a particular time or through the collection of a sample over a time period, sometimes described as passive or automatic sampling. The first approach rapidly obtains a sample that can be analysed whereas the second has the advantage of providing a degree of evening out of temporal variations in pathogen concentration. Monitoring of raw water sources is commonly done through grab sampling at different points, collecting perhaps 10L. *Cryptosporidium* testing, with the US EPA 1623 method, described in the previous chapter, is an example of sampling over time, where 1000L of finished drinking water is filtered over 24h before being subjected to further analysis. Both of the above methods provide batch samples.

Periodic, manual grab sampling is considered a relatively inefficient methodology for capturing mean concentrations of water subjected to highly variable loads [15]. The alternative is continuous, or online, monitoring. There are some examples of this for chemical monitoring of water quality, with fewer examples for microbiological testing [16]. One example is the PREDECT system, developed by a Swedish company. The technology is based on laser scanning of the water to detect particles as they pass by, triggering the capture of a sample if pathogen presence is suspected, and is designed to be located on the pipeline after the water treatment process.

There are two different types of collection regimes. The first and most commonly used is time-controlled collection, where samples are taken at regular time intervals, e.g. every 24h. The second approach, which might be more suitable during heavy rainfall, or flood events, is flow controlled collection where samples are taken at regular intervals, according to the flow rate.

If in case of outbreak investigations, the time of sampling coincides with the peak of contamination, small volumes up to 1 L may be adequate for the detection of the causative agent(s). For quality control or risk assessment purposes, larger volumes may be required as many kinds of human pathogens (in particular human viruses) exist at different levels in the environment where they can retain their infectivity for a long duration and be transmitted by the water route.

In general, the first approach in the above box is not applicable for waterborne pathogens, especially those present at low concentrations. Therefore, for both field testing and laboratory analysis, sample processing of the collected sample is essential to concentrate it to a level where detection can be satisfactorily performed. Different concentration techniques and how they can be applied to the various classes of waterborne pathogen is discussed in the next section.

4.3. Concentration techniques

4.3.1 Ultrafiltration

In ultrafiltration, size-exclusion filters with a pore size in the range of 10–70 kDa are used. Due to the small pore size, all types of microorganisms including viruses, but also macromolecules such as proteins, are retained in the filter. Most commonly, the filters consist of hollow fibres made from materials such as polysulfone, cellulose acetate or polyacrylonitrile and with surface areas of several m^2, enabling filtration of large volumes also when turbid surface waters are sampled. As the water sample is pumped into the fibres, a counter-pressure forces water and particles/molecules smaller than the pore size through the membrane (permeate or filtrate) while microorganisms are retained inside the fibres (retentate). These are collected directly, or by elution or back-flushing of the filter, after filtration is complete. The volume of the eluate will differ depending on the inner volume of the filter and elution technique, but usually secondary concentration is required before the detection of the microorganisms.

Two different setups are used: (1) tangential flow or recirculating filtration and (2) dead-end filtration (DEUF) (Fig. 4.2). In tangential flow filtration, the water sample is continually circulated through the filter until the volume has been reduced to a few 100 mL which is collected with or without an elution step. This setup requires a pump which means that the water sample has to be collected in containers and either brought to the laboratory for processing, or a portable pump with power supply has to be used. However, the continuous scouring of the filter is believed to reduce filter fouling and prevent adhesion of microbes to the filter and this is the technique most commonly used today. DEUF is more similar to membrane filtration where the sample passes the filter once. This technique is easier to perform for the nonexperienced operator. It has also been automated [17, 18] and a portable system is commercially available (Portable Multiuse Automated Concentration System (PMACS), IntelliSense Design, Inc., Tampa, FL, USA). In DEUF, the filter may be coupled directly to a tap or water inlet which may be useful in outbreak situations where samples may need to be collected in areas far from the laboratory, and transporting large volumes of water under cold conditions unrealistic. However, collecting water samples in containers has the advantage that various chemicals can be added to the water to improve the dispersal of microorganisms and prevent adhesion to the filter.

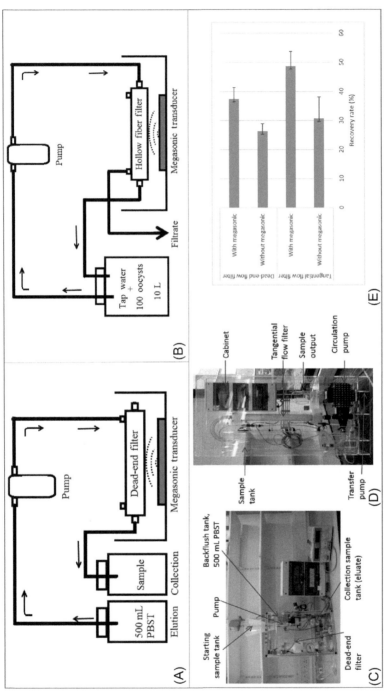

FIG. 4.2 Schematic and images of filtration set-ups. (A) Dead end ultrafiltration. (B) Tangential flow ultrafiltration where the water is recirculated through the filter while the tightened screw clamp forces a part of water to cross the filter and out through the upper port. (C) Automated DEUF set-up. (D) Automated tangential flow set-up. (E) Data reporting enhancement of performance with megasonic elution. *(Reproduced with permission from H. Bridle, A. Kerrouche, M.P.Y. Desmulliez, Megasonic elution of waterborne protozoa enhances recovery rates, Matters (2018).)*

Many different ultrafilters (UFs) are commercially available, differing in the filter material, pore size, and filter area. Holowecky et al. compared five different UFs (APS Series Dialyzer and Rexeed (Asahi Kasei), Exeltra Plus 210 (Baxter Corp), F200NR (Fresenius Medical Co.), and HPH 1400 (Minntech Corp.)) with cutoffs ranging from 20 to 70 kDa using tangential flow filtration combined with back-flushing [19]. The test included three different finished drinking waters seeded with vegetative bacteria, bacterial endospores, bacteriophages, and *Cryptosporidium* oocysts but no additives. Before tangential flow filtration, the entire system was flushed with 0.1% bovine calf serum albumin to reduce nonspecific binding. None of the filters performed better or worse than the others for any type of microorganism or water sample.

Automated systems have also been developed (Fig. 4.2), with some aiming to offer inline monitoring solutions, coupling the filtration step to secondary concentration and detection protocols [20]. The HOLM system reported 42% and 70% recovery, respectively, of bacteriophage MS2, for the tangential and dead-end operation of an UF concentrating from 1000 to 20 L. The Aquavalens project also developed automated UF applied to all three kingdoms for simultaneous concentration, using a Rexeed filter. Tangential UF was reported to offer 66%–95% recoveries from a range of viruses, bacteria, and protozoan species, with the study noting little correlation between water quality parameters and UF performance [21]. Relatively high, in comparison to existing literature, *Cryptosporidium* recoveries were reported, suggesting UF could potentially be a useful alternative to traditional EPA1.623 procedures [21], with protocol details being recently published [22]. Automated UF was evaluated as the most successful approach in a study comparing UF, DEUF, NanoCeram, and glass wool for all three kingdoms, and was shown to operate well with up to 200 L lake water samples [21, 23].

To prevent interactions between the filter and microorganisms, different filter pretreatments, alone, or in combination with chemical additives in the water sample, may be used. The filters can be coated with proteinaceous solutions to prevent electrostatic or hydrophobic interactions. Repulsion between negatively charged microorganisms and the filter surface may be increased by pretreatment of the filter with sodium polyphosphate (NaPP) and interactions with hydrophobic surfaces can be prevented by pretreatment with surfactants like Tween 80. The latter two can also be added to the water before filtration and also to the elution/backflush solution to improve efficiency.

Winona et al. investigated the effect of different proteinaceous pretreatments (BSA, beef extract, fetal bovine serum (FBS) and nutrient broth) on the recovery of three viruses in different types of water buffered with phosphate-buffered saline using tangential flow filtration [24]. They concluded that all treatments reduced viral adsorption to the filter compared to unblocked filters with the overall highest recovery for all three viruses for 1% FBS. FBS is a commonly used blocking agent but accidental contamination during blocking may result in microbial growth during storage and transport in the presence of FBS. Therefore, Hill et al.

investigated the use of NaPP as an alternative blocking agent for the recovery of various vegetative bacteria, bacterial endospores, viruses, and *Cryptosporidium* oocysts and no statistically significant difference was observed between filters coated with 5% FBS and 0.1% NaPP when the water and the backflush solution was amended with 0.1% NaPP [25]. Decreasing the NaPP concentration to 0.01% gave recoveries not significantly different from those with 0.1%. In the same study, it was also shown that the addition of 0.01% Tween 80 and Tween 20 to the back-flush solution increased microbial recovery while the addition of Tween 80 to the water sample increased fouling of the filter which decreased the permeate flow, thus resulting in longer filtration times. Whenever Tween was used, also 0.001% Antifoam A was added to prevent foaming. Worth noting is that it also was shown that no inhibition of PCR, or reverse-transcription PCR, was observed when a solution of 0.01% NaPP, 0.1% NaPP, or 0.01% Tween 80 was seeded with two different viruses, suggesting that the additives would not impair molecular detection.

Polaczyk et al. investigated the effect of seed level and NaPP in the water and compared the direct collection of the retentate with elution and backflushing, respectively, for recovery of vegetative bacteria, endospores, bacteriophages, *Cryptosporidium* oocysts and microspheres [8]. Seeding with 100 and 1000 cfu/pfu/oocysts/particles gave very similar results for all organisms. The addition of 0.01% NaPP to the water increased the mean recovery for all organisms from 46% to 71% but a significant difference was seen only for *Bacillus atrophaeus* endospores. For no single microorganism did NaPP decrease recovery. The difference in average mean recovery for all organisms between backflushing and elution was small and not significant but backflush increased recoveries of MS2-phages, *Salmonella enterica*, *B. atrophaeus* endospores, and *Cryptosporidium parvum* oocysts. However, in some cases elution decreased recovery compared to direct collection of the retentate while backflushing always resulted in an increase. An alternative approach to enhance elution from filters is the utilisation of megasonic energy or ultrasound, both of which enabled elution into smaller volumes, thus facilitating a reduction in the subsequent number of processing steps required [26, 27]. The megasonic approach was also applied to automated DEUF, with a Rexeed filter, and automated tangential UF, with a Fresenius filter, for *Cryptosporidium*, improving recoveries by 50% compared to operation without megasonic (Fig. 4.2E) [28].

Comparisons of recovery rates between studies are very difficult since so many parameters differ, and methods determining recovery efficiencies [29] and limits of detection vary [30]. Filters (brands and sizes), water quality, seeding organisms (levels and detection methods), blocking conditions, filtration method and sample amendments, elution techniques, and buffer composition, can be combined in an endless number of ways. Some examples of recovery rates from seeding experiments using different types of water and filtration/elution techniques are given in Table 4.1.

4.3.2   NanoCeram

A recently available NanoCeram cartridge filter (Argonide Corp.), containing nano alumina fibres, has the capacity to adsorb nanoparticles, latex microspheres, bacteria, viruses, and bacteriophages from water over a wide range of pH, turbidity, and salinity conditions [38]. NanoCeram cartridge filters have been demonstrated to be useful as a primary concentration step for enteroviruses and noroviruses from source waters, having similar or better efficiency in the recovery of those viruses as the commonly used electropositive 1MDS cartridge filters [7]. Furthermore, the unit cost of the NanoCeram cartridge filter is 1/10th of 1MDS cartridge filter [13].

In general, studies consistently show RNA viruses to have a higher recovery rate as compared to DNA viruses, [6, 7, 13] regardless of different eluting solutions. In 2011, Lee evaluated various elution buffers to release noroviruses (MNV/NoV) from the NanoCeram micro filters and found the most efficient recoveries using beef extract—Tween 80 (0.01%) (23.4/85.7%) opposed to beef extract—glycine (0.05 M) (18.3/26.5%) [39]. Further, in 2010, Gibbons et al. found that the elution of AdV/NoV could be increased from 1.0/88% to 1.4/119% if the concentration of Tween 80 was increased from 0.01% to 0.1% while using the beef extract—glycine buffer, a second rinse for 15 min improved poliovirus recoveries from 48% to 77% [7].

Using organic flocculation (sodium phosphate pH 9.0–9.5) as a secondary concentration, Lee found a higher recovery rate of MNV/NoV GII-4 (42.0/86.6%) opposed to precipitation with PEG 8000 (18.6/73.6%) and 6000 (12.5/11.1) [39].

A reliable reproducibility in the recovery of RNA and DNA viruses both as infectious viral particles and viral genome has been reported [13]. This consistence in the recovery of viral nucleic acids from different enteric viruses and enteroviruses was also found by Karim et al. and Ikner et al. [6, 7], while differences in recoveries of viral nucleic acids (0.4%–20%) and infectious viruses (21%–182%) have been reported [40].

The NanoCeram filter is capable of adsorbing virus in a wide range of turbidity and salinity conditions as well as pH [38], which was confirmed in studies showing no significant difference in recoveries of AdV from the deionised, tap and river water samples [13] or poliovirus from tap water with a pH range of 6–9.5 [7]. In contrast, based on NoV detection using NanoCeram and 1MDS filter in surface and groundwater where the results differed in 5 out of 10 samples, Lee et al. suggested the sensitivity of NoV recovery appeared to depend on the types and conditions of environmental water [39]. This was indeed confirmed in the study of Pang who observed a general better recovery rate for RNA viruses in deionised water than in raw water and tap water [13]. NanoCeram has been successfully applied in drinking water [41], surface water [42], and wastewater [43, 44] (Figs 4.3 and 4.4).

TABLE 4.1 Some examples of recovery rates in percent ± SD for different filtration setup and elution conditions for various microorganisms.

Filter	Filtration-technique	Elution-technique	Elution buffer	Seeding level[a]	Water type/volume
Rexeed 25-S	Dead end	Backflush	0.01% NaPP + 0.5% Tween 80 + 0.001% Y-30 Antifoam	MS2: $0.39–1.0 \times 10^5$ pfu E. faecalis $0.9–1.0 \times 10^3$ cfu C. perfringens: $1.5–3.76 \times 10^3$ endospores C. parvum: $0.9–1.4 \times 10^6$ oocysts	Mix[b] (0.29 NTU)/100 L Mix[b] (1.5 NTU)/100 L Mix[b] (4.3 NTU)/100 L
Fresenius F80A	Tangential	Backflushing	0.01% NaPP + 0.01%Tween 80 + 0.01% Tween 20 0.001% Antifoam A 0.01% NaPP + 0.5%– 1% Tween 80 + 0.001% Antifoam A	All 10^6 pfu/cfu/ oocysts/ Endospores	Tap/10 L
Fresenius F200NR	Tangential	Back-flushing	0.01% NaPP + 0.5% Tween 80 + 0.001% Antifoam A	All 10^6 pfu/cfu/ oocysts/ Endospores	Tap/100 L
		Elution	0.01% NaPP + 0.1% Tween 80 + 0.001% Antifoam A		
Fresenius F200NR	Tangential	Direct collection		MS2: 6300 ± 3400 pfu C. perfringens: 2100 ± 770 endospores E. coli: $24,000 \pm 16,000$ cfu C. parvum: $150,000 \pm 52,000$ oocysts	Reclaimed/10 L
		Elution	0.01% NaPP + 0.01%Tween 80 + 0.001% Antifoam Y-30		
Exeltra Plus 210 (blocked with 0.1% NaPP)	Tangential	Elution	0.01% NaPP + 0.1% Tween 80 + 0.001% Antifoam A; (Re-concentration: Centricon Plus-70)	High; Low (1–10 pfu or cfu/100 mL)	Surface/100 L Drinking/100 L

Surrogate viruses[a]	Endogenous human enteric viruses[a]	E. faecalis	Salmonella	E. coli	C. perfringens endospores	Bacillus endospores	C. parvum oocysts	Reference
MS2: 57±7.7		93±16			94±22		87±18	[9]
MS2: 82±14		71±11			60±8.1		63±21	
MS2: 73±13		78±12			57±21		83±12	
MS2: 65±35	EV: 97±58	79±20	74±15			70±19[c]	97±20	[25]
MS2: 91±33	EV: 49±47	83±13	70±13			84±17[c]	83±17	
MS2: 84±12	EV: 37±10		64±10			49±19[d]	79±23	[8]
MS2: 53±13	EV: 49±15		69±22			45±17[d]	89±31	
MS2: 110±18				52±18	110±37		92±26	[31]
MS2: 120±20				74±17	120±26		100±20	
MS2: 48±25; 34 PRD1: 43±32; ND MNV: ND; 74	PV: 40; 16	56±26; 183		68±46;71	30; ND			[32]
MS2: 58±10, 48±12 PRD1: 80; 7±16 MNV: 41±34, ND		ND; 53±8		51±5; 48±11	79±12; 64±8			

Continued

TABLE 4.1 Some examples of recovery rates in percent $\pm$ SD for different filtration setup and elution conditions for various microorganisms—cont'd

Filter	Filtration-technique	Elution-technique	Elution buffer	Seeding level[a]	Water type/volume
NanoCeram		Elution	1.0% NaPP, 0.05 M glycine (Re-concentration: Centricon Plus-70)	AdV: MS2: 10^4 and 10^8 PFU PV: EV: CV:	Tap/20 L
		Elution	3% beef extract, 0.1 M glycine pH 9.5	AdV: 10^9 PCR units Qβ: 10^{12} PCR units	Sea/40 L
					Source/40 L
					Finished/40 L
		Elution	1.5% beef extract pH 9.75. (Re-concentration: Flocculation in 0.1% FeCl3 followed by centrifugation).	EV: 10^2 IU AdV: 10^3 IU NoV GII: 10^6 GC	Pure/10 L
					Tap/10 L
					Raw/10 l
		Elution	1.5% beef extract, 0.05 M glycine pH 9.5. (Re-concentration: celite elution)	NoV: 10^6 MPN RT-PCR units	River/10–100 L
					Tap/10–100 L
Glass-cellulose filter, Zeta Plus Virosorp, 1 MDS		Elution	1.5% beef extract 0.05 M glycine buffer, pH 9.5. (Re-concentration: celite adsorption/elution)	NoV: 10^6 MPN RT-PCR units	River/10–100 L
					Tap/10–100 L

Surrogate viruses[a]	Endogenous human enteric viruses[a]	E. faecalis	Salmonella	E. coli	C. perfringens endospores	Bacillus endospores	C. parvum oocysts	Reference
MS2: 45±15 56±9	AdV 14±4 PV: 66±6 EV. 83±14 CV. 77±11							[6]
Qβ: 96±52	AdV: 2.5±2.3							[33]
Qβ: 34±16	AdV: 2.4±0.5							
Qβ: 36±20	AdV: 1.4±0.6							
	EV: 47±11 AdV: 19±2 NoV GII: 42±8							[34]
	EV: 42±37 AdV: 21±3 NoV GII: 29±15							
	EV: 78±26 AdV: 19±3 NoV GII: 18±3							
	NoV GI: 12±16 PV 1: 65±22 (10 L) 38±35 (100 L)							[7]
	CV: 27±17 EV: 32±8 NoV GI: 4±1 PV: 277±22 (10 L), 54±8 and 51±26 (100 L)							
	NoV GI: 0.4±2 PV: 30±11 (10 L), 36±21 (100 L)							[7]
	NoV GI: 1±1 PV 1: 44±9 (10 L), 67±6 (100 L)							

Continued

TABLE 4.1 Some examples of recovery rates in percent ± SD for different filtration setup and elution conditions for various microorganisms — cont'd

Filter	Filtration-technique	Elution-technique	Elution buffer	Seeding level[a]	Water type/volume
Glass wool	(Plain) filtration	Elution	3% beef extract, 0.05 M glycine buffer, pH 9.5. (Re-concentration: flocculation PEG 8000 + NaCl, centrifugation)	AdV: 10^1–10^3 GC; NoV GII: 10^3 GC; PV: 10^5 GC	Tap/10 L (PV), 20 L (NoV GII), 10–1500 L (AdV)
				AdV: 10^3 GC, CV: 10^6 GC EV: 18: 10^6 GC NoV GI: 10^7 GC NoV GII: 10^5–10^6 GC PV: 10^4–10^7 GC	Ground 1, 2/20 L (AdV, CV, EV, NoV GI), 10–20 L NoV GII, 20–1597 L (PV)
	Preacidification (pH = 3.5) prior to (Plain) filtration	Elution	3% beef extract, 0.05 M glycine buffer, pH 9.5. (Re-concentration: flocculation pH 3.5, centrifugation)		Tap/10 L
Monolithic chromatographic columns	Chromatography	Elution	Phosphate buffer + NaCl	HAV: 10^3–10^5 TCID$_{50}$ FCV: 10^2–10^5 TCID$_{50}$	Bottled/1.5 L

[a] The following abbreviations are used for the viruses: AdV, adenovirus 41; CV, coxsackievirus B5; EV, echovirus; FCV, feline calicivirus; HAV, hepatitis A virus; MNV, murine norovirus; MS2, coliphage MS2; NoV, norovirus (G, genogroup); PRD1, bacterial virus; PV, poliovirus; Qβ, coliphage Qβ.
[b] Tap water mixed with lake water to different turbidities.
[c] B. globigii.
[d] B. atrophaeus.

Surrogate virusesᵃ	Endogenous human enteric virusesᵃ	E. faecalis	Salmonella	E. coli	C. perfringens endospores	Bacillus endospores	C. parvum oocysts	Reference
	AdV: 28±14 NoV GII: 30±27 PV: 98±24							[3]
	AdV: 22 ± 16, 8 ± 4 CV: B5 12 ± 4, 15 ± 9 EV: 15 ± 11, 24 ± 19 NoV GI: 33 ± 1, 45 ± 0.5 NoV GII: 32 ± 14, 16 ± 3 PV: 56 ± 45, 31 ± 14							[3]
	NoV GII: 3 (range 0–19)							[35]
FCV: 6.6±6	HAV: 30–38							[36, 37]

FIG. 4.3 Nanoalumina fibres. *(Photograph by R. Ristau, IMS, UCONN.)*

NanoCeram is an example of what is known as the VIRADEL approach to concentrating water, i.e. virus adsorption and elution [45]. This method is widely used and has recently been adapted to allow for larger volume recoveries, e.g. 40L from river water and 1000L of tap water [46], as well as approaches to facilitate field use via application of the cations required for binding of negatively charged viruses directly to the membrane before use [45], or as an inline addition immediately prior to sample processing [46, 47]. The use of mixed cellulose ester membranes for lake, river, and groundwater demonstrated that a combination of VIRADEL for virus capture and size exclusion for protozoa capture was possible [45].

4.3.3 Glass wool

Glass wool or glass powder is an alternative adsorptive material for virus concentration. Sodocalcic glass wool held together by a binding agent and coated with mineral oil, contains both hydrophobic and electropositive sites on the surfaces capable of adsorbing viruses at near-neutral pH and like positively charged filters glass wool requires no conditioning of the water with exception of pH adjustments under some circumstances [3]. When a virus suspension flows through the pore space of the packed material, the fibre surface is able to attract and retain negatively charged virus particles at near-neutral pH [48].

Glass wool has been used in virus monitoring studies involving wastewater [49], drinking water [50–52], groundwater [53–55], and river water [52, 56].

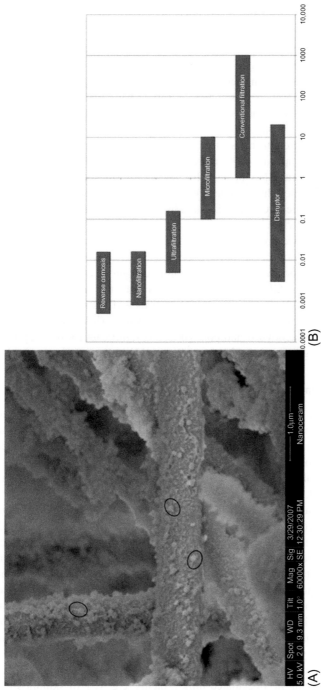

FIG. 4.4 Image and schematic of the functioning of the Disruptor fibre. (A) T4-Phages are seen attached to the Disruptor fibre. For the preparation of the raster electron microscopy (REM) image, the filter material was placed for 4 h in a highly concentrated T4-phage solution and dried afterward. Three phages are *encircled* as examples, where the icosahedral head and the tail are easily visible. (B) Filtration spectrum of Disruptor. *((A) Photo courtesy of E. Helmke, Alfred Wegener Institute, Germany.)*

Glass wool has been successfully applied in screening studies of a variety of viruses, adenoviruses, astroviruses, hepatitis A viruses, rotaviruses, and sapoviruses [4, 54, 57–59].

Using glass wool for recovery of enteric viruses has several advantages. Glass wool filtration is suitable for large-volume samples (1000L) collected at high filtration rates (4L/min), and its low cost makes it advantageous for studies requiring large numbers of samples [3, 60]. Other advantages were shown in studies where passing water samples through glass wool appeared to diminish PCR inhibition [3, 52]. Glass wool was found to have sufficient adsorptive capacity and strength to be used in long-term continuous sampling [3]. Finally, glass wool appears to be more effective in retaining infectious intact virus than naked viral RNA or partially degraded virions that have lost their infectivity. This is thought possible if the factors affecting adsorptive strength, such as charge, isoelectric point, size, and hydrophobicity, are more favourable for infectious intact virus [3].

However, the packaging of the glass wool into columns requires experience and variability in recovery has been reported using glass wool [3]. When comparisons of virus types were performed within categories of water matrix and vice versa, norovirus GII and poliovirus had statistically significant different recovery rates only for tap water, and adenovirus and norovirus GI had such differences only for one of two types of groundwater, while the difference in recovery between adenovirus and poliovirus was significant for both tap water and two types of groundwater [3].

4.3.4 Flocculation

Flocculation is a process by which a chemical coagulant added to the water acts to facilitate bonding between particles, creating larger aggregates that are easier to separate. The method is widely used in water treatment plants and can also be applied to sample processing for monitoring applications. A review of flocculation for bacteria using polymers also provides a good overview of the mechanisms [61].

In 2006, Ferrari and colleagues presented a calcium carbonate flocculation method for the concentration of pathogens from industrial wastewaters [62]. They stressed the need for internal controls given the highly variable recovery rates observed. Additionally, the authors considered it is essential to improve our understanding of the factors which lead to poor recovery rates when adapting techniques developed for drinking water to more complex matrices. Previous studies on the effect of turbidity on recovery rates have been contradictory. Recently, it has been shown that the addition of small silica particles improves *Cryptosporidium* oocyst recovery whereas particles greater than around 50 μm reduce oocyst recovery.

In 2008, Calgua presented a method for a one-step concentration of viruses from seawater, using preflocculated milk proteins (e.g. 10g of skimmed milk powder dissolved in the 1 L sample) [63]. The technique was applied to a study of adenoviruses in seawater samples, though the method could be adapted for other water types.

In 2012, Zhang et al. reported the use of the new coagulant for the concentration of *Escherichia coli* and *Helicobacter pylori* from raw and finished waters [64]. The study built on earlier work which had shown that lanthanum chloride was more effective as a sample concentration method than iron chloride or aluminium sulphate together with a microrespirometric detection method. However, this technique was not quantifiable for specific bacteria and so the process was adapted for molecular detection. The disadvantage of using iron coagulants is that the iron is known to interfere with DNA extraction, lowering the yield. This was not found for lanthanum. Ethylenediaminetetraacetic acid (EDTA) was employed as an elution agent, operating as a metal ion chelator and successfully dissolved lanthanum flocs in less than a minute, though performed less well with flocs of the other coagulant chemicals. A concentration of 10 times was reported which could be improved to 500 times by performing a 2-step process and including a centrifugation stage.

The use of nonmetal coagulants has been attracting increasing attention over the last few years. Polymers and natural compounds have been the main focus of attention and a schematic of the process of polymer flocculation is shown in Fig. 4.5. One example is the natural compound extracted from the *Opuntia ficus-indica* cactus as a flocculation and concentration agent for bacteria suspended in water [65]. Another, presented in 2012, is the use of *Moringa oleifera* seeds to flocculate *Giardia* and *Cryptosporidium* though the focus of the study was water treatment [66]. Alternatively, see the 2009 review by Renault et al. for more information on the uses of chitosan as a

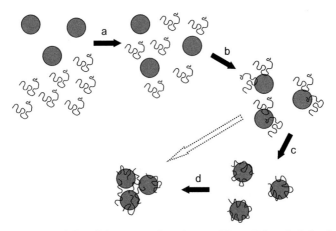

FIG. 4.5 Schematic of flocculation process by polymers. *(Figure 8 from B. Bolto, J. Gregory, Organic polyelectrolytes in water treatment, Water Res. 41(11) (2007) 2301–2324. Reprinted with permission.)*

coagulant [67]. Another recent study utilised celite for viruses, aiming at the secondary concentration step following UF (400–30 mL) and found that virus recoveries were reliable and ranged from 60% to 95% depending on the target virus [68]. Formalin-ether flocculation was also proposed as a low-cost alternative for parasites, and demonstrated successful field application in Colombia [69].

Polysaccharide polymers of microbial origin offer a unique and interesting class on account of their structural plasticity, stability, and functionality. Moreover, using metabolic engineering the turnover of such extracellular polymers from producer bacterial strains may be increased significantly. Ghosh and co-workers have applied such polymers for the flocculation of bacteria [70] and protozoan pathogens [71, 72].

4.3.5 Filtration specific for parasites

Concentration and detection of *Cryptosporidium* oocysts and *Giardia* cysts are standardised and described in for example "ISO 15553 Water quality—Isolation and identification of *Cryptosporidium* and *Giardia* cysts from water" and "Method 1623: *Cryptosporidium* and *Giardia* in Water by Filtration/IMS/FA" [73]. The methods consist of concentration by capsule filtration followed by a secondary concentration step, isolation using immunomagnetic separation and detection by microscopy (see below). The ISO-method also describes alternative concentration methods that can be used for small volumes of water (< 10L); carbonate flocculation, iron (II) sulphate flocculation, and membrane filtration.

Three different filters are included in the protocols (Table 4.2), all of which typically recover > 70% of the target organisms according to the manufacturers.

The main difference between the two Envirochek capsules is the maximum operating pressure tolerated, 2.1 bar for Envirochek, and 4.2 bar for Envirochek HV which affects the flow rate and the water volumes that can be filtered. It is

TABLE 4.2 Overview of parasite filters.

Filter	Capacity (L)	Filter media
Pall Envirochek		Supor membrane, hydrophilic polyethersulfone, 1300 cm^2, pore size 1 µm
Pall Envirochek HV	Up to 1000 L or more of treated water or up to 50 L of source water	Polyester, hydrophilic membrane, 1300 cm^2, pore size 1 µm
IDEXX Filta-Max	Up to 1000 L or more finished water or up to 50 L raw water	Compressed open-cell foam discs that give a nominal porosity of 1 µm

recommended that a pressure gauge is included in the filtration setup to ensure that the recommended pressure is not exceeded. After filtration is completed, the filter is emptied by draining. Parasites and particular matter are eluted in two steps with a Tris/EDTA/Laureth-12/Antifoam A-buffer using a wrist shaker. For Envirochek HV-filters, a pretreatment step with sodium polyphosphate and a rinse with deionised water precede the elution. The eluate normally exceeds 200 mL, and centrifugation is required for further concentration.

The compressed Filta-Max foam filter module is inserted into a reusable housing and connected to a water supply. After filtration, the filter is removed from the housing, and the filter washed twice in 600 mL PBS/Tween 20 using either a manual or an automatic wash station. During this procedure, the filter is decompressed and the parasites/particular matter released. Thereafter, the eluted material is concentrated by centrifugation or using a Filta-Max concentrator apparatus.

Filtration can either be achieved using a pumping system, in the laboratory or in field, or a pressurised source such as a tap. Although designed for *Cryptosporidium* oocysts and *Giardia* cysts, the filters have been used for the concentration of other parasites from drinking and environmental water, for example, *Toxoplasma* [74–76] and microsporidia [77].

Ultrafilters have been used as a more cost-effective alternative to capsule filters, with recent work recommending their use [21, 22]. In 2001, Simmons III et al. compared the concentration and detection of *Cryptosporidium* oocysts in seeded 10 L samples (100–150 oocysts) of reagent water and surface water using Envirocheck capsules and Hemoflow F80A ultrafilters (MWCO 80 kDa, Fresenius) [78]. They concluded that for reagent water, the capsule filter gave a somewhat higher recovery (46%) than the UF (42%) but the difference was not significant. For seeded surface water samples, the recovery remained at 42% for UFs but was only 15% for the capsule filter. Hill et al. [79] compared Envirochek HV capsules with F200NR ultrafilters (Fresenius) using tap water (100 L) and two different seeding levels, 150 (low) and 10^5 (high), of *Cryptosporidium* oocysts and *Giardia* cysts. For both seeding levels, recovery was higher with the ultrafilter for *Cryptosporidium* with the biggest difference observed for the low seeding level ($51 \pm 18\%$ vs $3.9 \pm 1.7\%$). For *Giardia*, recoveries were similar quite similar at both seeding levels. It should be noted that for the high seeding level, no secondary concentration (centrifugation and IMS) was required and thus, only the effect of the filtration step was seen. A comparison of recovery of *Toxoplasma gondii* was performed using a reusable hollow fibre filter (Microza membrane, Pall) and Envirochek HV capsules and different types of water (10 L freshwater and seawater) with different seeding levels (100, 1000, and 10,000 oocysts) with similar results for both filter types [75].

Another alternative for smaller volumes is flat membrane filtration. The ISO-standard 15553:2006 contains a protocol for the concentration of *Cryptosporidium* oocysts and *Giardia* cysts using 142 mm cellulose acetate membrane (pore size $\leq 2\,\mu m$) where, after filtration, particulate matter is released from the membrane by gentle rubbing of the filter in the presence of 0.1% Tween 80. For more information on membrane filtration, see Section 4.4.3.

Depending on the downstream application, factors other than the recovery rate may be worth considering when choosing a filter. The smaller pore size of ultrafilters will increase the amount of debris in the concentrated material. This will affect the size of the pellet which is, in turn, affects the IMS step in the protocols for *Cryptosporidium* and *Giardia* detection as several separate IMS reactions may have to be performed. It will also increase the background of fluorescent material in the microscopic analysis make the correct identification of oocyst and cysts more difficult. If molecular detection is used, it is likely to increase the amount of PCR inhibitory substances for example humic acids will be higher if ultrafiltration is used. The fraction of target DNA in the total DNA preparation will decrease if no further purification, such as IMS, is performed.

4.3.6 Continuous flow centrifugation

Centrifugation as a technique is discussed further under the secondary concentration techniques as it normally processes a smaller volume of the solution after one of the above primary techniques. However, the development of continuous flow centrifugation allows for the processing of larger volumes.

In 2002, Borchard and Spencer presented a continuous separation channel centrifugation, adapting existing blood cell separators for the purpose of waterborne pathogen concentration [80]. The advantages of this approach were the simultaneous concentration of many pathogens, suitability for a wide range of matrices, and high reproducibility. The main disadvantage was related to the achievable flow rate, with the authors finding maximums of a few hundred millimetres per minute depending upon the design. This method was applied to *Cryptosporidium*, *Giardia*, and *E. coli*.

In 2006 Zuckerman and Tzipori developed a portable continuous flow centrifuge system capable of simultaneous recovery of *C. parvum*, *Giardia intestinalis*, and *Encephalitozoon intestinalis*, with average recovery rates of 67%, 47%, and 72%, respectively [81]. However, there was a high degree of variability in the reported recovery rates. The authors also tested 1000 L of tap water spiked *C. parvum* with achieving a 35% average recovery and 10 L oocyst spiked source water samples, reporting 58% recovery. Zuckerman and Tzipori conclude that these recovery rates exceed what is acceptable in Method 1623, and compare favourably to filter results. Above, we saw that manufacturers claim over 70% recovery rates though for their filter systems. However, evidence suggests recovery rates do not always reach these levels [82] as does communication with water utilities (personal communication). This variation in recovery rate is observed for all methods as the recovery depends upon many factors, including the secondary concentration efficiency, operator experience, water quality, etc. whichever technique is used.

4.4. Secondary concentration techniques

Although after filtration, flocculation, or centrifugation, the sample volume has been reduced substantially, further concentration is normally needed. This usually includes a centrifugation step that pellets bacteria, parasites, and any debris present in the samples, while viruses remain in the supernatant. For parasites, depending on water quality and the detection method used, most often microscopic or PCR-based techniques, additional purification techniques may be required, e.g. to decrease the amount of debris and autofluorescing algae that may impair microscopy, or remove PCR inhibitory compounds.

After centrifugation, viruses remaining in the supernatant can be further concentrated using ultrafilter centrifugation, PEG-precipitation, flocculation, or ultracentrifugation, techniques which are presented in the following sections. Here we also present centrifugation, density gradient separation, and membrane filtration which can all be utilised for the further concentration of bacteria and parasites. Some discrimination between different pathogen types is possible at this stage by choice of conditions. However, for specific enrichment and isolation of a particular pathogen immunomagnetic separation is required. This is discussed in Section 4.3.4. An alternative means of pathogen isolation is the use of electrical separation techniques, which are discussed further in Chapter 6; such methods are not widely adopted for the manipulation of waterborne pathogens.

4.4.1 Centrifugation

For pelleting of *Cryptosporidium* oocyst and *Giardia* cysts, the ISO 15553:2006 recommends centrifugation at $1100 \times g$ and EPA Method 1623 $1500 \times g$ for 15 min, both without braking during deceleration. Several other centrifugal forces have been reported, e.g. $3000 \times g$ for 10 min after flocculation [83], $3300 \times g$ for 30 min [84] and $4000 \times g$ for 30 min [8] to recover also bacteria and bacterial endospores after ultrafiltration. Further, Stine et al. found that centrifugation at $1500 \times g$ followed by staining with Calcofluor white, resulted in 52.5% recovery of *E. intestinalis* spores while increasing the force to $2000 \times g$ decreased the fraction recovered to only 12.8% [85]. However, Dowd et al. detected both *E. intestinalis* and *Cyclospora cayetanensis* with PCR after centrifugation at $2000 \times g$ for 10 min [86]. For the concentration of bacteria after ultrafiltration, centrifugal forces of $3300-4200 \times g$ are commonly used [8, 31, 84, 87]. When deceleration conditions are mentioned, brakes are not applied [88].

Thus, no standard conditions for centrifugation can be given and conditions should be chosen based on the organisms studied and downstream processing. For example, higher forces are likely to pellet more bacteria and parasites but may decrease viability and damage surface antigens important for IMS. If PCR-based detection is used, these factors are not important while the effect on PCR-inhibitors needs to be considered.

4.4.2 Density gradient separation

Density gradient separation is mainly used for purification of parasites from faeces but may be applicable also to the isolation of parasites from water concentrates and some examples are given below. The technique is based on gradients formed by centrifugation of solutions of known gravity that either float parasites and pellet debris, or more rarely, float lighter debris and pellet the parasites. Commonly used solutions are percoll, caesium chloride, sucrose, and zinc sulphate.

Taghi Kilani and Sekla placed a *Cryptosporidium* oocyst suspension on top of a discontinuous gradient consisting of three layers of CsCl (1.40, 1.10, and 1.05 g/mL) [89]. After centrifugation for 1 h at $16,000 \times g$ at 4°C, a distinct band of very clean oocysts was observed and could be collected. Stine et al. purified *E. intestinalis* spores by placing a spore suspension on top of a Percoll-sucrose solution with a density of 1.2 g/L (created by diluting Percoll with 2.5 M sucrose) and centrifuged for 10 min at $1500 \times g$ [85]. The top layer and interface were collected and washed once in PBS. This purification technique was applied to tap water samples concentrated by Envirocheck-filtration and recoveries compared to centrifugation alone. Recoveries were very similar, but this extra purification step may be useful for turbid surface water samples. Percoll-sucrose gradients have also been used after filtration of surface water from reservoirs fed from wells or springs to separate protozoa from denser particular matter [86]. Both CsCl and Percoll are fairly costly and potassium bromide has been investigated as a cheaper alternative [90]. A faecal suspension containing *Cryptosporidium* oocysts was placed on top of three layers of KBr (28%, 16%, and 6%) in Tris-EDTA. After centrifugation at $3000 \times g$ or $16,000 \times g$ for 1 h at 4°C, the oocysts visible as a white band in the gradient that could be collected using a Pasteur pipette.

4.4.3 Membrane filtration

For tap water and other low turbidity waters, membrane filtration can be used for the secondary concentration of bacteria and parasites. In this context, the term membrane filtration is used to describe filtration of samples through flat, thin-sheet membranes. These membranes are often polymeric with pore sizes of 0.22 or 0.45 μm for bacteria and <2 μm for parasites and all material larger than the pre size will be retained on the filter (Fig. 4.6).

For bacteria, the filter is processed according to standard methods for the detection of bacterial pathogens in water, either placed on selective medium for quantitative detection or in enrichment broth qualitative detection. Parasites that are filtered onto membranes can then subjected to direct microscopic examination.

Warkiani et al. highlight some problems with traditional membrane filtration filters, including tortuous pore path, low pore density, overlapped pores,

FIG. 4.6 Image of flat-sheet membrane process The setup used for membrane filtration using a vacuum source or water suction. The funnel to the left is used for larger volumes (litres) and has a 90 mm filter on which the microorganisms are retained. The smaller funnel is used with 47 mm filters. Other types of filter housings for larger filters are available.

and a high coefficient of variation which all lead to relatively low sample throughput and cell recovery rate [91]. The solution proposed in 2011 by these authors is microfabricated filters to obtain membranes containing pores of uniform size and shape. The advantages of this approach are claimed to be the ability to use a larger (but controlled) pore size increasing throughput, the narrow pore size distribution ensuring passage of all smaller unwanted particles, and a straight pore path preventing the accumulation of particles within the pores. Warkiani fabricated filters using a 'dissolving mold technique' and report a recovery rate during elution from the filter of 97% for tap water spiked samples. A high flow rate was maintained throughout the operation and the filters could potentially be reusable. No cost comparison was presented in the paper, which could be a barrier to uptake of this type of technology, though if genuinely reusable a higher price might not be such a problem. A microfabricated refining system has also been reported, successfully concentrating 10 L of river water and achieving recovery rates exceeding 80% for both *Cryptosporidium* and *Giardia* [92, 93]. A palladium-coated nickel filter, reported to contain 36,300 pores of size $3 \pm 0.5\,\mu m$, has also been utilised for parasites, with recovery rates of 84% for *Giardia* and 70% for *Cryptosporidium*, when combined with ultrasonic elution; an advantage of this approach is the relentless nature and ability to elute into small volumes [27]. Alternative microfabricated solutions include microfluidic devices, discussed in more detail in Chapter 10, such as inertial focusing systems, which have reported high recoveries of parasites (Fig. 4.7) [94].

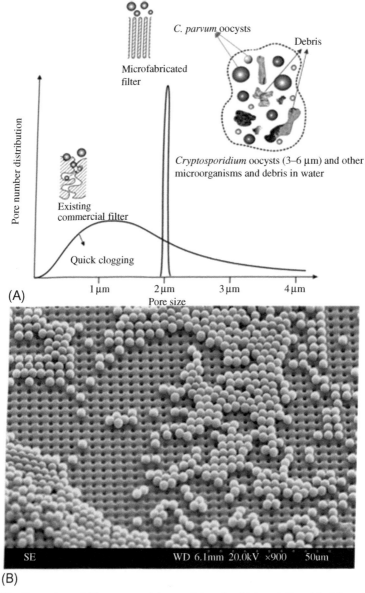

(A)

(B)

FIG. 4.7 Comparison of filter types and image of the microfabricated filter. (A) Schematic of pore size distribution in the commercial filters and the micro-fabricated filters. (B) SEM photo of a microfabricated filter with captured microbeads on the surface. *(Figures 2 and 9 from M.E. Warkiani, et al., Capturing and recovering of Cryptosporidium parvum oocysts with polymeric micro-fabricated filter, J. Membr. Sci. 369(1–2) (2011) 560–568.)*

4.4.4 Immunomagnetic separation

Immunomagnetic separation (IMS) both purifies and concentrates microorganisms. It is an established technique used for recovery of microorganisms, other types of cells, proteins, etc. using specific antibodies immobilised on paramagnetic beads. The process works by incubating the beads and sample, allowing time for the reaction to occur between the antibodies and the pathogen of interest. Mixing enhances this binding step. Subsequently, the beads are pulled to one side of the container using a magnet, isolating the pathogen of interest from the rest of the sample, which is removed and discarded. This bead-microorganism complex can be used for direct plating on selective media or used for the enrichment of a bacterium or for nucleic acid isolation. However, many protocols include dissolution of the bead-pathogen complex, often accomplished by adding acid. Next, the now dissociated beads are removed from the pathogen sample by using the magnet to once again concentrate them at the side of the container, and removing the solution now containing the isolated pathogen. IMS offers high recovery rates, though the drawbacks are the expensive reagents required, the time necessary for binding and unbinding and problems with unspecific binding or aggregation of nontarget cells within the bead-pathogen clusters.

IMS is used in the above mentioned standard methods for isolation of *Cryptosporidium* oocysts and *Giardia* cysts, and are commercially available from several suppliers. This protocol includes several IMS steps where the first reduces the sample volume from 10 to 1 mL and is followed by two rinses in 1 mL buffer. Thereafter, the parasites are eluted in HCl for 10 min and transferred to a microscope slide with NaOH for neutralisation and subsequent immobilisation and straining. Manufacturers report high recovery rates for IMS isolation of (oo)cysts, quoting over 95%. Acid dissociation has been found to be more successful than heat initiated dissociation [95].

To the best of our knowledge, beads conjugated with beads against other parasites are not available today. During method development, Hoffman et al. recovered *E intestinalis* spores, labelled with fluorescein isothiocyanate (FITC)-conjugated rabbit polyclonal antibodies using two different commercially available magnetic beads conjugated with antirabbit immunoglobulin antibodies [96]. A similar approach was used by Dunètre and Dardé for *T. gondii* using commercially available goat anti-mouse IgM coated magnetic beads and an in house produced monoclonal antibody [97]. However, this antibody lacked in specificity and a more specific antibody was later used by the same authors (2007) for tap and surface water (10 − 20L) concentrates seeded with approx. 100 sporulated oocysts with recoveries of 74.5% from drinking water and 30.6 and 37.1% from two different surface waters. For this antibody, acid elution was not sufficient enough but sonication was required to release the sporocysts. This antibody cross-reacted with sporocysts of *Hammondia hammondi*, *Hammondia heydorni*, and *Neospora caninum* and obtaining antibodies specific enough to isolate a microorganism from a complex background that also may contain closely related microorganisms is one of the major challenges in the development of IMS methods.

IMS is also used for recovery of bacteria, mainly from enrichment broths, in the analysis of pathogens in food, environmental, and clinical samples. Magnetic beads with antibodies against a common pathogen that can be encountered in the water, such as EPEC, *E. coli* O157, and *Salmonella* are commercially available from several suppliers. Cationic beads for recovery of viruses are also available and IMS for viruses has been reported in several research studies [98]. Automated IMS systems such as Pathatrix Auto System (ABI) will probably increase the number of pathogens against which antibodies are available.

Flow-through IMS techniques have been developed, with Ramadan and colleagues reporting a continuous flow magnetic separation system for *Cryptosporidium* and *Giardia* isolation and concentration [99]. Incubation of the protozoan pathogens with the IMS beads occurs as prescribed in US EPA Method 1623. Subsequently, the bead-pathogen complexes enter the flow-through system. Here, as the magnetic particulate matter passes through the channel it is repeatedly captured and released by the rotation of an external permanent magnet. Finally, the concentrated sample is captured by another magnet at the end of the channel. This is illustrated in Fig. 4.8. The aim of the multiple stages performed away from the wall was to avoid the problems of aggregation sometimes observed in IMS. Due to the relatively high magnetic particle concentrations, large aggregates are formed in which impurities might be trapped. This system was reported to concentrate samples of 50 mL down to 1 mL, with efficiencies comparable to the existing method, performing IMS in smaller volumes in a tube, for both tap (Fig. 4.9) and secondary effluent water.

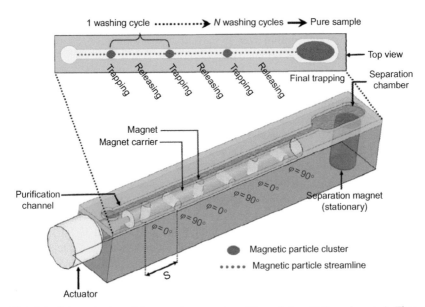

FIG. 4.8 Schematic view of the magnet array system. *(Figure 2 from Q. Ramadan, et al., Flow-through immunomagnetic separation system for waterborne pathogen isolation and detection: application to Giardia and Cryptosporidium cell isolation, Anal. Chim. Acta 673(1) (2010) 101–108.)*

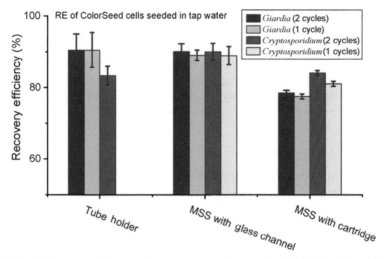

FIG. 4.9 Performance of the magnetic system compared to a standard tube holder. *(Figure 5 from Q. Ramadan, et al., Flow-through immunomagnetic separation system for waterborne pathogen isolation and detection: application to Giardia and Cryptosporidium cell isolation, Anal. Chim. Acta 673(1) (2010) 101–108.)*

When testing tap water samples, Ramandan et al. found that a glass channel flow system performed better than a disposable plastic moulded cartridge, which they speculate was due to rougher walls in the cartridge offering sites where pathogens could become trapped. The authors state their work is a step towards creating automated sample processing for protozoa, reducing the number of steps and human intervention required. See Chapter 10, Section 10.2.1, for further discussion of how microfluidics is applied to sample processing.

4.4.5 Ultrafilter centrifugation

Ultrafilter centrifugation is an alternative approach for the concentration of microbes from water that requires no preconditioning of the sample. Ultrafilter centrifugation is mainly used for the secondary concentration of viruses but will also concentrate bacteria and parasites if present in the sample. Spin-filter is available with different pore size, for examples Centricon Plus-70 (Millipore, 5-100 kDa) and Jumbsep Centrifugal Devices (Pall, 3–300 kDa). During centrifugation, water and molecules smaller than the cut-off will pass through the filter, and viruses and remaining particles/microorganisms/molecules are concentrated. Several parameters influence the efficiency of the ultrafiltration including membrane modules and material used, operational conditions, microbe types as well as water source and quality. Due to the small pore size, more turbid surface water samples may take a long time to process and filters are liable to clog during concentrating waters rich in solid matter. Therefore such samples are often subjected to a centrifugation step prior to loading of the supernatant onto the filter. The filters mentioned above have loading capacities of 70 and 60 mL, respectively, and concentrates down to 0.35 mL and 3.5–4 mL according to the manufacturers. A centrifugal ultrafiltration one-step method has been used to concentrate enteric viruses from 100 mL wastewater offering advantages of short processing time, ease of use and potential for the collection of bacteria and protozoa simultaneously (data not shown for this suggested application) though with the issues of potential clogging at high turbidity and potential lack of suitability to water types containing lower virus loads, e.g. drinking water [88].

4.4.6 PEG precipitation

As mentioned previously, polyethylene glycol (PEG) is commonly used for the precipitation of viruses. PEG is a chemically inert, water-soluble, synthetic polymer. PEG reduces the solubility of protein compounds by excluding proteins from the solvent, effectively increasing their concentration until solubility is exceeded and protein precipitation or crystallisation occurs [100]. More highly charged or hydrophobic proteins should be more easily precipitated than those of lesser charge [101]. As enteric viruses are charged due to their surface proteins, their precipitation will be favoured although not exclusively in this manner. PEG is nontoxic but the disadvantage of using PEG upstream a molecular detection is the risk to co-precipitate substances that can act as inhibitors.

4.5. Nuclei acid extraction for molecular detection

Based on the type of water concentrate, a variety of methods to extract RNA and DNA from virus, bacteria, and parasites while concurrently reducing the level of PCR inhibitors has been reported [4, 13, 32, 33, 102, 103] and the efficiency of these methods all depends on the characteristics of the prior water concentration.

Total nucleic acid can be released from concentrated water samples solely by heating at 99°C for 5 min as shown in several studies [7, 103]. Chemical nucleic acid extractions may include the use of Proteinase K [104] or the chaotropic guanidinium thiocyanate (GuSCN) [33, 105] in conjunction with a lysis buffer to degrade proteins and thereby disrupt the viral capsids and cell walls of bacteria and parasites to release, e.g. nucleic acids. These chemical principles have been adapted into commercial kits for nucleic acid purification as has bead beating as a means for releasing nucleic acids from hard to lyse-structures such as bacterial endospores and (oo)cysts. The released nucleic acid is bound specifically to silica, magnetic or paramagnetic beads or membranes and following a few effective washing steps, the nucleic acids are eluted from the binding matrix with an elution buffer or molecular-grade water. Spin columns to remove inhibitors such as HiBind RNA minicolumn may be applied in addition [33]. Most of these and similar kits can be used in manual, semiautomated, and fully automated formats (e.g. Magtration System (Precision System Science), miniMAG magnetic system (bioMérieux)), and are thus suitable for small and large laboratories with varied throughput requirements. More evaluation of kits has been undertaken with faecal samples compared to water samples [106]. Freeze-thaw cycles are typically employed as an alternative to bead-beating for parasites DNA extraction, although recent studies suggest that oocyst boiling and proteinase K treatment may be more effective [107].

DNA for the detection of bacteria and protozoa can be extracted from material pelleted by centrifugation or collected on membrane filters. However, this pellet is likely to contain a lot of material besides bacteria and protozoa, especially if surface waters are being analysed. This includes for example algae, debris, etc. that will release DNA during the extraction process. Thus, the target DNA will only constitute a small part of the total DNA. Further, it also will contain PCR-inhibitors, e.g. humic substances and phenolic that will hamper the subsequent PCR-based detection [108]. Various kits and methods for DNA preparation, use of postlysis additives such as polyvinylpyrrolidone (PVP), and Chelex-100 or PCR facilitators such as PVP, BSA, and dimethylsulphoxide to the PCR-reaction, or combinations thereof, have been investigated [12, 109] but no universal method has been presented. Francy et al. successfully used the PowerSoil DNA extraction kit (MO BIO Laboratories, Inc., Carlsbad, California) with a slightly modified protocol to extract DNA after ultrafiltration of drinking water and raw water (ground and surface) spiked with of various bacteria, bacterial endospores and *Cryptosporidium* oocysts [110]. We have tested also other kits intended for extraction of DNA from the soil for extraction of water concentrates after ultrafiltration of surface water and surface raw water. Although some PCR-inhibition was observed in all DNA preparations, the bacteria used for spiking were detected in 10 or 100× diluted DNA using a standard qPCR-kit and in undiluted DNA using a kit designed for environmental samples (unpublished). Dineen et al. evaluated a number of commercial DNA extraction kits for isolation of bacterial endospore DNA from different types of soil, a study that may constitute a good basis for the choice of kits also for DNA extraction from water concentrates [111].

Hill et al. [25] used membrane filtration for secondary concentration after ultrafiltration of spiked tap water samples and extracted DNA by bead beating of filters in PBS after which a noncommercial guanidine thiocyanate based lysis buffer was added and the lysate purified on a silica spin column. They concluded that tap water quality can affect the analytical performance in qPCR assays although no specific water quality parameter could be identified as responsible for the reduced performance. In this study, vegetative bacteria and endospores were targeted, but later the same technique has been used also for *C. parvum* oocysts [8].

Recently there has been growing interest in developing simplified approaches for nucleic acid extraction. A mechanical microfluidic approach was reported for bacteria DNA and tested with six waterborne bacteria, though not in actual water samples [112]. Another microfluidic solution incorporated an in-built corona discharge for ozone generation, subsequently employing the ozone for bacterial analysis [107]. In a 2019 study, ionic liquids were found to offer a rapid (5 min), low-cost and simplified approach for bacterial lysis, avoiding the need for complex clean-up steps, though there is some concern over the limit of detection achievable [113].

In conclusion, more work is needed to develop a DNA extraction method that can be used for inhibitor-free template DNA from bacteria and protozoa from different types of water.

4.6. Analytical controls

Whenever possible, analytical controls should be included in the sample processing [45]. For water analysis this may be more difficult than, for e.g. food. For instance, in an outbreak situation, no noncontaminated water sample from the same source will be available to serve as a negative sample process control, or for spiking with the target organisms, e.g. for detection of a bacterial pathogen by cultivation. Furthermore, process controls of the same volume as the water sample should be included which may not be possible to include when large water volumes are analysed. Below we give two examples of analytical controls that can be used in water sample processing.

4.6.1 Analytical controls in molecular detection

Whenever possible, analytical controls should be included in the sample processing. For molecular detection, these normally include process controls and amplification controls (Table 4.3). This is also illustrated in Fig. 4.10. using viruses as an example. A control panel of model organisms representing bacteria, viruses, and parasites, can be used to monitor the overall efficiency to recover the microorganisms from the three kingdoms from different types of waters (treated as well as untreated ground and surface waters) when using a specific combination of processing steps (concentration, elution, and extraction). Positive process controls are used to measure the recovery of target pathogen during the whole extraction and test procedure, thus evaluating the efficiency of the method as well as the error in human performance [114]. When choosing a heterologous microbe as process control, it is important that such microbes simulate the recovery of the target organism from the water sample as much as possible during sample treatment but at the same time are unlikely to be naturally present [115]. For example, for the detection of enteric viruses, a number of nonenveloped positive-sense single-stranded RNA viruses have been suggested as process controls, such as coliphage MS2 [116], FCV [117], mengovirus strain (MC_0) [118], and MNV [119].

4.6.1.1 Process controls

Process controls are used to control both sample treatment and test procedure and should therefore ideally be included at the beginning of the actual sample treatment. However, in on-line sampling, or sampling directly from the tap in

TABLE 4.3 Analytical controls for real-time detection of microorganisms in food.

Process controls	
Sample process control (SPC):	A sample portion spiked with a specified amount of a model organism and processed throughout the entire protocol. A positive signal should be obtained indicating that the entire process was correctly performed
Sample process control. Negative control:	A nonsample spiked with a specified amount of the model organism and processed throughout the entire protocol. A negative signal should be obtained indicating the lack of contamination throughout the entire process.
Target-negative process control (TNPC):	A test control free of sample and model organism and processed throughout the entire protocol. A negative signal should be obtained indicating that no cross contamination has taken place.
Amplification controls	
Positive amplification control:	A template known to contain the target sequence. Positive amplification indicates that amplification was performed correctly.
Negative amplification control (or No Template Control):	Including all reagents used in the amplification except the template nucleic acids. Usually, water is added instead of the template. A negative signal indicates the absence of specific contamination in the amplification assay
External amplification control (EAC):	An aliquot of a solution of control DNA, containing a defined quantity or copy number, added to an aliquot of the nucleic acid of the extracted sample and analysed in a separate reaction tube. A positive signal indicates that the sample nucleic acid extract did not contain inhibitory substances in a degree to completely prevent amplification
Internal amplification Control (IAC):	Chimerical nontarget nucleic acid added to the master mix to be co-amplified with the same primer set as the target but yet distinguishable from the target amplicon. The amplification of the IAC both in the presence and in the absence of the target indicates that the amplification conditions are adequate

Modified from (or based on) A. Bosch, et al., Analytical methods for virus detection in water and food, Food Anal. Methods 4 (2011) 4–12; D. Rodriguez-Lazaro, et al., Virus hazards from food, water and other contaminated environments. FEMS Microbiol. Rev. (2011) 1–20 for real-time RT pPCR detection of viruses in food.

an outbreak situation, it is not possible to add controls before sampling. Instead, they may be added after sample elution.

Sample Process Control (SPC) is added to every test sample and to the Target-Negative Process Control sample (TNPC, see below), ideally at the start of analysis. It verifies if sample treatment has functioned correctly, and

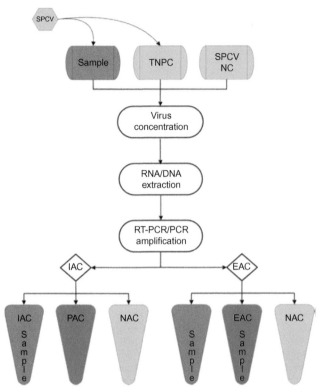

FIG. 4.10 Suite of controls for nucleic acid amplification-based methods for the detection of enteric viruses. Abbreviations: *EAC*, external amplification control; *IAC*, internal amplification control; *NAC*, negative amplification control; *PAC*, positive amplification control; *SPCV*, sample process control virus; *SPCVNC*, sample process control virus—negative control; *TNPC*, target-negative process control [115].

identifies those samples in which sample treatment has failed. The SPC must be detected in every sample. If the SPC cannot be added to the sample before concentration, it will only verify the step following its addition.

Target-Negative Process Control (TNPC) is a 'sample' in which the matrix is replaced with the same weight/volume of sterile distilled water or phosphate-buffered saline. This control should be included in every batch of samples analysed and is processed identical as a test sample (including the addition of the SPC). This control verifies that the sample treatment reagents and/or equipment are not contaminated with the target organism or its amplicon. The SPC must be detected but there should be no target signal obtained from the TNPC. For large sample volumes, including a TNPC may be unrealistic.

Sample Process Control—Negative Control is a 'sample' in which both sample matrix and SPC are replaced by sterile distilled water or equivalent. This control should contain all reagents used in test sample analysis and verifies that the sample treatment reagents and/or equipment are not contaminated with sample process control organism or its amplicon. Neither target nor SPC must be detected in this sample.

4.6.1.2   Amplification controls

Amplification inhibition is caused by the presence of co-extracted (RT)-PCR inhibitors. The purity of the nucleic extracts can be estimated using optical density measurements, and the amplifiability of the preparation can be tested by (RT)-PCR. The optical density measures the protein contamination and it is not necessarily correlated to the success of (RT)-qPCR [120], as other substances can interfere with this method as well. Thus, (RT)-qPCR provides a better estimate of purity and even if the nucleic acid is considered 'dirty' by, e.g. visual inspection the enzymes may well be able to amplify the target efficiently [121]. One approach to assess for—or reduce—inhibition from potential co-extracted inhibitors is to test for the target organism in undiluted and diluted nucleic acid extracts. However, this approach also affects the assay sensitivity since only a smaller amount of the sample is analysed. Another way to control for inhibitors present in the nucleic acid preparation is to compare the amplification efficiency of a target similar but yet distinguishable template added to the test material extract and a control reaction [114]. The amplification product from the amplification control could be distinguishable from the target template by for example an internal sequence modified to contain recognition sites for restriction enzymes [122] or an alternative probe sequence [102].

Amplification controls should be included in the qPCR reagent mixture, or additional reactions performed, depending on their character.

Internal Amplification Control (IAC) is a nucleic acid sequence present in every reaction. An IAC verifies if amplification reactions have functioned correctly, and identify those which have failed. An IAC is flanked by target primers to enable simultaneous co-amplification with the target sequence during (RT)-qPCR detection of the target but is distinguishable from the target sequence by an IAC-specific [102]. Ideally, an IAC should be included for both the target organism and the SPC. A successful amplified IAC, signals 'less than complete' inhibition in the matrix extract and that absence of target signal in the test reaction can be considered as a true negative sample under the test conditions. In contrast, the failure to amplify both IAC and target indicates inhibition.

External Amplification Control (EAC) is an alternative to the IAC. The EAC is included in a reaction separate to the target test reaction. An EAC may be identical to the target nucleic acid sequence and thus include both primer and probe sequences. Thereby it aids in the identification and evaluation of possible reaction inhibition [118, 122]. A successful amplified EAC, signals 'less than complete' inhibition in the matrix extract and that absence of target signal in the test reaction can be considered as a true negative sample under the given test conditions. In contrast, failure to amplify both EAC and targets indicates inhibition.

The advantage of an *IAC* is that the control for inhibition is performed in the same reaction as the detection of the target sequence. However, if the *IAC* concentration is not correctly optimised, this may lead to competition between amplification of target sequence and *IAC*. This competition is not a problem if an *EAC* is used; however, using an *EAC* does not control for some situations such as individual pipetting errors, and using an *EAC* increases the number of reactions that are required for each sample [115].

Negative Amplification Control (Fig. 4.10) is a separate amplification reaction in which the nucleic acid extracted from the matrix is replaced with water (also called a nontemplate control). This control enables to verify that the (RT)-PCR master mix reagents are not contaminated with the target organism, SPC, or its amplicon.

Positive Amplification Control verifies the amplification performance of the target sequence in a reaction, in which a known amount of nucleic acid from the target organism is included and the sample nucleic acid is replaced with water. A target signal must be obtained from the positive amplification control.

4.6.2 An analytical control in detection of *Cryptosporidium* and *Giardia* analysis

For analysis of *Cryptosporidium* and *Giardia*, an internal quality control consisting of vials containing a known number of (oo)cysts stained red are commercially available (ColorSeed, BTF, www.btf.com). The (oo)cysts are added to the water sample to be filtered or, if the sample has been filtered elsewhere and the filter sent to the lab or the filter is connected to a tap, they can be added to the eluate. If so, this will not show how well the filtration step has worked but only the subsequent concentration step. Staining is carried out, as usual, using a positive control (oo(cyst)-suspension) and negative control (PBS). During the subsequent microscopic analysis, the ColorSeed (oo)cysts will fluorescence in green under FITC conditions and in red under Texas Red conditions and thus are easily differentiated from naturally occurring (oo) cysts. The fraction of red (oo)cysts recovery shows the efficiency in the part of the method in which they were included.

4.7. Summary

This chapter has reviewed a stage critical to the detection of waterborne pathogens, namely that of sample preparation. The techniques described in this chapter are all critical to the overall success of the detection technologies described in the next chapters. Many of the methods in Chapters 5–8 operate with small sample volumes and require purified samples for success. In this summary, we provide a review of the sample processing techniques considering each class of pathogen in turn and finish by drawing conclusions upon the existing state of the art in sample preparation and expected future directions.

Sampling and analytical procedures for the microbiological analysis of water are well documented. A variety of techniques have been described for the recovery of viruses, bacteria and parasites from water, each with their own advantages and disadvantages as the physicochemical quality of the water, including but not limited to the pH, conductivity, turbidity, presence of particulate matter and organic acids can all affect the efficiency of recovery of microorganisms. It is important to acknowledge that no single method may universally be recognised as superior: efficiency, the constancy of performance, robustness, cost, and complexity are all factors to be considered for each method, and performance characteristics must be continually monitored.

Different methods of sample processing are also likely to be required depending upon the type of monitoring undertaken. For operational monitoring the main requirement is likely to be rapid detection and therefore simple, quick methods are preferable. During investigative monitoring of an outbreak, pathogens may be present in higher numbers such that large samples and high recoveries rates are less critical and it is more important to ensure rapid and efficient removal of interferents to enable accurate, detailed information on pathogen species being obtained. Finally, for surveillance monitoring, high recovery rates are likely to be a key requirement to detect potentially low numbers of pathogens.

A vision of networked in situ sensors is a particular challenge for waterborne pathogen monitoring and often sample processing would be required for concentration and enrichment before useful detection could take place. Such in-situ or online monitoring places a great demand on sample processing and requires its automation. Chapter 10 presents some microfluidic approaches to automation of sample processing, though systems will be needed to cope with larger volumes of water and for which clogging is less of a challenge.

Viral recovery and concentration techniques include ultrafiltration, adsorption-elution using filters or membranes [123], NanoCeram filters, glass wool [4] or glass powder, two-phase separation with polymers, flocculation, the use of monolithic chromatographic columns [124] ultrafilter centrifugation, IMS and PEG precipitation. The use of the glass wool adsorption-elution procedure for the recovery of enteric viruses from large volumes of water has proven to be a cost-effective method and has successfully been applied for the routine recovery of human enteric viruses from large volumes of water [4]. The same applies to NanoCeram filters. While many different solutions are available for concentrating viruses from water [45], the recovery rate generally does not exceed 60% [74], except in particular cases [61].

Bacteria are commonly processed using ultrafiltration, centrifugation, and membrane filtration. The use of membrane filtration is popular as the membranes can be immediately employed in a traditional culture-based detection techniques. Flocculation, IMS and density gradient centrifugation are also techniques applied to bacteria and there have been initial studies looking at the use of NanoCeram [38] and glass wool [1]. With a move towards more direct detection of bacteria, including addressing the concern that some bacteria exist in viable but nonculturable states, sample processing techniques will need to develop further to accommodate, e.g. issues of removal of interferents for molecular methods. In general recovery rates for bacteria appear to be higher than for viruses (Table 4.1) though there is a high degree of variability observed depending upon factors such as water quality, operator skill, secondary concentration efficiency, etc.

Parasites can be concentrated using ultrafiltration, flocculation, centrifugation, membrane filtration as well as isolated using IMS. Recovery rates are often comparable to those achieved for bacteria, though depending on the water type and it has been suggested that aiming to find one overarching sample processing solution is a flawed strategy [29]. Commercially available tools available for the specific isolation of parasites, other than *Cryptosporidium* and *Giardia* are lacking, such that for other parasites the detection options are either microscopic examination of, or molecular detection, in highly complex samples. To differentiate between parasites of different species under the microscope requires experience missing at most water laboratories. Molecular identification is hampered by the low concentrations of parasites normally present in water samples. For example, during the outbreak of cryptosporidiosis in Östersund, Sweden, in 2010–11, the highest number of confirmed oocysts per 10 L of drinking water ever isolated was 0.05 [125].

Flocculation has been applied to all types of microorganisms and new materials are emerging. Polymers using biomaterial (e.g., chitosan and lysine) can be used due to their affinity (e.g., the weak interaction (e.g., electrostatic, hydrogen bond)) with microbiological particles (containing electrostatic charges and hydrophobic groups expressed variably as a function of pH, chemical, and organic components of the water). However, to be compatible with rapid, on-site applications, they must be packaged in an appropriate form (i.e. powder, magnetic beads, hollow fibres, or foam) [15].

New materials are one area of development in sample processing, including materials/chemicals for flocculation, filters (e.g. NanoCeram or glass wool), elution protocols, and chemical additives or surface modifications to prevent adsorption as well as novel bacterial lysis methods. More work is required to optimise the use of these materials, to fully understand their interactions with pathogens, and perhaps to work towards cheaper, greener materials. Another novel approach is physical methods to produce uniform pore size filters [91, 92] or utilise, e.g. megasonic or ultrasound elution [26, 27] to enhance recovery rates, and such technologies should be further explored.

Microfluidic and mechanical/chemical lysis methods have also been applied to bacteria. Given microfluidic lysis has received reasonable attention, for a range of cell types, further investigation of adaptions of existing approaches to waterborne pathogens would be sensible. Other simplified lysis protocols have been recently reported and further studies are recommended both to develop new protocols and to benchmark and compare existing methods. There is considerable variation in protocols and little direct evaluation of the effectiveness reported; hopefully, future work will soon rectify the situation.

Automation of systems and integration of different stages of the process are also being demonstrated [21, 28]; this is challenging since the research evidence demonstrates that single methods are usually not capable of concentrating all species equally but offers many potential advantages and is a key area of future development.

Demands of new detection technologies impact on sample processing, with molecular methods being a key example of this. In viral detection, the advent of molecular methods has driven a lot of work into the development of procedures for the isolation of virus samples free from interferents. The traditional detection means identifying bacteria by cultivation and parasites by microscopic observation means less focus has been placed on sample processing for other end uses though this is changing with recent work reflecting developments in this area. However, replacing the use of traditional techniques for the detection of bacteria and parasites by molecular detection only will require the development of methods that also show that the organisms are indeed viable and thus may be infectious as DNA is isolated also from nonviable cells/(oo)cysts. This could be in the detection stages, as discussed in subsequent chapters, or perhaps in the development of sample processing techniques to distinguish viable and nonviable pathogens.

Sample processing has not received as much research attention as detection technologies yet is a key part of the process often determining the overall success of monitoring. Therefore it is essential to continue research in this area, particularly in characterising the performance of different methods, development of new materials and processes, and also in automation of processes, bringing the advantages of reduced reliance on operator skill, reduced variability, and reduced risk of cross-contamination of samples. Analytical controls are also essential to validate and quality assure the sample concentration and enrichment processes [45] and is another important area of future work.

References

[1] H.T. Millen, et al., Glass wool filters for concentrating waterborne viruses and agricultural zoonotic pathogens, J. Vis. Exp. 61 (2012) 3930.

[2] N. Deboosere, et al., Development and validation of a concentration method for the detection of influenza a viruses from large volumes of surface water, Appl. Environ. Microbiol. 77 (2011) 3802–3808.

[3] E. Lambertini, et al., Concentration of enteroviruses, adenoviruses, and noroviruses from drinking water by use of glass wool filters, Appl. Environ. Microbiol. 74 (2008) 2990–2996.

[4] D. Sano, et al., Quantification and genotyping of human sapoviruses in the Llobregat river catchment, Spain, Appl. Environ. Microbiol. 77 (2011) 1111–1114.

[5] V.K. Upadhyayula, S. Deng, G.B. Smith, M.C. Mitchell, Adsorption of *Bacillus subtilis* on single-walled carbon nanotube aggregates, activated carbon and NanoCeram, Water Res. 43 (2009) 148–156.

[6] L.A. Ikner, M. Soto-Beltran, K.R. Bright, New method using a positively charged microporous filter and ultrafiltration for concentration of viruses from tap water, Appl. Environ. Microbiol. **77** (2011) 3500–3506.

[7] M.R. Karim, et al., New electropositive filter for concentrating enteroviruses and noroviruses from large volumes of water, Appl. Environ. Microbiol. 75 (2009) 2393–2399.

[8] A.L. Polaczyk, J. Narayanan, T.L. Cromeans, D. Hahn, J.M. Roberts, J.E. Amburgey, V.R. Hill, Ultrafiltration-based techniques for rapid and simultaneous concentration of multiple microbe classes from 100 L tap water samples, J. Appl. Microbiol. 73 (2008) 92–99.

[9] C.M. Smith, V.R. Hill, Dead-end hollow-fiber ultrafiltration for recovery of diverse microbes from water, Appl. Environ. Microbiol. 75 (2009) 5284–5289.

[10] S.D. Leskinen, et al., Hollow-fiber ultrafiltration and PCR detection of human-associated genetic markers from various types of surface water in Florida, Appl. Environ. Microbiol. 76 (2010) 4116–4117.

[11] B. Mull, V.R. Hill, Recovery and detection of *Escherichia coli* O157:H7 in surface water, using ultrafiltration and real-time PCR, Appl. Environ. Microbiol. 75 (2009) 3593–3597.

[12] J. Jiang, et al., Development of procedures for direct extraction of Cryptosporidium DNA from water concentrates and for relief of PCR inhibitors, Appl. Environ. Microbiol. 71 (3) (2005) 1135–1141.

[13] X.L. Pang, et al., Pre-analytical and analytical procedures for the detection of enteric viruses and enterovirus in water samples, J. Virol. Methods 184 (2012) 77–83.

[14] I.J. Allan, et al., Strategic monitoring for the European water framework directive, Trends Anal. Chem. (2006) **25**(7).

[15] B. Roig, et al., Analytical issues in monitoring drinking-water contamination related to short-term, heavy rainfall events, Trends Anal. Chem. 30 (8) (2011) 1243–1252.

[16] I.J. Allan, et al., A toolbox for biological and chemical monitoring requirements for the European Union's Water Framework Directive, Talanta 69 (2) (2006) 302–322.

[17] E.A. Kearns, S. Magana, D.V. Lim, Automated concentration and recovery of micro-organisms from drinking water using dead-end ultrafiltration, J. Appl. Microbiol. 105 (2008) 432–442.

[18] S.D. Leskinen, et al., Automated dead-end ultrafiltration of large volume water samples to enable detection of low-level targets and reduce sample variability, J. Appl. Microbiol. 113 (2012) 351–360.

[19] P.M. Holowecky, et al., Evaluation of ultrafiltration cartridges for a water sampling apparatus, J. Appl. Microbiol. (2009) 738–747.

[20] D. Karthe, et al., Modular development of an inline monitoring system for waterborne pathogens in raw and drinking water, Environ. Earth Sci. 75 (23) (2016) 1481.

[21] M.A. Kahler, et al., Evaluation of an ultrafiltration-based procedure for simultaneous recovery of diverse microbes in source waters, Water (2015) **7**(3).

[22] A.M. Kahler, V.R. Hill, Detection of cryptosporidium recovered from large-volume water samples using dead-end ultrafiltration, in: J.R. Mead, M.J. Arrowood (Eds.), Cryptosporidium: Methods and Protocols, Springer New York, New York, NY, 2020, pp. 23–41.

[23] D.S. Francy, et al., Comparison of filters for concentrating microbial indicators and pathogens in lake water samples, Appl. Environ. Microbiol. 79 (4) (2013) 1342.

[24] L.J. Winona, A.W. Ommani, J. Olszewski, J.B. Nuzzo, K.H. Oshima, Efficient and predictable recovery of viruses from water by small scale ultrafiltration systems, Can. J. Microbiol. 47 (2001) 1033–1041.

[25] V.R. Hill, A.L. Polaczyk, D. Hanh, J. Narayanan, T.L. Cromeans, J.M. Roberts, J.E. Amburgey, Development of a rapid method for simultaneous recovery of diverse microbes in

drinking water by ultrafiltration with sodium polyphosphates and surfactants, Appl. Environ. Microbiol. 71 (2005) 6878–6884.

[26] A. Kerrouche, M.P.Y. Desmulliez, H. Bridle, Megasonic sonication for cost-effective and automatable elution of Cryptosporidium from filters and membranes, J. Microbiol. Methods 118 (2015) 123–127.

[27] M.N.S. Al-Sabi, et al., New filtration system for efficient recovery of waterborne Cryptosporidium oocysts and Giardia cysts, J. Appl. Microbiol. 119 (3) (2015) 894–903.

[28] H. Bridle, A. Kerrouche, M.P.Y. Desmulliez, Megasonic elution of waterborne protozoa enhances recovery rates, Matters (2018), https://doi.org/10.19185/matters.201712000007.

[29] A. Efstratiou, J. Ongerth, P. Karanis, Evolution of monitoring for Giardia and Cryptosporidium in water, Water Res. 123 (2017) 96–112.

[30] J.P. Stokdyk, et al., Determining the 95% limit of detection for waterborne pathogen analyses from primary concentration to qPCR, Water Res. 96 (2016) 105–113.

[31] P. Liu, V.R. Hill, D. Hahn, T.B. Johnson, Y. Pan, N. Jothikumar, C.L. Moe, Hollow-fiber ultrafiltration for simultaneous recovery of viruses, bacteria and parasites from reclaimed water, J. Microbiol. Methods 88 (2012) 155–161.

[32] K.E. Gibson, K.J. Schwab, Tangential-flow ultrafiltration with integrated inhibition detection for recovery of surrogates and human pathogens from large-volume source water and finished drinking water, Appl. Environ. Microbiol. 77 (2011) 385–391.

[33] C.D. Gibbons, et al., Evaluation of positively charged alumina nanofibre cartridge filters for the primary concentration of noroviruses, adenoviruses and male-specific coliphages from seawater, J. Appl. Microbiol. 109 (2010) 635–641.

[34] X.L. Pang, B.E. Lee, K. Pabbaraju, S. Gabos, S. Craik, P. Payment, N. Neumann, Pre-analytical and analytical procedures for the detection of enteric viruses and enterovirus in water samples, J. Virol. Methods 184 (1–2) (2012) 77–83.

[35] N. Albinana-Gimenez, P. Clemente-Casares, B. Calgua, J.M. Huguet, S. Courtois, R. Girones, Comparison of methods for concentrating human adenoviruses, polyomavirus JC and noroviruses in source waters and drinking water using quantitative PCR, J. Virol. Methods 158 (1–2) (2009) 104–109.

[36] K. Kovač, I. Gutiérrez-Aguirre, M. Banjac, M. Peterka, M. Poljšak-Prijatelj, M. Ravnikar, J.Z. Mijovski, A.C. Schultz, P. Raspor, A novel method for concentrating hepatitis A virus and caliciviruses from bottled water, J. Virol. Methods 162 (1–2) (2009) 272–275.

[37] A.C. Schultz, S. Perelle, S. Di Pasquale, K. Kovac, D. De Medici, P. Fach, H.M. Sommer, J. Hoorfar, Collaborative validation of a rapid method for efficient virus concentration in bottled water, Int. J. Food Microbiol. 145 (1) (2011) S158–S166.

[38] F. Tepper, L. Kaledin, Virus and Protein Separation Using Nano Alumina Fiber Media, 2007, p. 30-9-0012.

[39] H. Lee, et al., Evaluation of electropositive filtration for recovering norovirus in water, J. Water Health 9 (2011) 27–36.

[40] D. Li, H.C. Shi, S.C. Jiang, Concentration of viruses from environmental waters using nano-alumina fiber filters, J. Microbiol. Methods 81 (2010) 33–38.

[41] J.L. Cashdollar, et al., Development and evaluation of EPA method 1615 for detection of enterovirus and norovirus in water, Appl. Environ. Microbiol. 79 (1) (2013) 215.

[42] B. Prevost, et al., Large scale survey of enteric viruses in river and waste water underlines the health status of the local population, Environ. Int. 79 (2015) 42–50.

[43] R.M. Chaudhry, K.L. Nelson, J.E. Drewes, Mechanisms of pathogenic virus removal in a full-scale membrane bioreactor, Environ. Sci. Technol. 49 (5) (2015) 2815–2822.

[44] Y. Qiu, et al., Assessment of human virus removal during municipal wastewater treatment in Edmonton, Canada, J. Appl. Microbiol. 119 (6) (2015) 1729–1739.

[45] E. Haramoto, et al., A review on recent progress in the detection methods and prevalence of human enteric viruses in water, Water Res. 135 (2018) 168–186.

[46] A. Hata, et al., Concentration of enteric viruses in large volumes of water using a cartridge-type mixed cellulose ester membrane, Food Environ. Virol. 7 (1) (2015) 7–13.

[47] T. Asami, et al., Evaluation of virus removal efficiency of coagulation-sedimentation and rapid sand filtration processes in a drinking water treatment plant in Bangkok, Thailand, Water Res. 101 (2016) 84–94.

[48] N.G. Green, A. Ramos, A. González, A. Castellanos, H. Morgan, Electric field induced fluid flow on microelectrodes: the effect of illumination, J. Phys. D Appl. Phys. 33 (2) (2000) L13.

[49] C. Gantzer, et al., Enterovirus genomes in wastewater: concentration on glass wool and glass powder and detection by RT-PCR, J. Virol. Methods 65 (1997) 265–271.

[50] W.O. Grabow, M.B. Taylor, J.C. de Villiers, New methods for the detection of viruses: call for review of drinking water quality guidelines, Water Sci. Technol. 43 (2001) 1–8.

[51] J.C. Vivier, M.M. Ehlers, W.O. Grabow, Detection of enteroviruses in treated drinking water, Water Res. 38 (2004) 2699–2705.

[52] H.J. van Heerden, et al., Prevalence, quantification and typing of adenoviruses detected in river and treated drinking water in South Africa, J. Appl. Microbiol. 99 (2005) 234–242.

[53] K.L. Powell, et al., Microbial contamination of two urban sandstone aquifers in the UK, Water Res. 37 (2003) 339–352.

[54] W.B. van Zyl, et al., Application of a molecular method for the detection of group A rotaviruses in raw and treated water, Water Sci. Technol. 50 (2004) 223–228.

[55] M.M. Ehlers, W.O. Grabow, D.N. Pavlov, Detection of enteroviruses in untreated and treated drinking water supplies in South Africa, Water Res. 39 (2005) 2253–2258.

[56] D. Hot, et al., Detection of somatic phages, infectious enteroviruses and enterovirus genomes as indicators of human enteric viral pollution in surface water, Water Res. 37 (2003) 4703–4710.

[57] M.B. Taylor, et al., The occurrence of hepatitis A and astroviruses in selected river and dam waters in South Africa, Water Res. 35 (2001) 2653–2660.

[58] H.J. van Heerden, et al., Prevalence of human adenoviruses in raw and treated water, Water Sci. Technol. 50 (2004) 39–43.

[59] W.B. van Zyl, et al., Molecular epidemiology of group A rotaviruses in water sources and selected raw vegetables in southern Africa, Appl. Environ. Microbiol. 72 (2006) 4554–4560.

[60] A. Bosch, et al., Analytical methods for virus detection in water and food, Food Anal. Methods 4 (2011) 4–12.

[61] S. Barany, A. Szepesszentgyfrgyi, Flocculation of cellular suspensions by polyelectrolytes, Adv. Colloid Interf. Sci. 111 (2004) 117–129.

[62] B.C. Ferrari, K. Stoner, P.L. Bergquist, Applying fluorescence based technology to the recovery and isolation of Cryptosporidium and Giardia from industrial wastewater streams, Water Res. 40 (3) (2006) 541–548.

[63] B. Calgua, et al., Development and application of a one-step low cost procedure to concentrate viruses from seawater samples, J. Virol. Methods 153 (2) (2008) 79–83.

[64] Y. Zhang, et al., Determination of low-density *Escherichia coli* and *Helicobacter pylori* suspensions in water, Water Res. 46 (7) (2012) 2140–2148.

[65] A.L. Buttice, P.G. Stroot, N.A. Alcantar, Concentration and removal of waterborne bacteria for easy detection, Biophys. J. 98 (3) (2010) 408a.

[66] L. Nishi, et al., Application of hybrid process of coagulation/flocculation and membrane filtration for the removal of protozoan parasites from water, Procedia Eng. 42 (2012) 173–185.

[67] F. Renault, et al., Chitosan for coagulation/flocculation processes—an eco-friendly approach, Eur. Polym. J. 45 (2009) 1337–1348.

[68] E.R. Rhodes, et al., The evaluation of hollow-fiber ultrafiltration and celite concentration of enteroviruses, adenoviruses and bacteriophage from different water matrices, J. Virol. Methods 228 (2016) 31–38.

[69] F. Lora-Suarez, et al., Detection of protozoa in water samples by formalin/ether concentration method, Water Res. 100 (2016) 377–381.

[70] M. Ghosh, S. Pathak, A. Ganguli, Application of a novel biopolymer for removal of Salmonella from poultry waste water, Environ. Technol. 30 (2009) 337–344.

[71] M. Ghosh, S. Pathak, Application of microbial exopolymers for removal of protozoan parasites from potable water, Res. J. Biotechnol. 5 (1) (2010) 58–62.

[72] M. Ghosh, S. Pathak, A. Ganguli, Effective removal of cryptosporidium by a novel bioflocculant water, Environ. Res. 81 (2009) 123–128.

[73] Method 1623: Cryptosporidium and Giardia in Water by Filtration/IMS/FA, United States Environmental Protection Agency, 2005.

[74] I. Villena, et al., Evaluation of a strategy for *Toxoplasma gondii* oocyst detection in water, Appl. Environ. Microbiol. 70 (2004) 4035–4039.

[75] K. Shapiro, et al., Detection of *Toxoplasma gondii* oocysts and surrogate microspheres in water using ultrafiltration and capsule filtration, Water Res. 44 (2010) 893–903.

[76] D. Aubert, I. Villena, Detection of *Toxoplasma gondii* oocysts in water: proposition of a strategy and evaluation in Champagne-Ardenne Region, France, Mem. Inst. Oswaldo Cruz. 104 (2) (2009) 290–295.

[77] F. Izquierdo, et al., Detection of microsporidia in drinking water, wastewater and recreational rivers, Water Res. 45 (2011) 4837–4843.

[78] O.D. Simmons III, et al., Concentration and detection of Cryptosporidium oocysts in surface water samples by method 1622 using ultrafiltration and capsule filtration, Appl. Environ. Microbiol. 6 (2001) 1123–1127.

[79] V.R. Hill, A.L. Polaczyk, A.M. Kahler, T.L. Cromeans, D. Hahn, J.E. Amburgey, Comparison of hollow-fiber ultrafiltration to the USEPA VIRADEL technique and USEPA method 1623, J. Environ. Qual. 38 (2) (2009) 822–825, https://doi.org/10.2134/jeq2008.0152.

[80] M.A. Borchardt, S.K. Spencer, Concentration of Cryptosporidium, microsporidia and other water-borne pathogens by continuous separation channel centrifugation, J. Appl. Microbiol. 92 (2002) 649–656.

[81] U. Zuckerman, S. Tzipoli, Portable continuous flow centrifugation and method 1623 for monitoring of waterborne protozoa from large volumes of various water matrices, J. Appl. Microbiol. 100 (2006) 1220–1227.

[82] R. Das, Cryptosporidium Detection Through Antibody Immobilization on a Solid Surface, Bangladesh University of Engineering and Technology, 2007.

[83] G. Vesey, P. Hutton, A. Champion, N. Ashbolt, K.L. Williams, A. Warton, D. Veal, Application of flow cytometric methods for the routine detection of Cryptosporidium and Giardia in water, Cytometry 16 (1994) 1–6.

[84] H.D. Lindquist, et al., Using ultrafiltration to concentrate and detect *Bacillus anthracis*, *Bacillus atrophaeus* subspecies globigii, and *Cryptosporidium parvum* in 100-liter water samples, J. Microbiol. Methods 70 (2007) 484–492.

[85] S.W. Stine, F.D. Vladich, I.L. Pepper, C.P. Gerba, Development of a method for the concentration and recovery of microsporidia from tap water, J. Environ. Sci. Health A 40 (2005) 913–925.

[86] S.E. Dowd, et al., Confirmed detection of *Cyclospora cayetanesis, Encephalitozoon intestinalis* and *Cryptosporidium parvum* in water used for drinking, J. Water Health 1 (2003) 117–123.

[87] D.S. Francy, O.D. Simmons III, M.W. Ware, E.J. Granger, M.D. Sobsey, F.W. Schaefer III, Effects of seeding procedure and water quality on recovery of *Cryptosporidium parvum* by using U.S. Environmental Protection Agency Method 1623, Appl. Environ. Microbiol. 70 (2004) 4118–4128.

[88] Y. Qiu, et al., A one-step centrifugal ultrafiltration method to concentrate enteric viruses from wastewater, J. Virol. Methods 237 (2016) 150–153.

[89] R. Taghi Kilani, L. Sekla, Purification of Cryptosporidium oocysts and sporozoites by cesium chloride and percoll gradients, Am. J. Trop. Med. Hyg. 36 (1987) 505–508.

[90] E. Entrala, et al., *Cryptosporidium parvum*: oocysts purification using patssium bromide discontinuous gradient, Vet. Parasitol. 92 (2000) 223–226.

[91] M.E. Warkiani, et al., Capturing and recovering of *Cryptosporidium parvum* oocysts with polymeric micro-fabricated filter, J. Membr. Sci. 369 (1–2) (2011) 560–568.

[92] N.M.M. Pires, T. Dong, Recovery of Cryptosporidium and Giardia organisms from surface water by counter-flow refining microfiltration, Environ. Technol. 34 (17) (2013) 2541–2551.

[93] N.M.M. Pires, T. Dong, A cascade-like silicon filter for improved recovery of oocysts from environmental waters, Environ. Technol. 35 (6) (2014) 781–790.

[94] M. Jimenez, H. Bridle, Microfluidics for effective concentration and sorting of waterborne protozoan pathogens, J. Microbiol. Methods 126 (2016) 8–11.

[95] D. de Oliveira Pinto, L. Urbano, R.C. Neto, Immunomagnetic separation study applied to detection of Giardia spp. cysts and Cryptosporidium spp. oocysts in water samples, Water Supply 16 (1) (2015) 144–149.

[96] R.M. Hoffman, et al., Development of a method for the detection of waterborne microsporidia, J. Microbiol. Methods 70 (2007) 312–318.

[97] A. Dunètre, M.-L. Dardé, Immunomagnetic separation of *Toxoplasma gondii* oocysts using a monoclonal antibody directed against the oocyst wall, J. Microbiol. Methods 61 (2005) 209–217.

[98] I.A. Hamza, et al., Methods to detect infectious human enteric viruses in environmental water samples, Int. J. Hyg. Environ. Health 214 (2012) 426–436.

[99] Q. Ramadan, et al., Flow-through immunomagnetic separation system for waterborne pathogen isolation and detection: application to Giardia and Cryptosporidium cell isolation, Anal. Chim. Acta 673 (1) (2010) 101–108.

[100] D.H. Atha, K.C. Ingham, Mechanism of precipitation of proteins by polyethylene glycols. Analysis in terms of excluded volume, J. Biol. Chem 256 (1981) 12108–12117.

[101] J.C. Lee, L.L. Lee, Preferential solvent interactions between proteins and polyethylene glycols, J. Biol. Chem. 256 (1981) 625–631.

[102] A. Hata, et al., Validation of internal controls for extraction and amplification of nucleic acids from enteric viruses in water samples, Appl. Environ. Microbiol. 77 (2011) 4336–4343.

[103] G.S. Fout, et al., A multiplex reverse transcription-PCR method for detection of human enteric viruses in groundwater, Appl. Environ. Microbiol. 69 (2003) 3158–3164.

[104] D.L. Bronson, A.Y. Elliott, D. Ritzi, Concentration of Rous sarcoma virus from tissue culture fluids with polyethylene glycol, Appl. Microbiol. 30 (1975) 464–471.

[105] R. Boom, et al., Rapid and simple method for purification of nucleic acids, J. Clin. Microbiol. 28 (1990) 495–503.

[106] S. Paulos, et al., Evaluation of five commercial methods for the extraction and purification of DNA from human faecal samples for downstream molecular detection of the enteric protozoan parasites Cryptosporidium spp., *Giardia duodenalis*, and Entamoeba spp, J. Microbiol. Methods 127 (2016) 68–73.

[107] E.-H. Lee, et al., A disposable bacterial lysis cartridge (BLC) suitable for an in situ waterborne pathogen detection system, Analyst 140 (22) (2015) 7776–7783.

[108] I.G. Wilson, Inhibition and facilitation of nucleic acid amplification, Appl. Environ. Microbiol. 63 (1997) 3741–3751.

[109] J.A. Anceno, et al., IMS-free DNA extractionfor the pCR-based quantification of *Cryptosporidium parvum* and *Giardia lamblia* in surface and waster water, Int. J. Environ. Health Res. 17 (2007) 297–310.

[110] D.S. Francy, et al., Comparison of traditional and molecular analytical methods for detecting biological agents in raw and drinking water following ultrafiltration, J. Appl. Microbiol. 107 (2009) 1479–1491.

[111] S.M. Dineen, et al., An evaluation of commercial DNA extraction kits for the isolation of bacterial spore DNA from soil, J. Appl. Microbiol. 109 (2010) 1886–1896.

[112] V. Kamat, et al., A facile one-step method for cell lysis and DNA extraction of waterborne pathogens using a microchip, Biosens. Bioelectron. 99 (2018) 62–69.

[113] R. Martzy, et al., Simple lysis of bacterial cells for DNA-based diagnostics using hydrophilic ionic liquids, Sci. Rep. 9 (1) (2019) 13994.

[114] D. Lees, CEN/WG6/TAG4, International standardisation of a method for detection of human pathogenic viruses in molluscan shellfish, Food Environ. Virol. 2 (2010) 146–155.

[115] M. D'Agostino, et al., Nucleic acid amplification-based methods for detection of enteric viruses: definition of controls and interpretation of results, Food Environ. Virol. 2 (2011) 55–60.

[116] J. Dreier, M. Stormer, K. Kleesiek, Use of bacteriophage MS2 as an internal control in viral reverse transcription-PCR assays, J. Clin. Microbiol. 43 (2005) 4551–4557.

[117] J.A. Lowther, et al., Comparison between quantitative real-time reverse transcription PCR results for norovirus in oysters and self-reported gastroenteric illness in restaurant customers, J. Food Prot. 73 (2010) 305–311.

[118] M.I. Costafreda, A. Bosch, R.M. Pinto, Development, evaluation, and standardization of a real-time TaqMan reverse transcription-PCR assay for quantification of hepatitis A virus in clinical and shellfish samples, Appl. Environ. Microbiol. 72 (2006) 3846–3855.

[119] G.E. Greening, J. Hewitt, Norovirus detection in shellfish using a rapid, sensitive virus recovery and real-time RT-PCR detection protocol, Food Anal. Methods 1 (2008) 109–118.

[120] Y. Roussel, et al., Evaluation of DNA extraction methods from mouse stomachs for the quantification of *H. pylori* by real-time PCR, J. Microbiol. Methods 62 (2005) 71–81.

[121] R. Alaeddini, Forensic implications of PCR inhibition: a review, Forensic Sci. Int. Genet. 6 (2012) 297–305.

[122] F.S. Le Guyader, et al., Detection and quantification of noroviruses in shellfish, Appl. Environ. Microbiol. 75 (2009) 618–624.

[123] M. Gilgen, D. Germann, J. Lüthy, Ph. Hübner, Three-step isolation method for sensitive detection of enterovirus, rotavirus, hepatitis A virus, and small round structured viruses in water samples, Int. J. Food Microbiol. 37 (2–3) (1997) 189–199.

[124] I. Gutiérrez-Aguirre, M. Banjac, A. Steyer, M. Poljšak-Prijatelj, M. Peterka, A. Štrancar, M. Ravnikar, Concentrating rotaviruses from water samples using monolithic chromatographic supports, J. Chromatogr. A 1216 (13) (2009) 2700–2704.

[125] Cryptosporidium i Östersund, Swedish Institute for Communicable Disease Control, 2011. SMI rapport.

Section C

Detection

The main focus of this book is to review and evaluate the wide range of new and developing detection technologies for waterborne pathogens alternative to the traditional microbiological methods, described in Chapter 3. Some of these have been around for decades undergoing continual development and refinement, whereas as others are just emerging. For many, appropriate methods of sample preparation are critical, and advances in sample processing research are just as important as the development of new detection methods. Chapter 4 reviewed the state-of-the-art in sample preparation for viruses, bacteria and parasites.

The numerous detection technologies could be categorised in a range of different ways. There are many simple divisions: by limit of detection into those which have proven capability of single cell detection and those which have not; by the utilisable sample volume, for example, µL, mL, or L; by the ability to discriminate between viable and nonviable pathogens; by the ability to distinguish between different species; or by whether the technique has been proven to work in a range of complex sample matrices or not. All this information is essential to the evaluation of different detection methods, though not always reported in the literature, and we have included the above where available for all the reviewed technologies. Other characteristics important to the adoption of methods for waterborne pathogen monitoring include cost, robustness, reliability, portability and ease of use, even by untrained operators. In summarising and comparing techniques, we have included this information where available.

More complex classifications of the techniques could include by whether the method detects capture of the microorganism onto a surface, through an intrinsic property of the microorganism or via external labelling (or even a combination). Alternatively, the method of signal read-out (e.g. optical, gravimetric or electrical) could be used as a classification. The disadvantage with this approach, however, would be to divide the topic of molecular methods of detection up by the different final detection mode. We have opted to cover the important topic of molecular techniques within one chapter. Therefore, to create the chapters within this section we have chosen to divide up the different methods of detection as illustrated in the below schematic. This approach also means that only Chapter 8 considers the detection of nucleic acids, derived from the pathogens and all other chapters mainly look at the detection of intact pathogens.

The chapters in this section of the book will provide a general introduction to the basic science behind the detection methods they discuss followed by a review of the scientific literature, detailing how the particular methods have been applied to each of the taxa of waterborne pathogens. Each chapter concludes with a summary evaluating the state-of-the-art, highlighting the limitations for widespread application and suggesting future lines of enquiry to overcome some of these challenges and improve the technology.

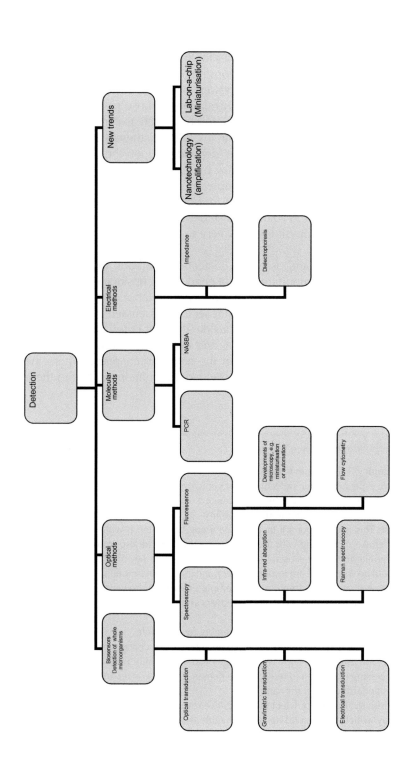

New rapid and sensitive approaches for the detection and enumeration of pathogens in water could have considerable value for the protection of public health and address weaknesses in relying on standard microbiological monitoring, particularly in relation to those pathogens which cannot be cultured or detected by microscopy, such as certain viruses or viable but nonculturable bacterial strains. Ideally, these systems should meet all the below criteria. However, depending upon the type of monitoring application certain features may be more important. For example, rapid multiplexed methods could enable microbial monitoring to be used in an operational sense. Alternatively, for both surveillance and investigative monitoring, the ability to distinguish between strains that pose a threat to human health from physiologically very similar strains that are not a risk would be very useful.

When reviewing, and assessing, detection methods for waterborne pathogens a range of important criteria need to be considered. Further detail regarding how a technology can progress from laboratory experiments to validation and adoption by the water sector is discussed in Chapter 11. Here, however, we present several considerations or criteria which should be remembered when reading about, and evaluating, the different types of detection technology presented in the next few chapters. Where the information was available we have provided information relating to these criteria.

Monitoring methods requirements

- Accuracy: False-positive and false-negative results must be low or preferably zero, especially when detecting pathogenic organisms.
- Assay time: The approach should produce a 'real-time' response, especially to achieve operational monitoring capability.
- Cost: This is an important factor determining how often it will be feasible to undertake monitoring with the technique, which could also impact the types of monitoring that it is useful for. Some settings where new detection methods would be very useful are resource-scarce and inexpensive monitoring is the aim.
- Recovery rate: More important for sample processing; however still essential here that the detection technique recovers and measures a high proportion of entering microorganisms.
- Reproducible: Each assay should be highly reproducible.
- Robust: The detection method should be able to resist changes in temperature, pH, ionic strength to cope with a range of source and finished waters. Sample processing may assist in reducing the demands placed on a particular method by standardisation of the sample.
- Sensitivity: Failure to detect false-negative results, lowers the sensitivity of the assay, which is not good for public health assessments. The LOD should aim towards single organism sensitivity.

- Size and portability: Portable systems avoid the requirement for shipping samples back to a central lab, reducing costs and speeding up detection.
- Specificity: The method should easily discriminate between the target organism and other organisms, and ideally between different species and strains.
- User friendly: The assay should be fully automated, or have the potential for future automation, and require minimal operator skills for routine detection.
- Validation: The assay should be evaluated against current standard techniques and a LOD obtained.
- Viability: The system should indicate whether detected pathogens are viable, and infectious, and ideally detect both viable and nonviable organisms.

(These characteristics are adapted from Table 2 in P. Leonard, et al., Advances in biosensors for detection of pathogens in food and water, Enzym. Microb. Technol. 32(1) (2003) 3–13.)

Chapter 5

Optical detection technologies for waterborne pathogens

Helen Bridle

Institute of Biological Chemistry, Biophysics and Bioengineering, Heriot-Watt University, Edinburgh, Scotland

This chapter will cover a range of optical detection technologies, which either detect the intrinsic properties of the microorganisms themselves or exploit external labels. Optical microscopy of viruses is not possible due to their small size, although it is possible that future developments in microscopy may enable this [1]. Bacteria and parasites can be observed using light microscopy, either with or without fluorescent labels. The use of fluorophores, coupled to specific antibodies or which react with specific intracellular components, eases identification, in some cases to the species level and/or providing viability determination. The existing fluorescence microscopy technology based on manual counting by lab technicians has been detailed in Chapter 3.

Here, an introduction is given to fluorescence and different applications and developments. For example, a brief mention of advances in fluorophore development and alternative optical strategies are given, though the use of nanomaterials for signal enhancement is covered in Chapter 9. Examples of miniaturisation of microscopes into portable systems, developed particularly for the detection of waterborne pathogens, are included as is work relating to the automation of optical imaging for this application. We have attempted to make a distinction between optical biosensors, covered in Chapter 7 and the optical detection methods presented here. We define an optical biosensor as a system where the act of biorecognition (capture of the analyte by a biological element) on a surface produces a change in the optical properties of the surface. Therefore, techniques such as photonic crystals, silicon microring resonators, and fibre optic evanescent field absorption are covered under Section 7.3 (and Sections 7.4 and 7.5 for application to viruses and bacteria) of Chapter 7 and all other optical methods are discussed here. However, at times this distinction has been unclear and some work that fits in this chapter is described by the authors as a biosensor as well as being included in biosensor reviews. This is

Waterborne Pathogens: Detection Methods and Applications. https://doi.org/10.1016/B978-0-444-64319-3.00005-8
117

particularly problematic for 'indirect' optical biosensors, where the detection occurs via a secondary label. Therefore, in some cases, work is referred to in both chapters.

The chapter first deals with optical detection techniques requiring sample labelling, mainly fluorescence (Section 5.1). A brief explanation of the fluorescence process, and different approaches, is provided. An alternative method for visualisation of pathogens is the use of enzymes to generate a coloured precipitate. In this section, we also describe attempts to automate the existing protocols, presented in Chapter 3, as well as general developments towards miniaturised microscopes.

Flow cytometry can utilise light scattering to discriminate between pathogens by size and shape, in a label-free way. However, the combination with fluorescence labelling offers greater selectivity and the ability to detect viability and even species; therefore, flow cytometry is presented under Section 5.1.

Section 5.2 deals with label-free technologies. In terms of intrinsic, label-free methods, vibrational spectroscopy techniques, including infra-red absorption, and Raman spectroscopy are described in detail.

5.1. Techniques using labelling

5.1.1 Fluorescence

The use of fluorescence is a popular method of detection. The Jablonski diagram below illustrates the fluorescence process. When a fluorophore is illuminated with light, the absorbance of a photon of sufficient energy results in excitation of an electron from the ground-state into higher electronic energy states. Various means are available to that electron to relax, losing the excess energy and returning to the ground state; one of these is fluorescence where a longer wavelength photon is remitted.

For a more detailed introduction to fluorescence and the different processes which can occur we recommend books, such as the Fundamentals of light microscope and electronic imaging, or for basic introduction websites with online tutorials, such as the Invitrogen website, are useful and have good animations (Fig. 5.1).

However, there are some drawbacks to fluorescence detection including the additional costs of the fluorescent label, problems of photobleaching, and the challenges to obtaining specificity. Photobleaching is a process by which the repeated cycling of a fluorophore between the ground and excited states eventually leads to molecular damage with a gradual reduction of fluorescence emission intensity from a sample over time. This limits the length of time for which a sample can be observed. Many fluorophore labels are conjugated to antibodies and therefore the specificity depends upon that of the antibody. Often, it is not possible to distinguish between different species, especially for protozoa where the cell membrane structure is generally highly conserved between species.

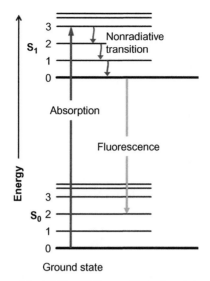

FIG. 5.1 Fluorescence explained: Jablonski diagram illustrating photon absorption, nonradiative internal conversion and finally fluorescence emission. *(From https://commons.wikimedia.org/wiki/ File:Jablonski_Diagram_of_Fluorescence_Only.png?uselang=en-gb.)*

For *Cryptosporidium* detection the fact the antibody binds all species is advantageous since the test thus determines whether any oocysts are present. The disadvantage is that no information is gained upon the human pathogenicity of detected oocysts. Another problem is that this antibody displays some cross-reactivity to algal cells, which are of a similar size to oocysts and this can complicate detection. Antibodies are discussed further in Chapter 7, as they are often utilised in biosensors to capture the pathogen of interest.

Fluorophores can also be conjugated to a wide variety of other molecules or incorporated into detection schemes where interaction with other molecules controls the production of fluorescence (Fig. 5.2). We have seen above how

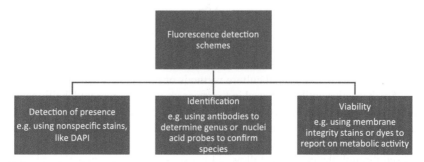

FIG. 5.2 Overview of fluorescent approaches to determining information about microorganisms.

antibody-based fluorophores can be utilised to bind to antigens on the cell surface. Alternatively, chemicals such as 4′,6-diamidino-2-phenylindole (DAPI) can be employed, which become fluorescent upon interaction with nucleic acids. DAPI can, therefore, be used for nonspecific detection of microorganisms or for an indication of viability. For *Cryptosporidium* DAPI is utilised to stain the nuclei in the sporozoites thus enabling confirmation of sporozoite presence, and therefore that the oocyst is unexcysted.

Finally, fluorescent stains can be employed to indicate viability, using for example propidium iodide (PI) or fluorescein diacetate (FDA). PI is membrane impermeable and therefore exclusion from a cell suggests that the membrane is intact. FDA is membrane permeable and is cleaved by enzymes in the cell to generate a fluorescent product, which is membrane impermeable. FDA, therefore, indicates both intracellular activity and membrane integrity. Further examples are given in [2].

An alternative means to detect viability is a fluorescent means of measuring enzyme redox activity. This was applied to protozoa in 2001 using 5-cyano-2,3-ditolyl tetrazolium chloride (CTC) to determine the redox activity potential of the respiratory electron transport system as an indicator of viability [3]. Waterborne Inc. was involved in the study but the findings do not yet seem to have translated into a commercial product.

There has been much development in superresolution imaging to identify smaller structures, in ways to image live cells and also in fluorophore development, all of which are outside the scope of this book though we recommend the following review articles for more information [4–7]. Some aspects of improved fluorophores are covered in Chapter 9, particularly quantum dots.

The use of fluorescence as a means of detection requires the availability of suitable probes. Fluorescence methods are being developed for emerging zoonotic pathogens such as leptospira [8]. Direct fluorescence analysis quicker than culturing techniques and easier to perform than molecular methods. The authors suggest polymerase chain reaction (PCR) is a complex and costly analysis for the underdeveloped countries that have the highest incidence of leptospirosis. Molecular methods are described in more detail in Chapter 8.

5.1.1.1 Bacteria

In 2003 a commercially available (Analyte 2000, Research International) fibre optic waveguide biosensor was applied using antibodies to capture the bacteria and fluorescent labels for detection. Positive samples were then subjected to further analysis by traditional culture methods or molecular methods.

An alternative antibody-based indirect sensor with fluorescent labelling was reported by Ho et al. in 2004. In this approach capture antibodies were immobilised on the interior surface of a microcapillary, through which the test sample subsequently flowed. Next, liposome secondary antibody conjugate was passed through the capillary, followed by a rinse to remove any unbound conjugate. The final detection step involved lysis of these liposomes to release the encapsulated fluorescent molecules, giving a limit of detection (LOD) of 360 cells/mL in 45 min. Liposomes can encapsulate 10^5–10^6 fluorescent molecules, thus offering a means of signal amplification, which resulted in the low LOD reported.

In 2005, Zhu et al. reported a LOD of 10^2 cells/mL in 2 h for *Escherichia coli* O157:H7 using a fluorescent sandwich immunoassay and fibre-optics for detection [9]. In 2011, U-bent optical fibres were adopted in an attempt to improve the LOD. However, during the 20 min allowed for detection, detection of concentrations lower than 10^2 cfu/mL was not possible due to diffusion-limited transport to the sensor surface [10]. Very recent work has applied bacteriophages on optical fibres, with a LOD similar to the previous reports of 10^3 cfu/mL [11].

5.1.1.2 Protozoa

Raptor plus, a portable optical sensor developed by Research International (Monroe, Washington, United States) was tested for *Cryptosporidium* detection [12]. With a detection mechanism similar to an immunofluorescent assay, described by the authors as a fibre-optic biosensor, target oocysts are anchored on the tip of an optical waveguide by antibodies binding, and then washed with reporter antibodies.

A laser diode is used to excite the fluorescence through an optical fibre inserted in a miniaturised optical set-up moulded in a disposable polystyrene chip. The light from the reporter antibodies is then coupled back into the waveguide and detected by a photodiode. A limit of detection of 10^6 oocysts per mL was obtained. However, when the oocysts were boiled prior to detection, a tenfold decrease in the LOD was observed (10^5 oocysts/mL). Although portable and highly integrated, this technique has several drawbacks. First, it relies on a heat treatment that destroys the oocyst viability, making a viability assessment difficult, although a comparison between preboiled and boiled samples might give an indication. Second, it necessitates a sample preparation, including concentration and heat treatment, as well as a labelling step, limiting continuous real-time operation. Third, being based on an antibody assay, the technique cannot provide information on the species of *Cryptosporidium*.

5.1.2 Alternatives to direct fluorescent labelling

Chemiluminescence is the emission of light as a result of a chemical reaction, and for detection schemes, enzymes are often incorporated to multiply the detectable product, with the aim of increasing sensitivity. The product can be fluorescent or colourimetric, i.e. a colour change discernable by the human eye. One problem is the discrimination between different colours or between shades of a colour indicating concentration differences, especially in variable ambient light conditions. Therefore, colour changes can be read-out using techniques like ultraviolet (UV)-vis spectroscopy. The advantages of colourimetric detection are that complex optics for excitation and detection are not required, especially for qualitative studies, and recent developments mean the results could even be read-out using a smartphone [13], enabling portable devices for use in the field.

The enzyme-chemiluminescence approach was incorporated into an optical biosensor set-up by Song and Kwon, who developed a photodiode array, using a secondary antibody conjugated to alkaline phosphatase, giving a LOD of 10^4 cells [14]. As signal detection depends upon the action of the enzyme-producing a blue precipitate, the absorbance of which is detected, this sensor is very sensitive to pH. pH sensitivity could be a problem for the monitoring of waterborne pathogens as the pH of raw and finished waters is variable.

Karsunke and colleagues performed chemiluminescence within a lab-on-a-chip system (see Chapter 10 for further discussion of microfluidics and device miniaturisation). Bacteria, such as *E. coli* O157:H7, *Salmonella typhimurium*, and *Legionella pneumophila*, were captured with polyclonal antibodies and labelled using antibodies labelled with biotin and horseradish peroxidase (HRP)-streptavidin conjugates. Detection then occurs on the addition of luminol and hydrogen peroxidase, following the enzymatic (HRP) generation of chemiluminescence, achieving LODs of 1.8×10^4 cells/mL, 2.0×10^7 cells/mL, and 7.9×10^4 cells/mL, respectively for the above bacterial pathogens.

A 2013 paper by Larsson et al. reports the paper-based colourimetric detection of viruses (Fig. 5.3). Polyelectrolyte modified paper was utilised for the filtration of viruses which were then detected using an HRP-based detection scheme. The reported LOD was 5×10^4 pfu/mL, said to be comparable to a similar enzyme linked immunosorbent assay (ELISA) HRP detection protocol.

ATP-based detection systems have also been reported for bacteria, and recently merged into automated, microfluidic systems [15].

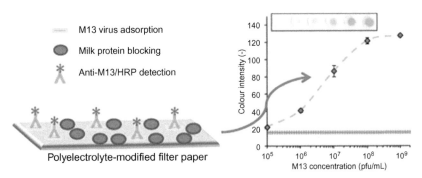

FIG. 5.3 Paper-based colourimetric detection of viruses. *Left*: Schematic of the detection scheme. *Right*: Range and detection limits of the assay using a sample volume of 0.50 mL at pH 5. Values are means of six discs given with 95% confidence limits. The *parallel lines* indicate background intensity and corresponding confidence limits. The *inset* shows scanned sample discs (10^5–10^9), including a background disc (no phage; far left). *(Graphical abstract from P.A. Larsson, et al., Filtration, adsorption and immunodetection of virus using polyelectrolyte multilayer-modified paper, Colloids Surf. B Biointerfaces 101 (2013) 205–209. Reproduced with permission.)*

Intrinsic fluorescence of four different bacterial species was analysed using multivariate analysis to enable discriminate between species, though the authors recognise there would be considerable challenges in real-world application, e.g. sample preparation for concentration and clean-up as well as dealing with variability due to growth phase [16].

5.1.3 Automation of existing methods

Automation of the testing process reduces costs, through a removing need for highly trained microbiologists, and reduces the possibility of contamination. Continuous, on-site operation also speeds up detection. Here we discuss automation of existing techniques based on optical means of detection. Developments in miniaturised microscopy are also briefly presented.

For traditional faecal indicator culture-based methods an automated system for on-site testing has been developed, through the detection of enzymatic activity with a 'Pathogen Detection System' that is capable of processing two 100 mL samples. The system uses a UV light source and charge coupled device (CCD) spectrometer and was found to perform as well as traditional microscopy and cell culture approaches [17]. The technology is now known as TECTA, part of ENDETEC, from Veolia Water Solutions and Technologies. It uses a patented polymer to sense the fluorescent indicator of enzyme activity. The company claims a dynamic range of detection from 1 to 10^6 CFU in 100 mL for *E. coli*, with the time required for detection inversely proportional to the level of contamination, i.e. 2 h for 10^6 and 18 h for 1 CFU.

A fully automated system to perform the US EPA 1623 method for the detection of *Cryptosporidium* was developed by the former company, Shaw Water Ltd., now no longer trading. The system comprised a filtration unit capable of pumping 1000 L within 24 h, complemented by a microfluidic chip known as the Crypto-Tect bioslide that enabled automated staining and counting of the *Cryptosporidium* oocysts. The Crypto-Tect bioslide is a 3-in. silicon wafer with 84 etched channels arranged in a circular way around a filter. Samples are introduced through the inlet and drawn by capillary action, or other means into the channel at the end of which oocysts are captured onto a filter membrane plug where they can be observed through a microscope. The company also developed automated image recognition software for the identification of the samples. Staining can be done by flowing the dye in the channel, the circular chip can be rotated to allow automated inspection of all channels and the filter membrane plug can be removed for further inspection.

Miniaturisation of optical microscopes also presents a way forward for cheaper and more portable instruments. In 2011, US researchers developed an integrated microscope weighing just 2 g and with a size slightly bigger than the end of a pencil [18], illustrated in Fig. 5.4. For waterborne pathogens, in 2010, Mudanyali et al. built a portable holographic microscope and developed a rapid image reconstruction algorithm, as well as an automated counting method. An incoherent

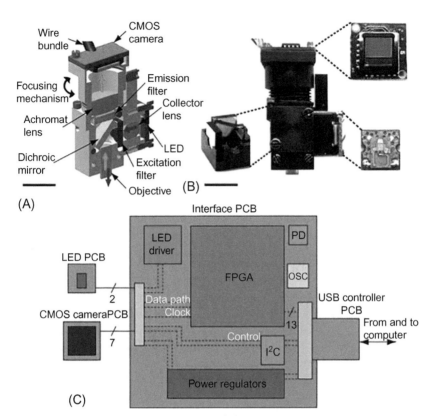

FIG. 5.4 Design of the integrated microscope. (A) Computer-assisted design of an integrated microscope, shown in cross-section. *Blue* and *green* arrows mark illumination and emission pathways, respectively. (B) Image of an assembled integrated microscope. Insets, filter cube holding dichroic mirror and excitation and emission filters *(bottom left)*, PCB holding the CMOS camera chip *(top right)* the LED illumination source *(bottom right)*. Scale bars: 5 mm. (C) Schematic of the electronics for real-time acquisition and control. *(Figure 1 from K.K. Ghosh, et al., Miniaturized integration of a fluorescence microscope, Nat. Methods 8(10) (2011) 871–878.)*

light source was used by the lightweight microscope to illuminate the sample of interest, while a Complementary Metal Oxide Semiconductor (CMOS) chip acquired holographic images of the sample in a 60 μL sample volume. A LOD of 380 cysts/mL was reported for *Giardia lamblia* and the automated counting algorithm proved capable of distinguishing between *Cryptosporidium parvum*, *G. lamblia*, microbeads, and dust particles. Without preconcentration, the system was incapable of accurately detecting 189 cysts/mL, though this could be improved by the use of sample processing techniques [19].

A 2012 paper showed how a cell phone camera could be used to detect *E. coli* to a LOD of ~ 5–10 cfu/mL in a fluorescent assay where the excitation light was provided by battery-powered inexpensive LEDs. Specificity was confirmed using *Salmonella* spiked samples as well as the detection in a complex matrix, like milk [20]. This device tested a 1 mL sample by pumping at a flow rate of 50 μL/min for 20 min. Another smartphone detection system employed a microfluidic microlens system, together with sample processing for immunoagglutination detection of bacteria [21]. An automated smartphone microscopy platform was used for *Giardia*, incorporating machine learning, and enabling field-use [22].

5.1.4 Flow cytometry

Flow cytometry has been proposed as detection technology for waterborne pathogens for several decades. Initially, the high cost and complexity of the instrumentation as well as the need for trained cytometrist proved off-putting, as described in a review by Veal et al. in 2000 [23]. This article covered some of the improvements undertaken in the 1990s rendering flow cytometry instruments easier to operate and improving detection. Despite the advantages of flow cytometry it has not become widely adopted in routine environmental monitoring applications. This is probably due to the challenge of standardising measurements for complex samples [23]. However, many water companies do use flow cytometry, especially to count, measure, and in some cases separate particles, including (oo)cysts. Flow cytometry is also used in the preparation of spike samples for analytical quality control.

The use of flow cytometry for waterborne pathogen detection was comprehensively reviewed in 2017 [24], with water companies interested in exploring the potential of this technique, and a UK workshop focussed on this approach was held in May 2018. The conclusion of the review paper was that a couple of years of data gathering and benchmarking would be necessary for each utility to understand the limits when using flow cytometry as a faecal indicator screen [24]. Further increases in adoption may occur as further smartphone-enabled flow cytometry platforms to emerge to enable lower-cost monitoring [25].

Flow cytometry measures the physicochemical properties of particles as they pass through an observation channel. As the particles enter the channel they are focussed using sheath flow to attempt to ensure single-particle passage. The light source employed is typically a laser and the resulting forward-scatter and side-scatter is measured (Figs 5.5 and 5.6). Fluorescence can also be employed for either auto-fluorescent particles or labelled particles.

Flow cytometers can be used for sorting a defined number of particles by either a mechanical (cheap, simple but slow) or droplet-based (fast, 10,000 cells/s but potentially risky with aerosol-forming pathogens) approach, to isolate target populations for further analysis or directly for detection.

In comparison with other methods flow cytometry often delivers an undercount compared to microscopy because sometimes clusters could be perceived by flow cytometer as one cell although it avoids false negatives in microscopy due to viewer fatigue [26]. Sonication has been suggested as a potential route to overcome this drawback, though finding optimal parameters is challenging

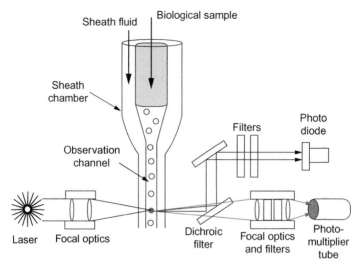

FIG. 5.5 Schematic of flow cytometry setup. *(Reproduced Figure 2 from C. Gruden, S. Skerlos, P. Adriaens, Flow cytometry for microbial sensing in environmental sustainability applications: current status and future prospects, FEMS Microbiol. Ecol. 49(1) (2004) 37–49.)*

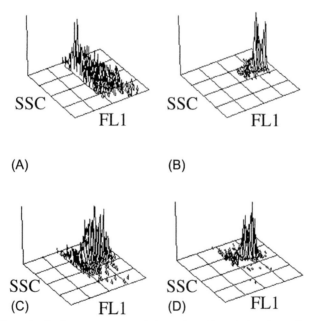

FIG. 5.6 Example of a flow cytometry result. Detection of bacteria in milk by flow cytometry. Samples were stained with SYTO BC and analysed using a FACSCalibur instrument. (A) Raw milk; (B) bacteria in phosphate-buffered saline; (C) raw milk cleared of lipids only; (D) raw milk cleared of proteins and lipids. FL1 wavelength range was 515–565 nm. *(Data were provided by T. Gunasekera. Figure 7 from D.A. Veal, et al., Fluorescence staining and flow cytometry for monitoring microbial cells, J. Immunol. Methods 243(1–2) (2000) 191–210.)*

[24]. Compared to culture-based methods flow cytometry often reports an over-count as 1 CFU could form from more than one initial cell [23].

The advantages of flow cytometry are that it is faster and more automated than culture-based techniques and that microorganisms are undisturbed by the analysis and therefore remain available for further testing, if required [24]. It is also compatible with many different fluorescent detection approaches, allowing for a range of information to be collected (Table 5.1).

Disadvantages of flow cytometry are that considerable processing is still needed and especially for those microorganisms with low detection limits flow cytometry may still need to be combined with microscopy. Flow cytometry is also challenged by the complex matrices of environmental samples, which may contain significant volumes of particulate matter of a similar size, some of which might auto-fluoresce, preventing the use of certain fluorophores.

In terms of sampling considerations for samples to be analysed by flow cytometry, the biggest challenge is to avoid aggregation of particles as they can cause blockage of the cytometer and also lead to undercounts. The use of resuspension in tetrasodium pyrophosphate has been suggested to reduce aggregation [23]. Immunomagnetic separation (IMS) is another useful option to reduce potential blockages.

Flow cytometry has been shown to distinguish between microorganisms on the basis scattering and auto-fluorescence alone, e.g. discrimination of yeast and bacteria using light scattering on the basis of size and shape. Changes in forward-scatter have also allowed for discrimination between excysted and

TABLE 5.1 Molecular methods for microbial sensing compatible with flow cytometry.

Assay type	Criteria	Specific examples
Nonspecific	Nucleic acid stains	Picogreen, SYBR-Green
	Cytoplasm stains	Rhodamine
Specific	Antigen-based	Immunoassays, fatty acid signatures
	Nucleic acid-based	DNA/RNA probes, FISH PCR, sequencing, ribotyping
Cell functioning	Membrane integrity	Propidium iodide
	Enzyme activity	Tetrazolium salts, fluorogenic enzyme assays, esterase assays
	Membrane activity	Rhodamine, Bis-oxonol

Modified from Table 2 in C. Gruden, S. Skerlos, P. Adriaens, Flow cytometry for microbial sensing in environmental sustainability applications: current status and future prospects, FEMS Microbiol. Ecol. 49(1) (2004) 37–49.

unexcysted *Cryptosporidium* oocysts. With fluorescent stains flow cytometry has also been utilised for detection of viruses [23]. For protozoa, a LOD of 100 parasites per litre was reported so preconcentration steps are needed [27].

Flow cytometry has been utilised for: the detection of *Cryptosporidium* (Vesey) and rotavirus [28] in drinking water; the detection of bacteria [29] and the protozoa *Cryptosporidium* and *Giardia* [30] in wastewater; the detection of marine viruses [31] and bacteria in surface water; the detection of legionella [32], *Cryptosporidium* [33] and bacteria [34] in groundwater; and the detection of *Giardia* [26, 35] in faeces. Table 5.2 summarises the above results, giving information upon the detection approach, LOD, and sample processing requirements. Flow cytometry has also been used to distinguish between viable and viable but nonculturable bacteria [36].

One drawback of flow cytometers is that they are large and expensive machines. However, despite high initial costs, a cost comparison study determined that automation and low consumables usage can actually result in lower costs per sample, particularly when processing large numbers [24]. There has been much development towards miniaturised systems though the typical 'portable' system reported is still somewhat large. A 2012 paper by Keserue and coworkers described the use of a portable flow cytometer for disaster relief [37]. However, this system was still relatively large, measuring $43 \times 37 \times 16$ cm and weighing 17 kg. Some progress has been made towards very small, microfluidic scale flow cytometry [23, 38] and this is discussed in detail in Chapter 10.

An alternative to flow cytometry is solid-phase cytometry (SPC). In SPC, the principles of epifluorescence microscopy and cytometry are combined. Microorganisms on a filter surface are fluorescently labelled and automatically counted by a laser scanning device. Subsequently, the data for each fluorescent spot are analysed by epifluorescence microscopy. The major advantages of this system are the short analysis time per filter and the low detection limit of only a few fluorescent particles. Viable bacteria in drinking and ultrapure water have been detected via standard plate count (SPC) [39]. Fluorescence in situ hybridisation (FISH) protocols for the quantification of *Vibrio cholerae* [40] and *Legionella* with SPC have also been developed for a platform to improve immunofluorescence based detection and enumeration of *Cryptosporidium* [41].

5.2. Spectroscopy

Spectroscopy is the interaction of radiated energy with matter, resulting in absorption, reflection or scattering (both elastic and inelastic), providing information about characteristics of the material. There are many different types of spectroscopy, including infrared (IR), Raman, multiwavelength UV/vis spectroscopy, and hyperspectral imaging.

Spectroscopy collects fingerprints of microorganisms across a range of wavelengths. For microbial samples, these spectral are most commonly DNA/ RNA, proteins, and membrane and cell wall amine- and fatty acid-containing

TABLE 5.2 Overview of waterborne pathogen detection by flow cytometry.

Pathogen	Detection approach	LOD	Sample processing	Reference
Marine viruses	Stained with nuclei acid dye SYBR green-I	Not reported	Small samples taken (1.5 mL), fixed and frozen	[25]
Rotavirus	Infection of cell lines, followed by fluorescein labelling	2–20	Concentrated from 500 L with MK II filters, eluted and organic flocculation to 50 mL	[23]
Bacteria (indigenous aquifer bacterium)	CFDA/SE stained, fluorescent microsphere used for calibration	1×10^3	1 mL from of 35 mL original sample	[27]
Legionella	FITC-labelled monoclonal antibody	10^2–10^3 cfu/L	Centrifugation to concentrate, IMS to isolate	[26]
Cryptosporidium	Oocyst wall fluorescently immunostained	Not confirmed but 20 oocysts/L detected	100 mL of wastewater concentrated by calcium carbonate flocculation and then centrifugation	[24, 29]
Giardia	Cyst wall fluorescently immunostained	Not confirmed but 40 oocysts/L detected	100 mL of wastewater concentrated by calcium carbonate flocculation and then centrifugation	[20, 24]

components. While this can provide extremely detailed information there is also a challenge involved in the identification of organisms from this data. There is the need to build up a spectral library, containing a number of representative spectra against which unknown samples can be compared. There is also the requirement to characterise the effects of the sample matrix on signals and to ensure the pathogen library is representative of a range of conditions, e.g. live, newly shed, environmentally aged, nonviable pathogens, for which the establishment of a database would be highly useful [42].

Other potential problems with spectroscopic techniques are interference from the substrate upon which the sample is immobilised and, in some set-ups, the necessity of long detection times to achieve good signal-to-noise ratios. However, the advantages of these methods are that they are low energy, noninvasive, and nondestructive with the potential to provide highly detailed information at the single organism level. A detailed review and comparison of the two main methods covered here, IR and Raman spectroscopy, was conducted in 2009 by Harz et al. [43] and more recently for food samples [44–46].

5.2.1 Infrared spectroscopy

IR spectroscopy is a technique where polychromatic light is applied to the sample, which absorbs a photon of this light whenever the frequency (energy) of the light is equal to the energy required for a particular bond to vibrate. All molecules undergo different forms of vibration, e.g. stretching, bending, at temperatures above absolute zero. IR spectroscopy works for bonds where the molecular dipole moment changes during the vibration. For more information we recommend the reader consult standard Chemistry textbooks. Different modes of IR spectroscopy are illustrated in Fig. 5.7.

An attenuated total reflectance (ATR) set up was selected in work aiming at developing automated, continuous monitoring for viruses [47]. The ATR crystal is made of geranium (Ge), which is a semiconductor and therefore also is operated to enable charge based virus collection. The collection efficiency depended upon the sample size, e.g. distance to be travelled to reach the surface. The technique only measures close to the surface thus preventing interference from other components in the solution through the sample collection is thus critical [47]. The experiments were shown to distinguish between poliovirus and MS2 bacteriophage even in variable tap water samples, with a LOD of 10^3 pfu/mL (Fig. 5.8)

IR spectroscopy has been applied for total bacterial counts [48], and to identify and differentiate microorganisms on both the species and strain levels [43]. Integrated microfluidic systems are also being developed, e.g. the 2018 work incorporating an ultrasonic trap in a microfluidic device with IR detection [49]. Further information can be deduced about the impact of environmental conditions over time. For example, Lu et al. studied the impact of storage at different

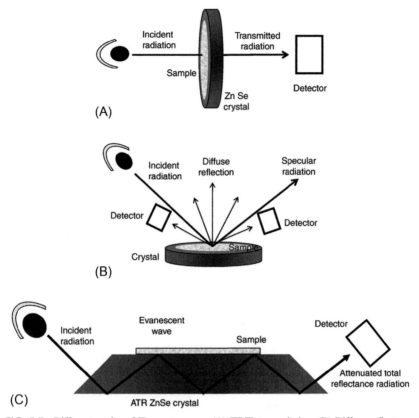

FIG. 5.7   Different modes of IR spectroscopy. (A) FT-IR transmission. (B) Diffuse reflectance FT-IR. (C) FT-IR-attenuated total reflectance. *(Reproduced from A. Alvarez-Ordonez, et al., Fourier transform infrared spectroscopy as a tool to characterize molecular composition and stress response in foodborne pathogenic bacteria, J. Microbiol. Methods 84(3) (2011) 369–378.)*

temperatures of the survival of the bacteria *E. coli* O157:H7, *Campylobacter*, and *Pseudomonas* [50] (see Fig. 5.9) While this sensitivity can be advantageous one potential difficulty with the widespread adoption of this method is that there may be considerable variation between spectra from the same species, depending upon the prior experienced environmental conditions.

The authors mention that an advantage of this technique over the traditional culture-based approaches is that it is has the ability to detect bacterial injury. For example, sublethally injured bacteria may not be detected by other means, e.g. culture-based test, but since they have the ability to repair themselves and resume growth under favourable conditions, a culture method may give a false negative result whereas IR spectroscopy would detect them. These authors employed Fourier-Transform-IR (FT-IR) spectroscopy in the mid-IR range of the electromagnetic spectrum (4000–400 cm^{-1}) and concentrated 50 mL of the sample using an aluminium oxide membrane filtration. In contrast to other

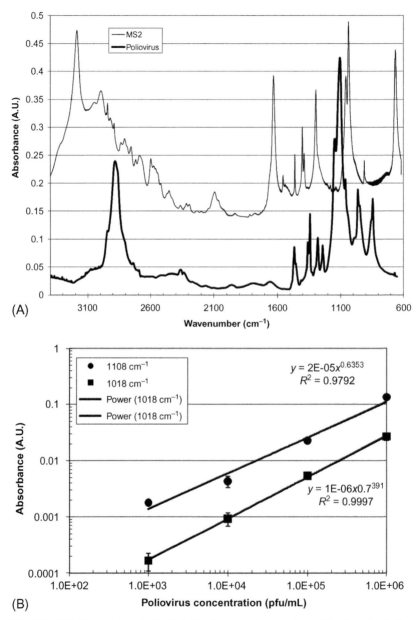

FIG. 5.8 ATR IR-spectroscopic discrimination between viruses and quantification of detection. (A) Infrared spectra of dried MS2 and poliovirus collected as transmission measurements through Ge chips. The spectrum of MS2 has been baseline shifted for ease of interpretation by adding 0.14 to all points in the MS2 spectrum. (B) Height of the sugar $(1108\,cm^{-1})$ and COO acetate $(1018\,cm^{-1})$ absorbance features for various concentrations of poliovirus deposited onto Ge chips. Lines represent power law best fits to the data. *(Reproduced from Figures 3 and 5 from C.A. Vargas, et al., Integrated capture and spectroscopic detection of viruses, Appl. Environ. Microbiol. 75(20) (2009) 6431–6440.)*

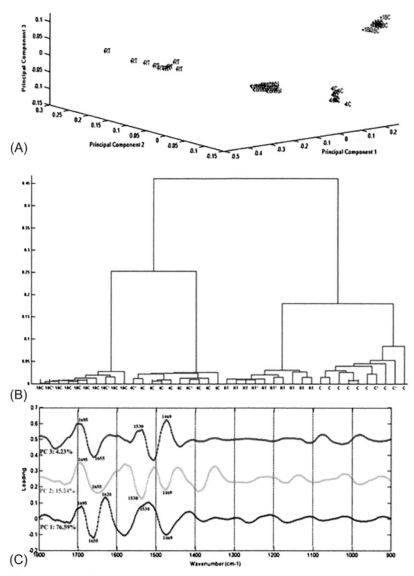

FIG. 5.9 Impact of storage time/age of bacteria on Raman results. (A). Principal component analysis of *E. coli* O157:H7 treated under room temperature for 5 days (RT), 4°C for 12 days (4C), −18°C for 20 days (−18C), and control (control). The cluster analysis model was validated by new treatments *(red colour in figure)*. (B). Hierarchical cluster analysis of *E. coli* O157:H7 treated under room temperature for 5 days (RT), 4°C for 12 days (4C), −18°C for 20 days (−18C), and control (C). The dendrogram function analysis model was validated by new treatments *(* in figure)*. (C). Loading plots of the first *(red)*, second *(green)*, and third *(blue)* principal component obtained from principal components-discriminant function analysis (PC-DFA) of FT-IR spectra of *E. coli* O157:H7 treated under room temperature for 5 days, 4°C for 12 days, −18°C for 20 days and control. *(Reproduced from X. Lu, et al., Using of infrared spectroscopy to study the survival and injury of Escherichia coli O157:H7, Campylobacter jejuni and Pseudomonas aeruginosa under cold stress in low nutrient media, Food Microbiol. 28(3) (2011) 537–546.)*

membranes, this aluminium oxide one contributes no spectral features between 4000 and 1000 cm^{-1} thus allowing direct entry to the FT-IR system. Better reproducibility with direct membrane capture, as opposed to prior centrifugation, is also obtained [50].

FT-IR has also been used to study the impact of stress on *Salmonella* as well as distinguish between viable and nonviable *E. coli* among other studies focussed on foodborne pathogens [51]. The approach could easily be translated to waterborne pathogens. IR has also been utilised for bacterial typing [52, 53]. to study heavy metal exposed bacteria [54], and identify antimicrobial resistance [52]. Bacterial IR detection has also been performed in bottled drinking water samples [55]. A recent study found transmittance mode coupled with a novel algorithm (random forest) for classification offered good discrimination of waterborne bacteria [56].

ATR-FT-IR has been applied to study the surface properties of *Cryptosporidium* oocysts. This article highlighted the large spectral differences between oocysts from different sources (see Figure 6 in the original article) [57]. The technique has also been applied to study surface adhesion of oocysts [58].

5.2.2 Raman spectroscopy

The Raman effect, the inelastic scattering of light, was discovered by Chandrashekhara Venkata Raman in 1928. When monochromatic light shines on a sample the majority of it passes through or is absorbed, depending upon the nature of the sample and the wavelength of the light. However, a small proportion (around 1%) of this light is scattered, either elastically, with the same frequency as the incident light (Rayleigh scattering), or inelastically, at a different frequency (Raman scattering) (see Fig. 5.10). The sample gains (or loses) some energy from the light and thus the frequency shift corresponds to vibrational energy shifts in molecules within the sample. Raman works for vibrations where a polarisability change occurs. The resulting Raman spectrum is a 'fingerprint' of vibrational modes with a molecule, providing a uniquely identifiable 'signature' (see Fig. 5.11). Raman can also be used for imaging and an excellent review of this area is provided in Ref. [59].

This inelastic scattering accounts for only 1 in 10$^{6-7}$ photons incident upon the sample, and therefore the Raman signal is obviously very weak. Various other means of enhancing Raman measurements have been developed including resonance Raman spectroscopy, stimulated Raman spectroscopy, coherent anti-Stokes Raman spectroscopy, and surface-enhanced Raman spectroscopy (SERS). Confocal Raman, Raman imaging, and mapping and Raman-activated cell-sorting have also been developed. All these methods and developments have been recently reviewed by Li et al. [42], Das et al. [60], and Stöckel et al. [61].

Raman has been applied to microbial analysis for both foodborne and waterborne pathogens. There is concern that cultural conditions can influence the

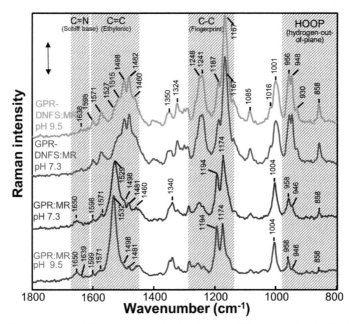

FIG. 5.10 Illustration of Raman spectra, showing how the different wavelength signals correspond to different chemical bonds. *(Figure reproduced under the Creative Commons Attribution Licence from Figure 4 in G. Mei, N. Mamaeva, S. Ganapathy, P. Wang, W.J. DeGrip, K.J. Rothschild, Raman spectroscopy of a near infrared absorbing proteorhodopsin: Similarities to the bacteriorhodopsin O photointermediate, PLoS ONE 13(12) (2018) e0209506. https://doi.org/10.1371/journal. pone.0209506.)*

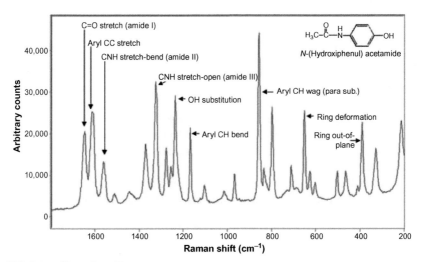

FIG. 5.11 Illustration of Raman spectra.

reproducibility and discrimination ability of Raman though one study has shown that even though there is some variation, identification accuracy and taxonomic resolution are maintained [62]. Tripathi and colleagues found that comparing distilled and tap water there was no significant effect on Raman spectra and that Raman discrimination performance was not influenced by sample ageing [63].

SERS has been used to distinguish between various viruses, including adeno-, noro-, and rotavirus, with a LOD of 10^2 viruses [64]. In addition to species discrimination, viability measurements have also been successfully demonstrated [65]. Confocal Raman has also been used for discrimination between bacteriophages and *E. coli*. The LOD was reported as 10^9 pfu/mL for bacteriophages and 10^6 cells/mL though this was considered to be due to the small sampling volume of the confocal system with signals actually resulting from around 100 cells [66]. In order to utilise small sampling areas, e.g. micro-Raman, Raman analysis was combined with an alternating current (AC) electrokinetically enhanced sampling method for virus detection [45].

Bacteria quantification was attempted by Escoriza et al. who trialled an approach measuring Raman signals directly on filter materials. The challenge with this method was the choice of filter material. The substrate is in general very important for Raman spectroscopy; given that the signal is so weak and complex to interpret the substrate should have a low background signal, low absorbance, and high optical reflectance as well as being nonreactive chemically. Although the membrane filters selected by these authors were expected to give low levels of background signal this was not the case and high levels of bacterial coverage (10^7 cells/13 mm membrane) were needed for detection. The authors propose that future work should concentrate on the development of new filter technologies that can overcome these problems [67]. The authors acknowledge that Raman of samples smeared onto microscope slides is simpler and faster (in terms of detection) but this obviously requires additional sample processing. One further challenge with using membranes is that the spectra can vary considerably between cells on membranes and those on slides (Fig. 5.12). Single bacteria/mL Raman detection was reported in 2017 based on dielectrophoresis on a microfluidic chip for sample concentration coupled with SERS [68]. Strain specification was reported in the previous work though it has been recognised elsewhere that this depends on the quality of the data used as a database or training set, with the use of training sets being suggested as most appropriate for environmental samples [69]. Meat-associated bacteria have been analysed and classified using Raman [70, 71], and transferability to waterborne pathogens should be possible.

Several Raman techniques have been applied to the detection of the protozoan, *Cryptosporidium*. The surface enhanced resonance Raman scattering (SERRS) method was reported to enable speciation between three species of *Cryptosporidium*: *C. parvum*, *C. hominis*, and *C. meleagridis* to the subspecies level, and it was observed that fresh oocysts (sample a few months old) and old oocysts (sample older than 12 months) had different fingerprints [72]

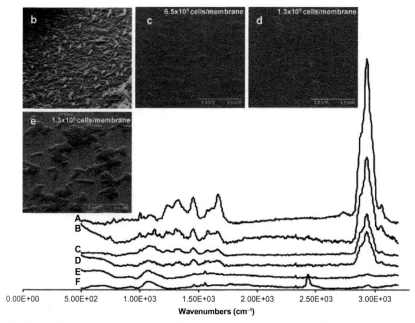

FIG. 5.12 *E. coli* Raman spectra. Part (b) was obtained from a colony of *E. coli* that was grown *on top of a membrane*. Parts (c), (d), and (e) have a decreasing concentration of cells (6.5×10^9 cells/membrane, 1.3×10^9 cells/membrane, and 1.3×10^8 cells/membrane, respectively) obtained through filtration. Even though there is a complete coverage of the membrane by cells in parts (b)–(d), spectrum B (corresponding to part b) is the only one that completely conserves bacterial signal. Thus, increased coverage does not seem to solve the issue except if a thick bacterial colony exists on the filter (b). This was unexpected since complete coverage is not needed for strong conserved bacterial signals on the aluminum slides. *E. coli* cells on Whatman membranes at different concentrations and their respective Raman spectrum. SEM pictures show a front view of the cells on the membranes. The spectra correspond to membranes on the parts with exception to parts (a) and (f) that correspond to *E. coli* applied on aluminum-slide and clean Whatman membrane, respectively. Raman spectra were acquired at 100×, 15 s acquisition time, 10 averages, and 200 mW. *(Reproduced from Figure 3 in M.F. Escoriza, et al., Raman spectroscopy and chemical imaging for quantification of filtered waterborne bacteria, J. Microbiol. Methods 66(1) (2006) 63–72.)*

(Fig. 5.13). A significant drawback of SERRS is the long data acquisition time, typically 15–20 min per oocyst depending on the range of wavelengths used. CARS offers acquisition of a *Cryptosporidium* oocyst in just a few seconds [73] and these authors suggest this would enable near real-time processing of oocysts in a water sample (assuming a preconcentration step). A patent has also been published relating to a Raman based method for assessing the occurrence of *Cryptosporidium* in a water sample and claiming the possibility of differentiating between viable and nonviable oocysts [74]. No LOD was reported. *Giardia* detection is also mentioned in the patent [74] as well as by Rule et al. [75].

To alleviate the bulkiness of current Raman spectroscopy instruments, researchers have reported successful attempts to miniaturise probes [76], and

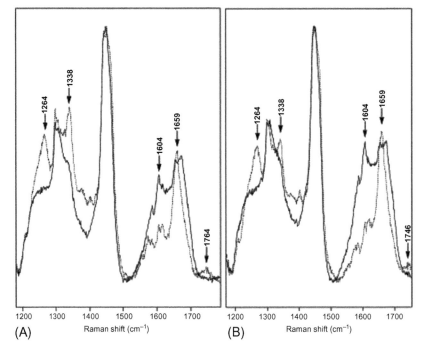

FIG. 5.13 *Cryptosporidium* Raman spectra. Illustration of the Raman technique. A window from the SERS spectra of two *C. parvum* genotype 2 strains: (A) a calf isolate from Iowa and (B) an isolate from a foodborne outbreak in Maine, comparing the fingerprints of recently passaged (- - -) and old (—) samples. Note that the old samples give similar fingerprints, whereas the recently passaged oocysts can be readily differentiated, e.g. by the ratios of the peak heights at $1659\,cm^{-1}$ and in the $1250-1350\,cm^{-1}$ range. Background corrected. *(Reproduced with the permission from A.E. Grow, et al., New biochip technology for label-free detection of pathogens and their toxins, J. Microbiol. Methods 53(2) (2003) 221–233.)*

handheld Raman spectrometers are commercialised by several companies such as HoribaScientific (2011), Intevac (2011) and Gammadata (2011). The development of portable systems is covered in a recent review, looking at forensic and homeland security applications [77].

5.2.3 Other methods

Multiwavelength UV/vis spectroscopy was presented in one conference paper showing significant spectral differences between three different pathogens (Fig. 5.14) [78]. Hyperspectral imaging is an emerging technique that has been applied to foodborne pathogens, finding more recognition and applications within the past 5 years [44, 79–82]. A recent review highlighted some outstanding challenges with the approach, e.g. long spectral acquisition times, spatial resolution and perhaps most challenging for adaption to water samples is the limit of detection, with the majority of existing studies concentrating on

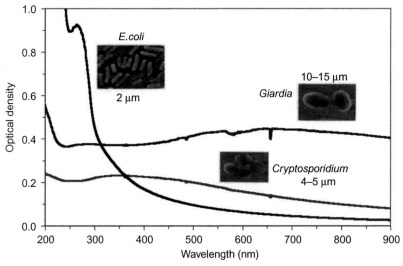

FIG. 5.14 Multiwavelength UV/vis spectroscopy for pathogen discrimination.

relatively high levels of spoilage organisms [81]. A digital holographic microscope has been combined with automated detection and classification for *Giardia* detection [83]. Laser scattering within a microfluidic device has also been applied to protozoa, with potential for adaptation to bacteria, though it was unclear how the approach would handle the numerous particulate matter present in water, given it was tested on reagent water [84]. Viruses have also been imaged using surface plasmon scattering allowing rapid imaging [85].

5.3. Summary

This chapter has covered a range of optical techniques and reviewed the position with regard to waterborne pathogens.

Firstly, fluorescence-based detection was discussed. While well-established as a means of detection, the need for reagents is a problem in terms of cost and storage, especially for portable applications or those in remote areas. New fluorescent probes offer greater stability but at an increased price. Further details of nanotechnology developments for signal enhancement are given in Chapter 9. However, the potential of automated and miniaturised systems with the wide-spread acceptance of fluorescent technologies bodes well for future applications. Lightweight, inexpensive microscopes seem very attractive for field-testing applications, and the authors of the paper using a mobile phone to read out the results stated that they those this method as many people in developing countries own cell phones. One challenge for such systems will be to integrate with miniaturised or simple sample-processing techniques to genuinely have a full, and portable, system. Some examples of integration are emerging but larger volumes are needed for effective sample processing and detection.

Automation of existing technology is well-advanced with the TECTA system and the Shaw Water system. The former automates the faecal indicator procedure enabling faster results and on-site testing. This perhaps moves microbial quality monitoring towards a position where results are obtained fast enough for operational, reactive monitoring. However, disadvantages remain with regard to the use of faecal indicator organisms and the lack of correlation to some pathogens. With the latter system, the company is no longer trading although the system would have offered a means of faster results and on-site testing. Discussion of some of the barriers to market entry and challenges companies face in developing new technology for the water utilities is presented in Chapter 11. However, it is clear that while both these automated approaches speed up monitoring and reduce technician time, the disadvantages of the traditional approaches remain, particularly the lack of information on species and viability.

Miniaturisation and/or automation is perhaps also the hope of flow cytometry for future applications, enabling wider access to such technology which at present is expensive and bulky. The ability to extract a wide range of information by appropriate staining procedures is very attractive. While very useful in cell sorting, rare cell detection remains challenging. Sample processing is again very important to prevent system blockages and also to prestain the pathogen of interest. SPC offers one potential solution.

Spectroscopic techniques avoiding the use of any labels have an immediate advantage in terms of simplified sample preparation (no need for any staining or rinsing steps) and reduced reagent cost. In some cases, relatively high LODs have been reported though spectra from single cells have been recorded, showing the potential for single-cell 'fingerprinting'. These noninvasive methods also for subsequent testing, if required. There is also the potential for these methods to provide highly detailed information allowing identification to the strain level as well as the ability to distinguish between viable, nonviable, and temporarily damaged bacteria. However, there are several major challenges for this technology.

Most importantly, a spectral library is needed characterising the results of numerous pathogen screens, enabling the identification of an unknown sample. More evidence is also required to determine the impact of environmental factors on the spectroscopic results. Some studies claim a high degree of sensitivity to, for example, age of oocysts, with this presented as evidence of the high-quality information delivered by this approach. One particularly exciting approach is the possibility to determine antimicrobial resistance (AMR), which could find important applications in understanding the spread of AMR in the environment. Others report robustness of detection between samples with different environmental histories, claiming this proves the technique can cope with a wide variety of input waters. Finally, instrument miniaturisation would also be of benefit to spectroscopic techniques as would the design of new instruments capable of faster sample scanning. With some set-ups acquiring a spectrum can be very time-consuming, and work is ongoing to reduce acquisition times. More work has been applied to foodborne pathogens and potential learning from recent developments in this sector could stimulate waterborne pathogen studies.

Overall, it seems as though there are several promising optical technologies, though for all of the instrument design and miniaturisation, as well as robust validation, is key to more widespread acceptance in the water quality monitoring arena.

References

[1]   Z. Wang, et al., Optical virtual imaging at 50 nm lateral resolution with a white-light nanoscope, Nat. Commun. 2 (2011) 218.

[2]   L.J. Robertson, B.K. Gjerde, Cryptosporidium oocysts: challenging adversaries? Trends Parasitol. 23 (8) (2007) 344–347.

[3]   R. Iturriaga, et al., Detection of respiratory enzyme activity in Giardia cysts and Cryptosporidium oocysts using redox dyes and immunofluoresce techniques, J. Microbiol. Methods 46 (2001) 19–28.

[4]   S. van de Linde, M. Heilemann, M. Sauer, Live-cell super-resolution imaging with synthetic fluorophores, Annu. Rev. Phys. Chem. 63 (2012) 519–540.

[5]   B. Hötzer, I.L. Medintz, N. Hildebrandt, Fluorescence in nanobiotechnology: sophisticated fluorophores for novel applications, Small 8 (15) (2012) 2297–2326.

[6]   J. Walker-Daniels, http://www.labome.com/method/Live-Cell-Imaging-Methods-Review.html.

[7]   W.H. Liu, G. Zheng, Photactivatable fluorophores and techniques for biological imaging applications, Photochem. Photobiol. Sci. 11 (3) (2012) 460–471.

[8]   S. Schreier, et al., Development of a magnetic bead fluorescence microscopy immunoassay to detect and quantify Leptospira in environmental water samples, Acta Trop. 122 (1) (2012) 119–125.

[9]   P. Zhu, et al., Detection of water-borne *E. coli* O157 using the integrating waveguide biosensor, Biosens. Bioelectron. 21 (4) (2005) 678–683.

[10]   R. Bharadwaj, et al., Evanescent wave absorbance based fiber optic biosensor for label-free detection of *E. coli* at 280 nm wavelength, Biosens. Bioelectron. 26 (7) (2011) 3367–3370.

[11]   S.M. Tripathi, et al., Long period grating based biosensor for the detection of *Escherichia coli* bacteria, Biosens. Bioelectron. 35 (1) (2012) 308–312.

[12]   M.F. Kramer, et al., Development of a Cryptosporidium oocyst assay using an automated fiber optic-based biosensor, J. Biol. Eng. 1 (3) (2007) 1–11.

[13]   L. Shen, J.A. Hagen, I. Papautsky, Point-of-care colorimetric detection with a smartphone, Lab Chip (2012).

[14]   J.M. Song, H.T. Kwon, Photodiode array on-cip biosensor for the detection of *E. coli* O157:H7 pathogenic bacteria, Methods Mol. Biol. 503 (2009) 325–355.

[15]   C.B. Hansen, et al., Monitoring of drinking water quality using automated ATP quantification, J. Microbiol. Methods 165 (2019) 105713.

[16]   H.-M. Fang, et al., Sensing water-borne pathogens by intrinsic fluorescence, in: M. Ghandehari (Ed.), Optical Phenomenology and Applications : Health Monitoring for Infrastructure Materials and the Environment, Springer International Publishing, Cham, 2018, pp. 133–147.

[17]   M. Habash, R. Johns, Comparison study of membrane filtration direct count and an automated coliform and *Escherichia coli* detection system for on-site water quality testing, J. Microbiol. Methods 79 (1) (2009) 128–130.

[18] K.K. Ghosh, et al., Miniaturized integration of a fluorescence microscope, Nat. Methods 8 (10) (2011) 871–878.

[19] O. Mudanyali, et al., Detection of waterborne parasites using field-portable and cost-effective lensfree microscopy, Lab Chip 10 (18) (2010) 2419–2423.

[20] H. Zhu, U. Sikora, A. Ozcan, Quantum dot enabled detection of *Escherichia coli* using a cell-phone, Analyst 137 (11) (2012) 2541–2544.

[21] T.-F. Wu, et al., Rapid waterborne pathogen detection with mobile electronics, Sensors (2017) **17**(6).

[22] H.C. Koydemir, et al., Automated detection and enumeration of waterborne pathogens using mobile phone microscopy and machine learning, in: 2017 Conference on Lasers and Electro-Optics (CLEO), 2017.

[23] D.A. Veal, et al., Fluorescence staining and flow cytometry for monitoring microbial cells, J. Immunol. Methods 243 (1–2) (2000) 191–210.

[24] S. Van Nevel, et al., Flow cytometric bacterial cell counts challenge conventional heterotrophic plate counts for routine microbiological drinking water monitoring, Water Res. 113 (2017) 191–206.

[25] Z. Li, S. Zhang, Q. Wei, Smartphone-based flow cytometry, in: J.-Y. Yoon (Ed.), Smartphone Based Medical Diagnostics, Academic Press, 2020, pp. 67–88 (Chapter 5).

[26] B.R. Dixon, et al., A comparison of conventional microscopy, immunofluorescence microscopy and flow cytometry in the detection of *Giardia lamblia* cysts in beaver fecal samples, J. Immunol. Methods 202 (1997) 27–33.

[27] M. Bouzid, D. Steverding, K.M. Tyler, Detection and surveillance of waterborne protozoan parasites, Curr. Opin. Biotechnol. 19 (3) (2008) 302–306.

[28] F.X. Abad, R.M. Pinto, A. Bosch, Flow cytometry detection of infectious rotaviruses in environmental and clinical samples, Appl. Environ. Microbiol. 64 (1998) 2392–2396.

[29] C. Gruden, S. Skerlos, P. Adriaens, Flow cytometry for microbial sensing in environmental sustainability applications: current status and future prospects, FEMS Microbiol. Ecol. 49 (1) (2004) 37–49.

[30] B.C. Ferrari, K. Stoner, P.L. Bergquist, Applying fluorescence based technology to the recovery and isolation of Cryptosporidium and Giardia from industrial wastewater streams, Water Res. 40 (3) (2006) 541–548.

[31] D. Marie, et al., Enumeration of marine viruses in culture and natural samples by flow cytometry, Appl. Environ. Microbiol. 65 (1999) 45–52.

[32] S. Riffard, et al., Occurrence of Legionella in groundwater: an ecological study, Water Sci. Technol. 43 (2001) 99–102.

[33] J.B. Rose, Environmental ecology of Cryptosporidium and public health implications, Annu. Rev. Public Health 18 (1997) 135–161.

[34] M.F. DeFlaun, et al., Comparison of methods for monitoring bacterial transport in the subsurface, J. Microbiol. Methods 47 (2001) 219–231.

[35] H.A. El-Nahas, et al., Giardia diagnostic methods in human fecal samples: a comparative study, Cytometry B Clin. Cytom. 84B (1) (2013) 44–49.

[36] M.M.T. Khan, B.H. Pyle, A.K. Camper, Specific and rapid enumeration of viable but nonculturable and viable-culturable gram-negative bacteria by using flow cytometry, Appl. Environ. Microbiol. 76 (2010) 5088–5096.

[37] H.A. Keserue, et al., Comparison of rapid methods for detection of Giardia spp. and Cryptosporidium spp. (oo)cysts using transportable instrumentation in a field deployment, Environ. Sci. Technol. 46 (2012) 8952–8959.

[38] Y.-C. Tung, et al., PDMS-based opto-fluidic micro flow cytometer with two-color, multi-angle fluorescence detection capability using PIN photodiodes, Sens. Actuators B 98 (2–3) (2004) 356–367.

[39] M. Riepl, et al., Applicability of solid phase cytometry and epifluorescence microscopy for rapid assessment of the microbiological quality of dialysis water, Nephrol. Dial. Transplant. 26 (11) (2011) 3640–3645.

[40] S. Schauer, R. Sommer, A. Kirschner, Rapid detection and enumeration of Vibrio cholerae with CARD-FISH combined with Solid-phase Cytometry, Appl. Environ. Microbiol. 78 (20) (2012) 7369–7375.

[41] M. Montemayor, et al., Comparative study between two laser scanning cytometers and epifluorescence microscopy for the detection of Cryptosporidium oocysts in water, Cytometry A 71 (2007) 163–169.

[42] M. Li, et al., Single cell Raman spectroscopy for cell sorting and imaging, Curr. Opin. Biotechnol. 23 (2012) 56–63.

[43] M. Harz, P. Rosch, J. Popp, Vibrational spectroscopy: a powerful tool for the rapid identification of microbial cells at the single-cell level, Cytometry A 75A (2009) 104–113.

[44] M. Kamruzzaman, Y. Makino, S. Oshita, Non-invasive analytical technology for the detection of contamination, adulteration, and authenticity of meat, poultry, and fish: a review, Anal. Chim. Acta 853 (2015) 19–29.

[45] M.R. Tomkins, D.S. Liao, A. Docoslis, Accelerated detection of viral particles by combining AC electric field effects and micro-Raman spectroscopy, Sensors (Basel, Switzerland) 15 (1) (2015) 1047–1059.

[46] S. Hameed, L. Xie, Y. Ying, Conventional and emerging detection techniques for pathogenic bacteria in food science: a review, Trends Food Sci. Technol. 81 (2018) 61–73.

[47] C.A. Vargas, et al., Integrated capture and spectroscopic detection of viruses, Appl. Environ. Microbiol. 75 (20) (2009) 6431–6440.

[48] S. Numthuam, et al., Method development for the analysis of total bacterial count in raw milk using near-infrared spectroscopy, J. Food Saf. 37 (3) (2017) e12335.

[49] S. Freitag, et al., Towards ultrasound enhanced mid-IR spectroscopy for sensing bacteria in aqueous solutions, in: SPIE BiOS, vol. 10491, SPIE, 2018.

[50] X. Lu, et al., Using of infrared spectroscopy to study the survival and injury of *Escherichia coli* O157:H7, *Campylobacter jejuni* and *Pseudomonas aeruginosa* under cold stress in low nutrient media, Food Microbiol. 28 (3) (2011) 537–546.

[51] A. Alvarez-Ordonez, et al., Fourier transform infrared spectroscopy as a tool to characterize molecular composition and stress response in foodborne pathogenic bacteria, J. Microbiol. Methods 84 (3) (2011) 369–378.

[52] U. Sharaha, et al., Using infrared spectroscopy and multivariate analysis to detect antibiotics' resistant *Escherichia coli* bacteria, Anal. Chem. 89 (17) (2017) 8782–8790.

[53] C. Quintelas, et al., An overview of the evolution of infrared spectroscopy applied to bacterial typing, Biotechnol. J. 13 (1) (2018) 1700449.

[54] R. Gurbanov, A.G. Gozen, F. Severcan, Rapid classification of heavy metal-exposed freshwater bacteria by infrared spectroscopy coupled with chemometrics using supervised method, Spectrochim. Acta A Mol. Biomol. Spectrosc. 189 (2018) 282–290.

[55] H.M. Al-Qadiri, et al., Rapid detection and identification of *Pseudomonas aeruginosa* and *Escherichia coli* as pure and mixed cultures in bottled drinking water using Fourier transform infrared spectroscopy and multivariate analysis, J. Agric. Food Chem. 54 (2006) 5749–5754.

[56] K.-X. Mu, et al., Near infrared spectroscopy for classification of bacterial pathogen strains based on spectral transforms and machine learning, Chemom. Intel. Lab. Syst. 179 (2018) 46–53.

[57] M.A. Butkus, J.T. Bays, M.P. Labare, Influence of surface characteristics on the stability of *Cryptosporidium parvum* oocysts, Appl. Environ. Microbiol. 69 (7) (2003) 3819–3825.

[58] X. Gao, J. Chorover, In-situ monitoring of *Cryptosporidium parvum* oocyst surface adhesion using ATR-FTIR spectroscopy, Colloids Surf. B Biointerfaces 71 (2) (2009) 169–176.

[59] S. Stewart, et al., Raman imaging, Annu. Rev. Anal. Chem. 5 (2012) 337–360.

[60] R.S. Das, Y.K. Agrawal, Raman spectroscopy: recent advancements, techniques and applications, Vib. Spectrosc. 57 (2011) 163–176.

[61] S. Stöckel, et al., The application of Raman spectroscopy for the detection and identification of microorganisms, J. Raman Spectrosc. 47 (1) (2016) 89–109.

[62] D. Hutsebaut, et al., Effect of culture conditions on the achievable taxonomic resolution of Raman spectroscopy disclosed by three Bacillus species, Anal. Chem. 76 (21) (2004) 6274–6281.

[63] A. Tripathi, et al., Waterborne pathogen detection using Raman spectroscopy, Appl. Spectrosc. 62 (1) (2008) 1–9.

[64] C. Fan, et al., Detecting food- and waterborne viruses by surface-enhanced Raman spectroscopy, J. Food Sci. 75 (5) (2010) 302–307.

[65] K. Rule Wigginton, Surface enhanced raman spectroscopy as a tool for waterborne pathogen testing, in: Civil and Environmental Engineering, Polytechnic Institute and State University, Virginia, 2008, p. 150.

[66] L.J. Goeller, M.R. Riley, Discrimination of bacteria and bacteriophages by Raman spectroscopy and surface-enhanced Raman spectroscopy, Appl. Spectrosc. 61 (7) (2007) 679–685.

[67] M.F. Escoriza, et al., Raman spectroscopy and chemical imaging for quantification of filtered waterborne bacteria, J. Microbiol. Methods 66 (1) (2006) 63–72.

[68] C. Wang, et al., Detection of extremely low concentration waterborne pathogen using a multiplexing self-referencing SERS microfluidic biosensor, J. Biol. Eng. 11 (1) (2017) 9.

[69] J.-C. Baritaux, et al., A study on identification of bacteria in environmental samples using single-cell Raman spectroscopy: feasibility and reference libraries, Environ. Sci. Pollut. Res. 23 (9) (2016) 8184–8191.

[70] S. Meisel, et al., Identification of meat-associated pathogens via Raman microspectroscopy, Food Microbiol. 38 (2014) 36–43.

[71] S. Pahlow, et al., Isolation and identification of bacteria by means of Raman spectroscopy, Adv. Drug Deliv. Rev. 89 (2015) 105–120.

[72] A.E. Grow, et al., New biochip technology for label-free detection of pathogens and their toxins, J. Microbiol. Methods 53 (2) (2003) 221–233.

[73] S. Murugkar, et al., Chemically specific imaging of cryptosporidium oocysts using coherent anti-Stokes Raman scattering (CARS) microscopy, J. Microsc. (Oxford) 233 (2) (2009) 244–250.

[74] S. Stewart, J.S. Maier, P.J. Treado, Water Quality Monitoring by Raman Spectral Analysis, 2005. US.

[75] K.L. Rule, P.J. Vikesland, Surface-enhanced resonance Raman spectroscopy for the rapid detection of *Cryptosporidium parvum* and *Giardia lamblia*, Environ. Sci. Technol. 43 (4) (2009) 1147–1152.

[76] H. Sato, et al., Biomedical applications of a new portable Raman imaging probe, J. Mol. Struct. 598 (1) (2001) 93–96.

[77] E.L. Izake, Forensic and homeland security applications of modern portable Raman spectroscopy, Forensic Sci. Int. 202 (2010) 1–8.

[78] Y.D. Mattley, L.H. Garcia-Rubio, Multiwavelength spectroscopy for the detection, identification and quantification of cells, in: SPIE National Meeting. East Boston, 2000.

[79] A.A. Gowen, et al., Hyperspectral imaging: an emerging process analytical tool for food quality and safety control, Trends Food Sci. Technol. 18 (12) (2007) 590–598.

[80] Y. Peng, et al., Potential prediction of the microbial spoilage of beef using spatially resolved hyperspectral scattering profiles, J. Food Eng. 102 (2) (2011) 163–169.

[81] A.A. Gowen, et al., Recent applications of hyperspectral imaging in microbiology, Talanta 137 (2015) 43–54.

[82] G. Foca, et al., The potential of spectral and hyperspectral-imaging techniques for bacterial detection in food: a case study on lactic acid bacteria, Talanta 153 (2016) 111–119.

[83] L. Nan, Z. Jiang, X. Wei, Emerging microfluidic devices for cell lysis: a review, Lab Chip 14 (6) (2014) 1060–1073.

[84] W. Huang, et al., Microfluidic multi-angle laser scattering system for rapid and label-free detection of waterborne parasites, Biomed. Opt. Express 9 (4) (2018) 1520–1530.

[85] X. Lu, et al., Fast detection to single nanoparticle and virus by using surface Plasmon scattering imaging, in: JSAP-OSA Joint Symposia 2016 Abstracts, Optical Society of America, Niigata, 2016.

Chapter 6

Electrochemical detection

Timothée Houssin[a,b], Helen Bridle[c], and Vincent Senez[a,b]
[a]*University Lille Nord de France, Villeneuve d'Ascq, France,* [b]*Institute of Electronics, Microelectronics and Nanotechnology, Lille, France,* [c]*Institute of Biological Chemistry, Biophysics and Bioengineering, Heriot-Watt University, Edinburgh, Scotland*

6.1. Introduction

The focus of this chapter is on analytical methods that use a measurement of electrical potential, charge, or current that can find applications in the detection of a pathogen and in the analysis of its potential effect on a living organism.

The contact between an electrical conductor and an ionic conductor induces several electrochemical phenomena occurring in series and/or in parallel. They involve mass transport and electrochemical reactions. Phenomenological models [1, 2] and microscopic theories [3] predict relatively well the observations in a large spectrum of experimental conditions. However, there is no general theory of charge transfer in these electrical-ionic systems [4]. Over the last century, various techniques have been developed to identify these various phenomena and when possible to develop practical applications. Recently, interest in electrochemistry has been renewed, driven by new nanotechnologies applications in energy conversion, water treatment, materials processing, and biotechnology.

The measurement of the electrical impedance changes of a biological tissue contaminated by pathogens (Fig. 6.1A) [5] or the measurement of the oxidation–reduction potential changes of a functionalised carbon nanotube by a captured pathogen (Fig. 6.1B) [6] are good examples to illustrate the challenges facing the development of electrochemical biosensor. In the first case, we want to detect the variation in charge transport far from the electrode due to the interaction between the pathogen and the cell tissue while in the second case, we want to detect the variation in the concentration of the redox species at the interface between the ionic and electronic conductors due to the presence of the pathogen. These systems are particularly complex and they involve various electrochemical phenomena that are characterised by different time constants.

So, when developing the application and thus having to choose the technique to monitor these changes, one has first to wonder if one of the phenomena has a much longer time constant that will impose its own rate on the overall process and second if the corresponding phenomena are indeed modified by the

Waterborne Pathogens: Detection Methods and Applications. https://doi.org/10.1016/B978-0-444-64319-3.00006-X
147

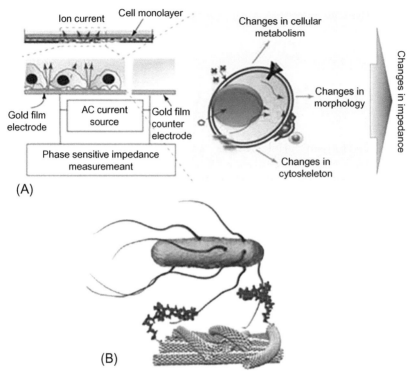

FIG. 6.1 (A) Schematic representation of an electric cell–substrate impedance sensing (ECIS) measurement system; (B) Schematic representation of the interaction between a target bacteria and a hybrid system aptamer-single-walled carbon nanotube. *((A) From I. Giaever, C.R. Keese, A morphological biosensor for mammalian cells, Nature 366 (1993) 591–592; (B) From G.A. Zelada-Guillen, S.V. Bhosale, J. Riu, F.X. Rius, Real-time potentiometric detection of bacteria in complex samples, Anal. Chem. 82 (2010) 9254–9260.)*

appearance of the pathogen in the system. Then, one can choose the monitoring technique among various possibilities where either the electrical voltage or the electrical current is imposed in (non)-steady state conditions and where the nonperturbed quantity (either current or voltage) is observed in time or frequency domains. One can also choose between large amplitude signal analysis and small amplitude signal analysis. The techniques which use small amplitude signals (i.e. a few mV peak-to-peak) on a polarisation point assume that the response is a linear function of the applied solicitation giving access to the electrochemical impedance of the system. In such conditions, any type of perturbing signal (i.e. sine wave, step, etc.) gives the same information. The advantage of nonsteady state conditions (i.e. variation of the amplitude or frequency of the solicitation) is their capability to separate the contributions of the various phenomena in the overall response of the system. Finally, a distinction can be made between static (i.e. no charge flow) and dynamic (i.e. charge flow) conditions.

Historically, many electroanalytical techniques have been developed and it may be difficult to understand their specificity for the newcomer in this field. However, they can be classified into four main categories: potentiometry, voltammetry, coulometry, and impedancemetry [7]. These techniques use two, three, or four electrodes depending on the specific measurement conditions. The electrode's material has to be chosen carefully since it has a direct effect on the electrochemical reactions at the interface. Potentiometric measurement is a static technique and is discussed in Section 6.2. The three other techniques are dynamic and can be carried out either in steady or nonsteady state conditions. In voltammetry (see Section 6.3), the current is measured as a function of a fixed or variable potential. In coulometry (see Section 6.4), the current is measured as a function of time. In impedancemetry (see Section 6.5), the current is measured as a function of voltage frequency. In parallel to these four main electrochemical techniques, the biosensing community is showing a growing interest in dielectrophoresis (see Section 6.6). This technique uses two, four, or eight electrodes to build a 3D, static or dynamic, nonuniform map of the electric field within the sample. The interaction of the electric field with the particle dipole induces a linear or rotational displacement that can be used to characterise the particle. Scaling down of the electrodes is a very active research field driven by the recent progress of miniaturisation technologies. However, the isomorphic reduction of the dimension of the electrode has a direct effect on the relative importance of the various physical phenomena that can be electrochemically monitored as discussed in Section 6.7. Section 6.8 gives a review of the literature on how the above techniques have been applied to pathogen detection, in particular, waterborne pathogens, and what are the current developments that would definitely impact electrochemical techniques in the near future. The chapter is briefly summarised in Section 6.9 and is concluded with perspectives on the potential use of these electrochemical technologies in waterborne pathogen detection.

6.2. Potentiometry

Potentiometry measures, through a high input impedance equipment (potentiometer), the oxidation–reduction potentials difference of a solution between a working or indicator electrode and a counter electrode under static conditions. Electrodes used for providing a reference potential usually include hydrogen, saturated calomel, or silver chloride (AgCl). However, for routine analytical work, a hydrogen electrode is rarely chosen due to difficulty in the preparation and use of the setup. The silver chloride electrode is preferred when the temperature of the system is not controlled. The working electrode measures the current generated by the system without any external excitation. In the modern arrangement, each electrode is placed in a separate solution and connected to the potentiometer instrument, while a salt bridge, containing an inert electrolyte, such as potassium chloride, is exposed to each sample, completing an

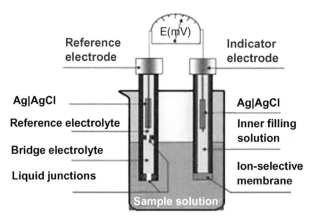

FIG. 6.2 Typical arrangement for a potentiometric sensor. It consists of an indicator and a reference electrode that are immersed in the sample solution and connected to the two terminals of a voltmeter. The working electrode is made of an ion-selective membrane. This membrane separates the sample solution and the inner filling solution of the working electrode. *(From E. Lindner, B.D. Pendley, A tutorial on the application of ion-selective electrode potentiometry: an analytical method with unique qualities, unexplored opportunities and potential pitfalls, Anal. Chim. Acta 762 (2013) 1–13.)*

electrical circuit (Fig. 6.2). Usually, the counter electrode is called the anode and the working electrode is referred to as the cathode.

Measurements are always made when no or very little current is present, so the composition of the substance measured is not altered, making quantitative analysis possible. In such conditions, the measured potential can be related to the concentration of one or more electroactive analytes present in the sample. Indeed, this potential is the difference between the reduction potentials for the redox reaction at the cathode and the anode, those reduction potential depends linearly with the logarithm of the ionic concentration according to the formulation introduced by Walther Nernst in 1889 (Eq. 6.1).

$$E = E° - (RT / nF) * \ln \left(a_{\text{Red}} / a_{\text{Ox}} \right) \tag{6.1}$$

where $E°$ is the standard-state potential, R is the gas constant, T is the temperature in Kelvins, n is the number of electrons in the redox reaction, F is Faraday's constant, and $a_{\text{Red}}/a_{\text{Ox}}$ is the reaction quotient (ratio between the reductant and oxidant electrochemical activities).

Indicator electrodes can be metallic types or membrane versions, which are also called ion-selective electrodes (ISE). Metal electrodes can be subdivided into four kinds. In the first kind, the metal electrode is in direct contact with the electrolyte. If ions of this metal are contained in the system, then the equilibrium is obtained at the metal surface depending on the concentration of the metal ions in the solution. Metal ions are accepted by the metal surface and simultaneously released into the electrolyte. However, this type of electrode suffers from poor selectivity since it can react to any more easily reduced cations. The metal electrode of the second kind consists of metal either coated with or immersed in one of its soluble salts (e.g. AgCl). This electrode reacts to the anions of the salt. The metal electrode of the third kind uses two equilibrium reactions to respond to a cation other than that of the metal electrode. If the electrolyte does not contain any ions of the corresponding metal then metal electrodes can still form an oxidation/reduction potential if a redox reaction occurs in the electrolyte. The electrode surface is inert to the redox reaction. No metal ions are released from the metal; in this case, the metal surface only acts as a catalyst for the electrons. Typically, gold or platinum are used for the metal indicator electrode as they are chemically inert (do not contribute to the reaction). This electrode is the fourth kind of metal electrodes.

Nowadays, potentiometry usually uses electrodes made selectively sensitive to the ion of interest, such as a fluoride-selective electrode. Membrane electrodes have a filling solution sealed inside. This solution contains ions to which the membrane is selective. If there is a difference in the activity of these ions on the two sides of the membrane, ions will enter the membrane from the side where the activity is higher, and they will exist the membrane on the other side. This ion flow alters the electrochemical properties of the membrane and causes a change in potential. A perfect selectivity to one type of ions is almost never possible. Most ion-sensitive electrodes often react with ions with similar chemical properties or a similar structure. These membranes can be made of either a solid-state material (e.g. a crystalline or polycrystalline inorganic salt like Ag_2S or LaF_3, or noncrystalline like glass or polymer) or a liquid material (e.g. an aqueous solution of Ca^{2+}). The glass-membrane electrode used in a pH meter belongs to this family of electrodes. It was discovered by Cremer in 1906. It measures the difference in the activity of hydronium ions (H_3O^+).

The working range for most ISE is from 0.1–1 M to 10^{-5}–10^{-11} M. This broad range is significantly greater than many other analytical techniques. There are several limitations to the determination of an analyte's activity by potentiometry. One problem is that standard-state potentials ($E°$) are temperature-dependent, and the values in reference tables usually are for a temperature of 25°C. A second problem is that standard-state potential depends on the chemical composition of the solution. A third problem is a fact that a small electrical current may pass through the cell. A final limitation is the appearance of junction potentials at the interface of two ionic solutions (i.e. at the interfaces between the sample and the salt bridge). Due to all of these uncertainties, accuracies of 1%–10% are usually observed depending on the type of ions. The precision is typically lower than ±0.8%, again depending on the type of analytes.

Perhaps the most frequent use of potentiometry is the determination of a solution's pH. It consists of determining the activity of hydrogen ions (H^+). When a glass surface is immersed in an aqueous solution then a thin solvated layer (gel layer) is formed on the glass surface in which the glass structure is softer. This applies to both the outside and inside of the glass membrane. The concentration of protons inside the membrane is constant (pH = 7), and the concentration outside is determined by the concentration, or activity, of the protons in the electrolyte. This concentration difference produces the potential difference that we measure with a pH meter. The measurement of pH is not trivial and needs careful calibration and procedure. Similarly, for clinical applications, most common analytes' activity, such as Na^+, K^+, Ca^{2+}, and Cl^-, can be measured by potentiometry.

Another application of a potentiometric sensor is gas-sensing. Indeed, a number of membrane electrodes respond to the concentration of dissolved gas. The basic arrangement consists of a thin membrane separating the sample from an inner solution containing an ISE. The membrane is permeable to the gaseous analyte. The gas molecules pass through the membrane where they react with the inner solution, producing a species whose concentration is monitored by the ISE. For example, in a CO_2 electrode, CO_2 diffuses across the membrane where it reacts in the inner solution to produce H_3O^+.

Potentiometric electrodes can also respond to a biochemically important species. The most common class of potentiometric biosensors are enzyme electrodes, in which an enzyme is immobilised at the surface of a potentiometric electrode. The analyte reacts with the enzyme and produces a product whose concentration is monitored by the potentiometric electrode. Potentiometric biosensors have also been designed around other biologically active species, including antibodies, bacterial particles, tissues, and hormone receptors. One example of an enzyme electrode is the glucose electrode, which is based on the catalytic hydrolysis of glucose by glucose oxidase.

6.3. Voltammetry

In modern voltammetry, a time-dependent potential is applied to a working electrode, changing its potential relative to a fixed potential of a reference electrode, and the resulting current, flowing between the working electrode and an auxiliary electrode, is measured as a function of the potential. This technique was first developed by Jaroslav Heyrovsky in 1922 for which he was awarded the Nobel Prize in Chemistry in 1959. The auxiliary electrode is generally made of platinum, the reference electrode can be a silver/silver chloride, calomel, or, more rarely, hydrogen electrode. For the working electrode, different materials are available including gold, silver, platinum, mercury, and carbon. The first studies on voltammetry (e.g. polarography) used a mercury working electrode. Mercury electrodes have evolved from classical dropping mercury electrode [8] through hanging mercury drop electrode, static mercury drop electrode, mercury film electrode, mercury amalgam electrodes [9]. Routine applications of mercury electrodes are not too frequent nowadays. This is caused by fast developments of modern spectrometric and separation techniques, by concerns about mercury toxicity and by the lack of properly validated methods. However, liquid mercury electrodes remain perfect sensors for voltammetric measurements since their renewable surface eliminates or reduces problems with surface fouling and their broad potential window enables reaching negative potentials up to -2.5 [10].

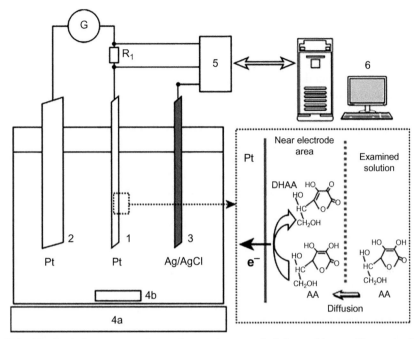

FIG. 6.3 Typical arrangement for a voltammetric sensor. 1, 2, 3—working, auxiliary and reference electrodes inside the cell; 4a and 4b—magnetic stirrer; 5—data acquisition system; 6—computer equipped with the program for recording, visualisation and archiving voltammograms, G—electronic generator; RI—measuring resistor, inside the rectangle on the right side: an example of an idea of the vitamin C oxidation on the working electrode. *(From J. Wawrzyniak, A. Ryniecki, M. Przybyt, Application of voltammetric and amperometric techniques to design enzymatic biosensors for food processing industry, Electron. J. Pol. Agric. Univ. 9 (2006).)*

A typical setup for a voltammetric sensor is given in Fig. 6.3. It measures the electrical current between the working electrode and the auxiliary electrode induced by oxidation–reduction reactions. More precisely, if the analyte is oxidised at the working electrode, the resulting electrons pass through the potentiostat to the auxiliary electrode, reducing the solvent or some other component of the solution. On the contrary, if the analyte is reduced at the working electrode, the current flows from the auxiliary electrode to the working electrode. This current is called a faradaic current. However, the current observed at a working electrode is not only due to redox reactions and other currents such as capacitive charging/discharging of the electrode double layer may occur. The plot of the current as a function of the applied voltage is called the voltammogram.

By convention, a faradaic current due to the analyte's reduction at the working electrode is a cathodic current and is positive. If the analyte is oxidised at the working electrode, then the charge flow is an anodic current and is negative. The relation between the potential at the working electrode and the concentration of an analyte and its reduced or oxidised form is given by the Nernst equation.

Since we impose the potential at the working electrode, the ratio between the different redox species has to be adjusted giving rise to the faradaic current. The magnitude of the current is determined by the rate of the oxidation or reduction reaction. It depends on the rate at which the reactants and products are transported to and from the electrode and the rate at which electrons pass between the electrode and the redox species in solution. The mass transport is the result of a diffusion mechanism due to the existence of a concentration gradient for the redox molecules. Electrodynamic effects can also be involved in mass transport. Additionally, hydrodynamic effects have to be taken into account if stirring is used or if the temperature gradient exists in the solution. This convection phenomenon has a direct impact on the shape of the voltammogram (i.e. existence of a peak current or limiting current). The electron transfer can exhibit two extreme cases. If it is fast, the redox reaction is at equilibrium and the reaction process is reversible. On the contrary, if it is slow, the Nernst equation is not verified and the reaction process is irreversible.

Apart from this faradaic current, a capacitive charging current can also be temporarily recorded due to the application of the potential on the electrode. Indeed, when applying the potential, we modify the charges on the electrode. In order to maintain electroneutrality at the solid–liquid interface, charges of opposite sign migrate from the solution to the interface and charges of the same sign migrate away from this interface. This flow of ions is called the capacitive charging current and cannot be distinguished from the electron flow. This phenomenon is responsible for the creation of the so-called electrical double layer.

Many different voltammetric techniques are depending on how the voltage is applied at the working electrode, when the current is measured and whether the solution is stirred. In polarography, a dropping mercury electrode is chosen as the working electrode. The current flowing through the system is measured while applying a linear potential ramp or a series of potential pulses for better sensitivity. Polarography is used for the analysis of metal ions, inorganic anions, such as NO_3 and some organic compounds (e.g. carboxylic acids). Hydrodynamic voltammetry is very similar to polarography (i.e. same applied voltage solicitations) except that the liquid working electrode is replaced by a solid-state electrode and the electrolyte is stirred to reproduced the distribution of electroactive species found in polarography at the liquid–liquid interface. Anodic stripping voltammetry involves the preconcentration of a metal phase onto a solid electrode surface or into Hg (liquid) at negative potentials through an electrolysis process. Metal deposition is usually enhanced by stirring. In a second step, the deposited metal is stripped from the electrode by selective oxidation of each metal phase species during a potential sweep towards more positive values. The detection limit of stripping voltammetry is much smaller than other electrochemical techniques (i.e. three orders of magnitude) thanks to the preconcentration step. Cathodic and adsorptive stripping voltammetry differs from anodic stripping voltammetry by their deposition mechanisms. In a cyclic voltammetry experiment, the working electrode potential is ramped linearly versus time and when it reaches a set potential, the working electrode's potential ramp is inverted. This inversion can happen multiple times during a single experiment. Cyclic voltammetry is carried out in an unstirred solution. Scanning the potential in both directions gives the opportunity to explore the electrochemical activity of species generated at the electrode. This is a major advantage of cyclic voltammetry over other voltammetric techniques. Finally, amperometry is a voltammetric technique where a constant potential is applied to the working electrode and one can measure a current as a function of time. One advantage of amperometry from other forms of voltammetry is that the current readings are averaged over time, thus bringing greater precision. In most voltammetric techniques, current readings must be considered independently at individual time intervals. One of the first amperometric sensors measured the dissolved O_2 in blood and was developed by Clark in 1956. An extension of single-potential amperometry is pulsed amperometry which prevent fouling of the working electrode. In pulsed amperometry, the potential is applied for a short time (usually a few hundred milliseconds), followed by higher or lower potentials that are used for cleaning the electrode.

Limit of detection in the micro-molar range is routine. For some analytes and for some voltammetric techniques, nano- and pico-molar levels can be reached, in particular with stripping voltammetry. The accuracy is limited by residual currents (i.e. charging of the electrode). At the micro-molar range, an accuracy of $\pm 1\%$–3% is routine. Precision is generally limited by the uncertainty in measuring the limiting current or the peak current. Under most conditions, a precision of $\pm 1\%$–3% is normal. Sensitivity can be improved by better control of the experimental conditions.

Voltammetry is used for the analysis of trace metals in environmental samples, including groundwater, rivers, and seawater. Level of dissolved oxygen and anionic surfactants in waters and wastewaters, as well as the concentration of CO_2, H_2SO_4, and NH_3 in atmospheric gases, have been measured by voltammetry technique. The concentration of trace metals in blood, urine, and tissue can be measured by voltammetry. Choline, ethanol, formaldehyde, glucose, lactate are other examples of molecules measured in the clinical sample by voltammetric techniques. These techniques find also applications in the food and pharmaceutical industries. On a more fundamental aspect, voltammetry is a technique of choice for the study of the electrochemical behaviours (i.e. equilibrium constant, reversibility, etc.) of complex electrolytes. Indeed, if the analytes behave independently, the voltammogram of a multicomponent solution is a sum of each individual voltammograms.

6.4. Coulometry

Coulometry uses either an applied current or potential to exhaustively convert an analyte from one oxidation state to another at the working electrode. In these experiments, the total current passed is measured directly or indirectly to determine the number of electrons passed. Knowing the number of electrons passed, we can extract the concentration of the analyte using faraday's law:

$$Q = nFN_A \tag{6.2}$$

where n is the number of electrons per mole of analyte, F is Faraday's constant ($96,487\,C\,mol^{-1}$), and N_A is the moles of the analyte. To obtain an accurate value for N_A, 100% of the current must be used to oxidise or reduce the analyte.

There are two forms of coulometry: controlled potential coulometry and controlled-current coulometry. A three-electrode potentiostat is used to set the potential in controlled potential coulometry. The working electrode is either platinum or liquid mercury electrode. The auxiliary electrode is very often made of platinum and is separated by a salt bridge from the analytical solution. A saturated calomel or silver chloride electrode serves as the reference electrode. To measure the total charge, the potentiostat monitors the current as a function of time and use electronic integration to calculate the corresponding charge as a function of time. During the electrolysis process, the analyte's concentration in the solution decreases, and as a consequence the electrolysis rate decreases. This situation leads to quite long analysis time (i.e. between 30 and 60 min) that can be problematic for industrial applications. Controlled-current coulometry has the advantage of a constant electrolysis current and thus a shorter analysis time (i.e. < 10 min). Furthermore, the calculation of the analyte concentration is straightforward since the charge is simply the product between the fixed current and the time. However, it is more difficult to ensure 100% current efficiency and to know when the electrolysis process is ended and refinements in the measurement procedure are required. Controlled-current coulometry is carried out using an amperostat composed of a working electrode, often made of platinum, and a counter electrode. This counter electrode can be isolated from the analyte by a salt bridge. The setup needs also an accurate clock for measuring the electrolysis time and a switch for starting and stopping the electrolysis (Fig. 6.4).

The smallest concentration of analyte that can be determined by coulometry depends on the capability to accurately determine the endpoint of electrolysis. At the best, it can nowadays reach the micromolar range. When using controlled-current coulometry, an accuracy of 0.1%–0.3% is feasible. In controlled-potential coulometry, accuracy better than 0.5% is achievable. Precisions of $\pm 0.1\%$–0.3% are routinely obtained in controlled-current coulometry and precisions of $\pm 0.5\%$ are typical for controlled-potential coulometry.

Controlled-potential coulometric analyses have been used to determine the concentration of more than 50 chemical elements, including trace metals and halides ions. It is frequently used for the determination of uranium and plutonium as well as for the measurement of oxygen content. However, controlled-current coulometry is more versatile and thus used in a wider range of applications. It can determine concentrations of water on the order of milligrams per litre in substances such as butter, sugar, cheese, paper, and petroleum. It is also used to measure the thickness of metallic coatings.

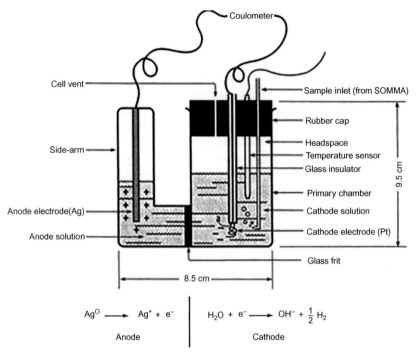

FIG. 6.4 Typical arrangement for a coulometric sensor made of a cathodic and anodic half-reactions for the coulometric titration of the H^+ from the acid formed by the reaction of CO_2 and ethanolamine. *(From K.M. Johnson, A.G. Dickson, G. Eischeid, C. Goyet, P. Guenther, R.M. Key, F.J. Millero, D. Purkerson, C.L. Sabine, R.G. Schottle, D.W.R. Wallace, R.J. Wilke, C.D. Winn, Coulometric total carbon dioxide analysis for marine studies: assessment of the quality of total inorganic carbon measurements made during the US Indian Ocean CO₂ Survey 1994–1996, Mar. Chem. 63 (1998) 21–37.)*

6.5.   Impedance spectroscopy

Small signal electrochemical impedance spectroscopy is a valuable tool for studying transport properties in electrolytes and electrochemical reactions at the interface between the solid and the electrolyte. The value of impedance spectroscopy derives from the effectiveness of the technique in isolating individual phenomena in a multistep process. Assuming that each phenomenon has a unique associated time constant, the various phenomena can be separated in the frequency domain. Although any kind of perturbation can be applied, most reported studies apply a small sinusoidal current perturbation to an equilibrium system and measure the corresponding voltage response.

Typically, alternative current impedance experiments are carried out over a wide range of frequencies (several millihertz to several megahertz), and the interpretation of the resulting spectra is performed by extracting an equivalent analog circuit made of resistors and capacitors [11]. In general, the equivalent circuits are not unique. One has to choose a physically plausible one containing a minimal number of components. From the value of each component, some meaningful insight into the system properties can be found.

The current flowing at an electrified interface due to an electrochemical reaction is made of faradaic and nonfaradaic components. When the charge transfer takes place at the interface, the mass transports of the reactant and product determines the rate of electron transfer. To eliminate or minimise the effects of capacitive currents during transient electrochemical experiments and identify the various transport processes, voltammetric (Section 6.3), and coulometric (6.4) have been developed [12]. However, even if they give good performance in terms of accuracy and rapidity, most often, they are not able to give a complete picture of the mechanisms involved in the electrochemical system. Only techniques involving impedance measurement are able to identify the various phenomena.

First measurements of full impedance spectra have been performed in the early 1970s when reliable equipment became available. Among the various methods of impedance measurements, the frequency response analyser (FRA) is nowadays a standard. FRA is a single-sine method in which a small potential of 5–15 mV of a given frequency is overlaid on a given bias potential and applied to a working electrode. The resulting electrical current is then measured between this working electrode and a counter electrode. This process is repeated by scanning the frequency and computing the impedances from the voltage and current data. Typically 5–10 measurements are performed for a decade change in frequency (Fig. 6.5).

This two-electrode configuration is most often used. However, it suffers from the parasitic influence of the electrode–electrolyte interface impedance which hides any signal detection in the low-frequency range. As a consequence, tetrapolar arrangements have also been proposed [13, 14]. The nature of the electrode's material depends on the phenomenon that is monitored. If it is the ionic transport that is under scrutinising, then any metal electrode can be used. On the contrary, if one wants to monitor a specific electrochemical reaction at the interface, then a careful choice has to be performed as for the previous techniques (i.e. potentiometry, voltammetry, and coulometry).

However, FRA measurement is viable only for a stable and reversible system in equilibrium, as the system's linearity and stability must be ensured. Due to the relatively long data-acquisition time of the single-sine FRA method, the validity of the method can be questioned for nonstationary systems. In efforts to reduce the measurement time, investigators have proposed methods in which no frequency scanning is employed. These methods use broadband perturbing signal (e.g. white noise) and a Fast Fourier transform algorithm to compute the impedance [15, 16]. This approach has been refined over years allowing now the exploration of complex problems encountered during the studies on biosensors and other electrochemical systems [17, 18].

The working range for impedance analyser is between a few microhertz for some equipment to hundred of megahertz for some others. Polarisation range is of the order of tens of volts. Accuracy on impedance measurement is typically

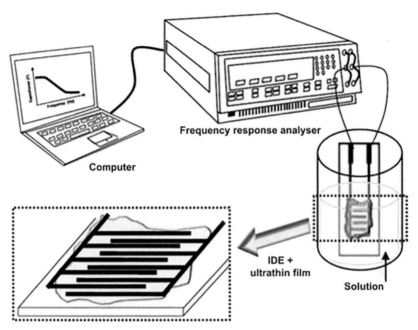

FIG. 6.5 Typical two-electrode arrangement for an impedimetric sensor. *(From A. Riul, C.A.R. Dantas, C.M. Miyazaki, O.N. Oliveira, Recent advances in electronic tongues, Analyst 135 (2010) 2481–2495.)*

around 0.1% and impedance dynamic is between few milli-ohms to hundred of megaohms. An advantage of the electrochemical impedance method compared with voltammetry or potentiometry is that labels are no longer necessary, thus simplifying sensor preparation. However, the detection limits of electrochemical impedance spectroscopy are still poor compared with traditional methods.

6.6. Dielectrophoresis

Dielectrophoresis is an electrokinetic phenomenon acting on polarisable particles in an inhomogeneous electric field. In 1971, H.A. Pohl [19] manipulated yeasts (*Saccharomyces cerevisiae*) thanks to a nonuniform electric field induced by the alternative current polarisation of two metallic electrodes. Few years before, he has established the relationship between the force acting on the yeasts and the frequency of the polarisation, the dielectric properties of the particle and surrounding medium, the amplitude, and gradient of the electric field (Eq. 6.3).

$$F_{\text{dep}} = 2\pi r^3 \varepsilon_{\text{m}} \text{CM}(\omega)\nabla E^2 \tag{6.3}$$

$$\text{CM}(\omega) = \left(\varepsilon_{\text{p}}^{*} - \varepsilon_{\text{m}}^{*}\right) / \left(\varepsilon_{\text{p}}^{*} + 2\varepsilon_{\text{m}}^{*}\right) \tag{6.4}$$

where $\varepsilon_p^* = \varepsilon_p - j\sigma_p / \omega$ and $\varepsilon_m^* = \varepsilon_m - j\sigma_m / \omega$, ε_p and ε_m are the absolute dielectric permittivity ($\varepsilon_r\varepsilon_0$) of the particle and the surrounding medium, respectively, σ_p and σ_m are the electrical conductivity of the particle and surrounding medium, respectively, ω is the angular frequency ($\omega = 2\pi f$) of the applied ac field, r is the particle radius, E is the amplitude (rms) of the electric field, ∇ represents the gradient operator and $CM(\omega)$ is the Clausius–Mossotti factor related to the effective polarisability of the particle. Indeed, dielectrophoresis relies upon the interaction of the electric field with an induced dipole in the particle. The direction of the resulting force depends upon whether the particle is more, or less polarisable than the medium, as expressed by the Clausius–Mossotti factor (Eq. 6.4). If the particle is more polarisable it will be attracted to areas of high electric field strength and vice versa.

Related techniques include travelling wave dielectrophoresis [20] and electrorotation [21] both of which exploit phase-shifted electric fields to achieve translational or rotational particle movement. Travelling wave dielectrophoresis can be formed using either a linear or a spiral set of electrodes to move particles, perpendicular to the electrode array (Fig. 6.6). The direction of particle movement depends upon the relative polarisability of the particle and the suspending medium. Electrorotation utilises a set of electrodes, around which the electric

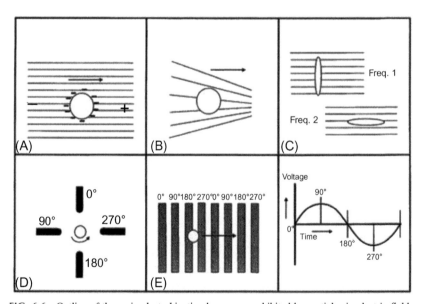

FIG. 6.6 Outline of the main electrokinetic phenomena exhibited by particles in electric fields. Electrophoresis occurs in DC electric fields, whereas AC electric fields can be used to observe the other effects. To observe electrorotation and particle movement in travelling wave devices, phase-shifted electric signals are applied to the electrodes. (A) Electrophoresis. (B) Dielectrophoresis. (C) Electroorientation. (D) Electrorotation. (E) Travelling waves. *(From G.H. Markx, C.L. Davey, The dielectric properties of biological cells at radiofrequencies: applications in biotechnology, Enzyme Microb. Technol. 25 (1999) 161–171.)*

field is cycled, to create a central area in which particles undergo rotation. Each particle type exhibits a near-unique profile of particle rotation rate against the applied electric field for given environmental conditions.

The technique offers potential selectivity according to the size, dielectric properties, and surface electrical charge of the particle. If we consider a biological cell at rest and without the electric field, its dielectric properties and surface charge depend on its viability and maybe some other physiological parameters. These cytoplasm and surface charges are counter-balanced by ions of the surrounding medium [22]. When applying the electric field, the spatial distributions of charges in the cytoplasm, in the surrounding medium and on the membrane surface are modified according to the dielectric properties of the system. This reorganisation induces a dipole on the particle which interacts with the nonuniform electrical field. If it is quite simple to characterise the dielectric properties of the surrounding medium, the characterisation of the dielectric properties (i.e. conductivity and permittivity) is rather complex. Literature shows very different values depending on the experimental arrangement. A key parameter for colloidal particles is the surface conductance K_s which appears in the equation giving the conductivity of the particle (Eq. 6.5).

$$\sigma_p = \sigma_v + 2K_s / r \qquad (6.5)$$

where σ_v is the conductivity of the bulk and r is the radius of the particle. The surface conductivity cannot be ignored for particle radius smaller than $10\,\mu m$. Three methods have been reported to measure the surface conductance. The first one measures the frequency of the polarisation for which the Clausius–Mossotti factor is null [23]. The accuracy depends on the capability to distinguish between dielectrophoretic and Brownian motion. A second method has been proposed by Morganti [24] which used electrorotation to extract the surface conductance from the electrorotation rate spectrum. The third technique has been introduced very recently by Honegger which measures the particle velocity in pure dielectrophoretic conditions [25].

The techniques used for the study and exploitation of dielectrophoresis have made many progress in recent years [26]. There have been significant advances in electrode design, the construction of microfluidic chambers, and with methods to monitor the collection of particles at electrodes depending on their dielectric properties. In total, more than 2000 scientific papers have been published on this topic during the last 10 years. Although developed after the other techniques presented in Sections 6.2–6.5, dielectrophoresis is now well established and theoretically understood allowing the development of accurate, robust, and reliable biosensors.

6.7. Scaling effect

Miniaturising machines and physical systems is an ongoing effort in human history. This effort has been increased recently as market demands for low-cost, robust, and multifunctional products, has become stronger than ever. The only solution to produce them is to package many modules (for parallelization and redundancy of operations) into the final product and thus to miniaturise each component. There are additional arguments in favour of miniaturisation. Small objects require less energy and material during the fabrication. Small sensors and actuators can be easily integrated with electronics for signal processing and communication. Miniaturised systems are less invasive and needless operating energy, opening the possibility to develop the 'internet of things'. A wide variety of environmental, industrial, and security applications exist where long-term remote (bio)-chemical analysis is either required or at least highly desirable [27]. As a result, the design of 'smart' miniaturised sensors, capable of performing stable and reliable measurements without operator intervention has become an area of increasing interest in instrument and sensor development. Further, more general, discussion of the application of miniaturised systems to the detection of waterborne pathogens is given in Chapter 10.

However, there are several physical consequences of scaling down many physical quantities. Miniaturisation is a lot more than just scaling down spatial dimensions. A careful analysis of the scaling laws is necessary to understand if a given system can be scaled down favourably of not. Depending on if intensive or extensive properties are involved in the static and/or dynamic behaviour of the system, the miniaturisation will have either no effect, a positive effect or negative effects on the performances. In summary, small dimensions of microsystems are crucial for some applications but lack of importance for other fields. In such a case, integration becomes more important than miniaturisation.

Focusing our analysis on scaling issues encountered in electrochemical sensors, one can wonder if electrical potential, current, and charge are intensive or extensive properties. Potential is an intensive property, it does not depend on the amount of metal present at the electrode, neither the size of the electrode, the concentration of ions in the solution, and as a consequence is scale-invariant. On the contrary, the electrical current and charge are extensive properties. They depend on the size of the system, only the current and charge densities are intensive. As a consequence, one cannot expect improved performances when scaling down a potentiometric technique while modifications will occur for voltammetric, coulometric, and impedimetric techniques where an electrical current flowing in the system is monitored.

Going back to cyclic voltammetry to illustrate the effect of scaling on an electrochemical technique, one can remember from Section 6.3 that the measured current is limited by analyte diffusion at the electrode surface. The electrode geometry dictates the mass transport to and from the electrode surface. The Cottrell equation describes how the current decays as a function of time. The current at the electrode is a sum of both planar and spherical diffusion and the magnitude of each depends on the time and size of the microelectrode. Planar diffusion predominates at short times while spherical diffusion is predominant at a sufficiently long time. Indeed, at short times, the size of the diffusion layer is smaller than that of the electrode, and planar diffusion dominates even if the electrode has been miniaturised. On the contrary, at longer times, the dimensions of the diffusion layer exceed those of the electrode, and the diffusion becomes hemispherical. In this configuration, the analytes diffusing to the electrode surface come from the hemispherical volume (of the reactant-depleted region) that increases with time. A miniaturised electrode can then be defined as an electrode that has a characteristic surface dimension smaller than the thickness of the diffusion layer (i.e. between hundred nanometres to several microns) on the timescale of the electrochemical experiment. This means that the time required to attain a steady state (i.e. plateau in cyclic voltammograms instead of peaks) strongly depends on the electrode dimensions. Miniaturised electrode enables the study of very fast kinetics. Furthermore, capacitive current (IC) decreases in proportion to the decreasing area of the electrode, while the steady-state faradaic current (IF) is proportional to its characteristic dimension. Therefore, the IF/IC ratio, which represents the signal to noise ratio of the technique, increases with the reciprocal of the characteristic length [28, 29]. Analysis of noise in electrochemical can be an extremely sensitive technique to monitor biosystems [30]. With a miniaturised electrode, the absolute current level is small. A first positive consequence is that the ohmic drop of potential is decreased and measurements in lower conductivity media are possible. A negative consequence is that the current is not measurable and it is then necessary to work with arrays of electrodes to improve the absolute current level. The small size of the electrodes permits measurements in a very limited volume or at a very short distance allowing high collection efficiency of analytes. Detailed description and demonstration of these scaling effects can be found in [31, 32] for integrated planar electrodes (voltammetry, coulometry, and impedancemetry) and in [33] for ion-sensitive field-effect transistors (ISFET) (potentiometry).

As a consequence, lots of effort has been made in recent years to apply electrochemical techniques to microscopic scales. All above-described techniques have been miniaturised. The well-established integrated circuit (IC) and the emerging microsystems (MEMS) technologies give the possibility to create a variety of miniaturised devices dedicated to the detection of (bio) chemical analytes and microorganisms. Potentiometric sensors have been reported using the metal-oxide sensitive field-effect transistor (MOSFET) [34], the light-addressable potentiometric sensor (LAPS) [35], ISFET [36, 37], and the ISE [38]. Amperometric sensors utilising either flow-by or stopped-flow analysis have been adapted to diverse applications including biosensing [39, 40], gas sensing [41], and chlorine [42]. Coulometric sensors have been constructed for the measurement of chemical oxygen in lake water [43]. The study of mediated enzyme reactions by rapid coulometric has also been reported [44]. More recently, it has been shown that the very precise release of calcium ions from ion-selective membrane can be controlled by electrical current [45]. A very robust implementation of a coulometric chip showing low deviation and high reproducibility has also been reported [46]. Many promising applications of impedance spectrometry have been reported, for example for counting and characterising suspended cells in a microchannel (Fig. 6.7) [47–49], or for

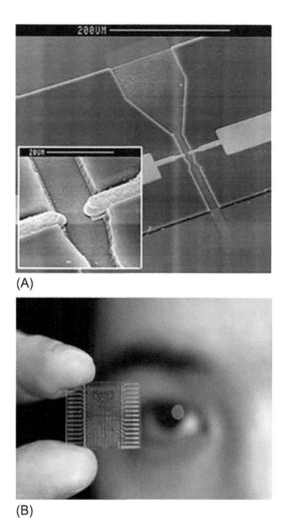

(A)

(B)

FIG. 6.7   (A) SEM photo of miniaturised electrodes embedded in a micromachined channel [49]. The inset is a zoomed-in image of microelectrodes; (B) Photo of an electrical flow cytometer within the inset a photo of the integrated coplanar electrodes [47].

monitoring physiological changes (migration, proliferation, apoptosis, etc), in cell cultures or small tissue samples [50–54]. Cerrioti took clever advantage of the benefits offered by these different techniques to realise an integrated system monitoring in parallel acidification rate, oxygen consumption, and cell adhesion respectively with potentiometric, voltammetric, and impedimetric biosensors [55]. Dielectrophoresis has been implemented in numerous microsystems to transport, separate, concentrate, or characterise micro and nanoparticles [56]. The electric field utilised in dielectrophoresis has either been generated by patterned external or internal electrodes, light patterns [57] or insulating structure patterns to create nonuniformities in a uniform electric field [58]. It has been used with biological cells including protozoa [59], bacteria [60], viruses [61], large deoxyribonucleic acid (DNA) strands [62], as well as chemicals, such as protein molecules [63] and pesticides.

6.8. Miniaturised detection of waterborne pathogens

Classifying miniaturised electrochemical biosensors for waterborne pathogen detection and analysis is a difficult task. One can either define different categories according to the type of bioreceptor, the type of electrical probe, the fabrication processes, the type of electrochemical transduction (see Section 6.2–6.6), or the type of pathogen (see Chapter 2 for an overview of waterborne pathogens).

6.8.1 Bioreceptors

Bioreceptors can be classified into five different categories including antibodies, enzymes, nucleic acids, cells or specific cells proteins, biomimetic structure, and bacteriophage. The choice of one of them will depend on the target analyte. It can be an antigen on the whole pathogen surface, protein production by the pathogen, a nucleic acid of the pathogen, a specific cellular response of a living microorganism in contact with the pathogen (i.e. antibody production, membrane deterioration, etc.). The two main factors important for every biorecognition element are affinity and specificity. New techniques allow the selection of more specific ligands. They are evolving rapidly and represent the future of recognition elements. Antibodies are the most common bioreceptor [64, 65]. They may be polyclonal, monoclonal, or recombinant depending on the way they are synthesised. They can bind with high affinity against a target protein, another antibody or even a whole pathogen. Biosensors using enzymes as the biorecognition elements have been reported [66–68]. Enzymes are chosen based on their specific binding capability and their catalytic activity, which provides electron transfer to the working electrode, is the key element for highly sensitive analysis. However, in most cases, enzymes are used to function as labels and not as bioreceptor. Recent advances in nucleic acid recognition have expanded the use of DNA as a bioreceptor [69, 70]. For pathogen detection, the identification of a target analyte's nucleic acid is achieved by hybridisation of the complementary base pairs. Since each organism has unique DNA sequences, any self-replicating microorganism can be easily identified. With cells based bioreceptors, biorecognition is based on a whole-cell that is capable to respond to stimuli [71, 72]. The transduction mechanism is a two steps process. First, the cell converts the effect of the pathogen into a cellular response. Second, the electrochemical sensor measures this cellular response. Using cells as bioreceptors is particularly interesting because cells provide (i) sensitivity to a wide range of biochemical stimuli, (ii) very low detection limits, and (iii) a functional assay for biochemical agents. Their major limitation is their need for a specific well-controlled environment to function normally. Proteins found within cells can also serve the purpose of bioreceptors. In particular, biological cell membranes are the host of a great variety of molecules acting as receptors in the interaction with the external environment. Being more robust than antibodies due to their function, they are of special interest for the development of novel biotechnological tools [73]. A receptor that is fabricated and designed to mimic a bioreceptor (antibody, enzyme, cell, or nucleic acids) is often termed a biomimetic receptor. There are several methods to produce these receptors such as genetic engineering (e.g. aptamer), artificial membrane fabrication, and molecular imprinting. Aptamers are similar to nucleic acids in that they are ligands of deoxyribonucleic or ribonucleic acids. They are specifically designed (using SELEX) to bind a target protein [74, 75]. Molecular imprinting produces artificial recognition sites by moulding a polymer around a molecule which can be used as a template [76, 77]. Recently, bacteriophages have been employed as biorecognition elements for the identification of various pathogenic microorganisms. They are viruses that bind to specific receptors on the bacterial surface in order to inject their genetic material inside the bacteria [78].

6.8.2 Electrode material

Integrated electrochemical biosensors use mainly inorganic material as semiconductor (doped silicon, silicon nanowire, carbon nanotube, boron-doped diamond, etc.), metal (gold, titanium, platinum, etc.) or dielectric (oxide). Nanomaterials open a new perspective for the development of electrochemical biosensors. Nanoscale materials have been used to achieve direct wiring of enzymes to electrode surfaces, to promote electrochemical reactions, to impose nanobarcodes on biomaterials, and to amplify the signal from biorecognition events. New materials are currently studied (i.e. conducting doped-polymers, graphene, and composite materials).

These last years, conducting doped-polymers have been extensively studied to provide flexible and cost-effective bio-compatible electrodes for electrochemical biosensing. Doped poly(3,4-ethylenedioxythiophene) (PEDOT) has been successfully micropatterned to fabricate integrated polymeric electrochemical biosensors by associating them with a polymeric substrate. For example, Kiilerich-Pedersen has made an impedimetric biosensor to assess in-vitro the viral infection of human foreskin fibroblasts with tosylate-doped PEDOT electrodes patterned on COC substrate [79]. In vivo neural recordings have been also carried out with poly(styrene sulfonate)-doped PEDOT microelectrodes deposited on the Parylene C substrate [80]. Owens has monitored the gastrointestinal epithelium disruption with an organic transistor [81]. Polyaniline has been also micropatterned [82] and was further used for conductivity detection

[83]. Finally, polydimethylsiloxane (PDMS) has also been mixed with carbon alone [84] or carbon and oil [85] (oil-reducing capacitive current) to form doped polymeric electrodes for bioapplications. In the latter case, microelectrodes have been patterned and thus used to detect catecholamine release from PC12 cells by voltammetry.

Graphene is a unique-atom-thick membrane of carbon atoms arranged as a honey-comb crystal lattice. Compared to zero-dimensional nanoobjects (nanoparticles and quantum dots) and one-dimensional nanostructures (CNTs and silicon nanowires (SiNWs)), graphene sheet could represent an interesting alternative thanks to the fact that it presents as well outstanding electrical and mechanical properties [86] and that it could interact directly with a much larger biological sample like cells and even tissues. For example, a graphene-based transistor has been used to monitor cardiomyocytes' electrical activity [87]. Concerning cell detection, it has been also used to detect single bacteria [88] and graphene functionalised with aptamer has been employed to detect suspended cancer cells by impedancemetry [89]. Integrated detection of cell biomolecules release can also be carried out: assemble of several layers of graphene sheet with peroxidase (for biomolecule selectivity) and extracellular matrix protein (for enhanced biocompatibility) have been able to detect hydrogen peroxide release from living cells by voltammetry [90] and patterned graphene oxide sheet integrated as the gate of a FET have monitored the catecholamine release from neuroendocrine cells [91]. Recently, graphene-based electrochemical sensors have been implanted in vivo, on teeth, to monitor wirelessly mouth bacterial activity [92]. Interestingly, by functionalising graphene with gold nanoparticles themselves functionalised with calcium ions, graphene sheets have been wrapped around single yeast cells [93]. Then, by interconnecting this graphene-cell complex with gold electrodes, cell volume change caused by exposure to alcohol has been electrochemically quantified. Various other chemical elements are progressively integrated as an electrode material, particularly for the reference electrode, which has been for a long time a technological bottleneck for fully integrated electrochemical devices [94, 95].

Composite materials can also be employed to fabricate flexible electrodes for disposable biosensors by combining the advantageous properties of different materials. For example, titanium dioxide nanoparticles have been associated with carbon paper with carbon nanotubes (CNTs) to enhance the biocompatible properties of these letters [96]. Thus, different types of leukaemia cells have been electrochemically discriminated against with a flexible biosensor. A 2D-assemble of gold nanoparticles has been integrated on a graphene sheet to further increase electron transport [97]. With this system, hydrogen peroxide release from living HepG2 cells has been monitored by conductometry. Other composite materials have been patterned on rigid electrodes for the label-free analysis of living cells: CNTs have been integrated with cellulose for leukaemia cells detection by impedancemetry as an example [98].

6.8.3 Fabrication process

So far, integrated electrochemical biosensors have been mainly fabricated on rigid substrates like silicon, or optically transparent substrate as glass or Pyrex. Much research now focuses on the use of flexible substrates for electrochemical biosensors to integer them in smart textiles, to improve interfacial contact with tissue in vivo and mostly to reduce biosensors fabrication cost. Interestingly, insulating polymers present a particularly cheap fabrication and processing cost. PDMS is extensively used to fabricate whole or partially polymeric devices (often associated with glass, Pyrex or other rigid substrate). Nevertheless, PDMS is more and more criticised for its noncompatibility with mass manufacturing, notably because of its protein adsorption, its permeability causing liquid evaporation, its natural hydrophobicity, and its ageing [99, 100]. Numerous other polymers are currently investigated for integrated biomedical sensors. Among them, polyimide, a flexible dielectric with excellent thermal stability and resistance to solvent, also employed for encapsulation purposes in microelectronics packaging [101], has been often exploited. Indeed, it has been used in vivo for the implantations of microelectrodes for the intra [101] and surficial [102] integrated monitoring of brain electrical activity by potentiometry. Muscle and nerve action potential measurements have been thus also monitored in vivo [103]. In vitro extracellular cardiomyocyte recording has been also carried out with an electrochemical biosensor made on a polyimide substrate [104]. Parylene has been also used as a flexible and dielectric substrate to study in vitro [105] and in vivo [106] neuron electrical activity with integrated devices. Cyclic olefin copolymer (COC) is also an interesting polymeric substrate which has already been used for integrated electrochemical cell analysis [79]. Other biocompatible polymeric substrate which has been already used for the electrochemical analysis of protein or DNA and could be also employed for pathogens analysis: Polyethylene terephthalate (PET) [107], polyethylene naphthalate (PEN) [108], and polystyrene [109]. Impedimetric biosensors for in vitro cell-based assay could be made from these types of polymeric substrates to reduce their consumable costs.

Academic biosensors use mainly electrodes fabricated by photolithography and thin-film deposition (i.e. chemical or physical vapour deposition), which are expensive methods issued from microelectronics. Printing techniques represents a seductive alternative since they enable to deposit electrodes with conductive ink (metal [110] or conductive polymers [111] or nanostructures (nanoparticles [112] or CNTs [113])) on a much larger surface, with more flexibility and reduced fabrication costs (around 10 times cheaper than the photolithographic process for screen-printing [114]) (Fig. 6.8A). For biosensing, printing techniques encompass mainly screen-printing and inkjet printing. Screen-printing is the most employed printing technique (by considering the number of iteration in scientific publications). This printing technique enabled previously to develop commercial disposable amperometric biosensors for home blood glucose tests [115] and could be so used for disposable cell analyses. Several electrochemical

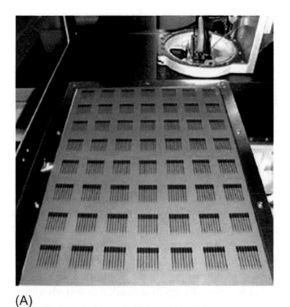

(A)

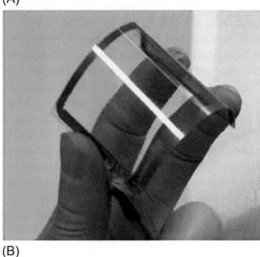

(B)

FIG. 6.8   (A) Photo of AuNP arrays screen-printed on a Kapton substrate [119]. (B) Photo of a flexible grapheme/PET touch panel fabricated by roll-to-roll process [124].

biosensors have been fabricated by screen-printing and employed for cellular analysis [116–118]. Integrated electrochemical biosensors have been fabricated by inkjet [119] but, as far as we know, they have not been applied for cell analysis yet. Resolution is considered as the main drawback of screen printing [120]: 30 µm maximum resolution can be achieved [121]. Interestingly, ink-jet printing enables to obtain submicrometric resolution [122]. The main advantage of exploiting flexible substrates and/or electrodes is they can be manufactured with roll-to-roll process which is one of the fastest and cheapest manufacturing techniques. Roll-to-roll process can be straightforwardly associated with printing techniques to further reduce manufacturing cost. Interestingly, graphene has been also recently successfully patterned by inkjet-printing for submicrometric resolution [123] and manufactured by roll-to-roll and screen-printing processes [124] (Fig. 6.8B). Reduced graphene sheets can be also large-scale micropatterned [91]. For nonplanar biosensors (with a microfluidic channel for example), hot embossing can also be used to manufacture whole polymeric electrochemical biosensors [125].

6.8.4 Pathogens

Waterborne pathogens can be classified into three different categories including viruses, bacteria, and protozoa. A careful survey of the scientific literature shows that few electrochemical techniques have been evaluated for the detection and analysis of waterborne pathogens. However, works performed on other types of pathogens (i.e. human pathogens from other sources or plant and animal pathogens) can be generally easily extended to waterborne microorganisms and viruses. That is why our review is not limited to waterborne pathogens.

Molecular imprinting technology was applied to design sensing elements for the detection by potentiometry of virions of poliovirus in a specific manner (no cross-reactivity to adenovirus, no response by a nonimprinted sensor) [126]. Their findings demonstrated the application of the principles of molecular imprinting to the development of a new method for the detection of viruses. Sensors have been developed with immobilised single-stranded DNA complementary to that of the viral DNA on a glassy carbon solid-phase transducer. Upon analyte binding, the newly formed double-stranded DNA transferred electrons, detected by differential pulse voltammetry and linear sweep voltammetry. This method has been used by several groups in the detection of the hepatitis B virus. Albers has reported the fabrication of Si-based biochip incorporating interdigitated ultramicroelectrodes of high spatial resolution [127]. A fully electrical array for voltammetric detection of redox molecules produced by enzyme-labelled affinity binding complexes has been shown by Nebling [128]. The electronic detection is based on ultramicroelectrode arrays manufactured in silicon technology. It requires an electroactive mediator, hydroxaniline, and has been used for the detection of virus DNA in real unpurified multiplex polymerase chain reaction (PCR) samples. Ding has implemented a similar label-free biosensor for the detection of oligonucleotides related to the hepatitis B virus sequence via the interactions of DNA with a redox-active complex [129]. It measures the flow of electrons directly through the double-stranded DNA helix, using only cobalt complexes which increased efficiency and sensitivity, the limit of detection being 1.94×10^{-8} M. [130]. Detection of antihuman immunodeficiency virus (HIV) antibodies in serum has been reported by Laczka [131]. The novelty of the work was the combination of allosteric enzymes and coulometry to yield a fast, simple, and reliable HIV indirect diagnostic method. Detection of whole viruses using carbon nanotube thin film field effect devices have been reported by Mandal [132]. Selective detection of M13-bacteriophage has been demonstrated using a simple two-terminal configuration. Chemical gating through the specific antibody-virus binding on carbon nanotube surface has been proposed to be the sensing mechanism. Compared to electrical impedance sensors with identical microelectrode dimensions, this sensor exhibits sensitivity 5 orders higher. Jarocka has reported the development of an immunosensor for the detection of Prunus necrotic ringspot virus extracts using electrochemical impedance spectroscopy [133]. The immunosensor uses glassy carbon electrodes and a polyclonal antibody. It was capable of discriminating between samples consisting of extracts from healthy plants and consisting of leaf extracts from infected plants diluted 10,000 times with extract from healthy plants. Important efforts have been carried out for detecting, concentrating, and separating the virus from a sample using dielectrophoresis. Morgan has studied cowpea mosaic virus [134] and the tobacco mosaic virus [135]. In 2007, Balasubramanian presented a microfluidic system based on electrophoretic transport and electrostatic trapping to study MS-2 virus and Echovirus 11 found in potable water [136]. Two gold electrodes were used for dielectrophoretic trapping. MS-2 virus was captured with 88%–99% efficiency and Echovirus showed capture efficiency above 70%. Grom reported the combination of electrohydrodynamic flow and dielectrophoretic forces to trap Hepatitis A virus [137]. They showed that the microsystem can be useful to accumulate viral particles from relatively large sample volumes.

In the past decade, only a few further examples of impedance and dielectrophoretic detection schemes for viruses have been developed. However, one such system was targeted specifically at waterborne viruses, including norovirus, rotavirus, and hepatitis A, and was tested across a range of pH and salinity [138]. The polymeric aptasensor using impedimetric detection and was shown to operate stably for 2 weeks under continuous flow. An impedimetric sensor using aptamers has also been developed to determine virus viability, which is a crucial parameter in monitoring and controlling waterborne viruses. [139]. Dielectrophoretic capture, and the resulting conductance increase, enabled quantification of norovirus with a LOD of 2.5 ng/mL of NoV capsid in 5 min [140].

Several works have been reported in the literature based on the monitoring of bacteria metabolism. For instance, oxygen consumption can be measured by amperometric technique [141]. Impedance biosensor chips were developed for the detection of *Escherichia coli* O157:H7 based on the surface immobilisation of affinity-purified antibodies onto indium tin oxide (ITO) electrode chips [142]. The biosensor can detect the target bacteria with a detection limit of 6×10^3 cells/mL. An interesting strategy to detect bacteria through their metabolic process was that based on the electrochemical detection of specific marker enzymes (i.e. β-D-galactosidase). Perez described a method to detect viable *E. coli* in water samples using an amperometric sensor for the detection of 4-aminophenol

(4-AP) after hydrolysis of the substrate 4-aminophenyl-β-D-galactopyranoside (4-APGal) by the bacterial enzyme β-D-galactosidase [143]. Impedance detection of bacterial growth has also been used for food samples, enabling more rapid results than traditional growth based processes [144]. Impedance has also been used as a readout for cantilever sensors, by the same group which applied them to waterborne pathogens [145].

Immunological detection with antibodies has been successfully employed for the detection of specific bacteria and their toxins. Brewster developed an assay for the detection of *E. coli* O157:H7 measuring the current produced by the oxidation of an electroactive enzyme product. The sensor had a detection limit of 5×10^3 CFU/mL with a 25-min analysis time [146]. *Salmonella typhimurium* was sandwiched between antibody-coated magnetic beads and an enzyme-conjugated antibody [147]. The beads were transported onto the surface of graphite electrodes with the aid of a magnet. Conversion of enzyme substrate to an electroactive product was measured giving a minimum detectable level of 8×10^3 cells/mL. The same group described the development of an immunoligand assay in conjunction with a light-addressable potentiometric sensor for the rapid detection of *E. coli* O157:H7 cells in buffer [148]. *Campylobacter jejuni* detection was performed with an enzyme-linked immunoassay coupled with a tyrosinase-modified enzyme electrode [149]. Recently, Li has reported about an amperometric immunosensor that uses gold nanoparticles to enhance the surface area. The device was tested with *E. coli* O157:H7 bacteria and have a working range, under optimal conditions, of 4.12×10^2–4.12×10^5 colony-forming units/mL [150]. Immunomagnetic separation was performed to isolate *C. jejuni* from the sample solution. This system was evaluated using *C. jejuni* pure culture and poultry samples inoculated with *C. jejuni*.

In recent years various kinds of electrochemical biosensors based on the identification of the bacterial nucleic acid have been developed. *Legionella pneumophila* [151] and Vibrio cholera [152] have been detected and quantified by combined PCR and differential pulse voltammetry. Wang developed genosensors for *Cryptosporidium*, *E. coli*, *Giardia*, and *Mycobacterium tuberculosis* based on the immobilisation of specific oligonucleotides onto a carbon paste electrode and chronopotentiometry for monitoring the hybridisation events [153]. Baeumner described the development of a field-usable RNA biosensor for the detection of viable *E. coli* in water. Bacteria were thermally lysed and mRNA was extracted, purified, amplified, and quantified. A detection limit of 5 fmol per sample was determined for a synthetic target sequence [154]. More recently, an LoD of 1×10^{-17} was reported using graphene nanoflakes to enhance the detection of *E. coli* O157:H7 [155].

Radke described a high-density microelectrode array biosensor for the detection of *E. coli* O157:H7 [156]. The biosensor was fabricated from (100) silicon with a 2-μm layer of thermal oxide as an insulating layer, an active area of 9.6 mm^2, and consisted of an interdigitated gold electrode array. The change in impedance caused by the bacteria attachment to the electrode surface was measured over a frequency range of 100 Hz–10 MHz. The biosensor was able to discriminate between bacterial concentrations of 10^4–10^7 CFU/mL in pure culture and inoculated food samples. Maalouf has shown more recently that the detection limit can be decreased to 10 CFU/mL [157]. Lectins have also been used as the recognition element for bacteria, particularly *E. coli* and *Salmonella*, in an impedimetric system, but more work is needed to optimise the designs and lower the LOD [158]. The development of a disposable immunochip system has been described by Li [159]. It used electrochemical impedance spectroscopy and fluorescence microscopy to study the detection of Legionellosis.

Electrochemical detection of pathogens was reviewed in 2015 [160], and Furst and Francis reviewed impedance-based detection for bacteria in 2019, concluding that though there was little commercial output at present the future looked highly promising [161]. Jiang et al. used antimicrobial peptide assisted impedance detection, conferring selectivity for *E. coli* with a LOD of 10^2 CFU/mL demonstrated in water samples between pH 7 and 9 [162]. Another Jiang achieved a slightly better LOD of 10 cells/mL using a smartphone as a platform for impedance detection that also incorporated a microfluidic preconcentrator [163]. Single-cell level detection is achieved using a microfluidic flow-through an impedance screening system that was successfully applied to *E. coli* spiked conductivity-adjusted drinking water samples [164].

Dielectrophoresis has been used in the concentration of microbial pathogens but also in combination with a variety of detection schemes to enhance performance, e.g. with immunocapture, impedance detection, Raman, and molecular methods [165]. With impedance approaches Kim et al. reached 300 CFU/mL with a flow rate of 1.5 mL/h [166] and Suehiro tried various approaches to further enhance DEP assisted impedance detection including the addition of antibodies and the use of electropermeabilisation [167]. A 2017 review of DEP applied in this way concluded that the main challenges in application to waterborne pathogens were the low flow rate and the need for integrated systems, while also highlighting the potential of the use of aptamers and the application to antimicrobial resistance determination [165]. In 2019, 10 cells/mL for a multiplexed bacterial pathogen detection system were reported, combining dielectrophoresis (DEP) concentration and impedance [168]. A similar performance was reported for a foodborne bacteria DEP and impedance sensor [169]. This approach also allows for viability determination [170].

DEP has been applied to the study of both *Cryptosporidium* and *Giardia* (oo)cysts. It has been shown that viable and nonviable oocysts electrorotate at different rates and in opposite directions, depending upon the field strength [59, 171]. Goater et al. designed a system in which travelling wave DEP was used to collect oocysts in the centre of a spiral electrode where electrorotation was applied for detection. In this paper, it was observed that, in the frequency window of 20–600 kHz, viable oocysts rotated faster than nonviable ones, at rates discernible to the human eye or an automated image recognition system. In 2010, a US patent was granted to Simmons et al. for the use of an insulating DEP microfluidic chip to capture *Cryptosporidium* [172]. The patent claims that the device could process 1–10 mL of water concentrating the sample to 25 μL for further study such as immunofluorescence. Potential clogging problems were addressed by the utilisation of an ultrafiltration membrane prior to sample entry into the dielectrophoretic segment. Recently, it has been shown that *Plasmodium falciparum* trophozoites modify the zeta potential of red blood cells, opening the way for characterisation of infectivity by measurement of zeta potential changes [173]. A capacitative approach for *Cryptosporidium* detection, using antibody capture, was successful down to 40 oocysts per mm [174].

Reybier has used electrochemical impedance spectroscopy to monitor the changes affecting murine macrophage cell lines in response to parasite infection by *Leishmania amazonensis* and have proposed a model describing the interaction of the parasite with the host cell metabolism [175]. Electrochemical impedance spectroscopy has been used to detect *Cryptosporidium* in water [176]. In this example, the release of ions from *Cryptosporidium* oocysts results in a change of conductivity in a water buffer. A chip consisting of an arrangement of four sensors with 4 μm wide interdigitated electrodes was manufactured by optical lithography and metal deposition on a Pyrex substrate. The limit of detection of the device was measured to be 10^4 oocysts/mL in the buffer. Additionally, it was found that nonviable oocysts show a 15% difference in impedance compared to viable oocysts at the same concentration of 1000 oocysts/mL. The same group also applied this technique to monitor cell culture infections [177], and used another electrical method, exploiting electro-hydrodynamic forces for the selective concentration of different *Cryptosporidium* species and *Giardia* into small droplets [178]. A technique known as impedance cytometry allows for rapid, noninvasive, and rapid (1000/s) single-cell characterisation via impedimetric measurements, and in 2017 this was applied to protozoan evaluation [179]. Different species of *Cryptosporidium* could be discriminated and *Cryptosporidium* and *Giardia* distinguished, more successfully than utilising flow cytometry (Fig. 6.9). Additionally, the approach allowed viability discrimination, with the data comparing untreated and heat-inactivated oocysts, and throughput could potentially be improved by increasing channel size [180]. Evaluation of six single-stranded DNA probes was performed by simulation and experiment as a biorecognition element of *Bonamia ostreae* and *Bonamia exitiosa* parasites showing that bioinformatics can be used to developed genosensors [181].

6.9. Summary and future outlook

Electrochemistry is issued from electrophysiology and has been developed by important pioneer works in the first part of the twentieth century using macroscopic electrodes. Electrochemistry moved into the micrometre scale world in

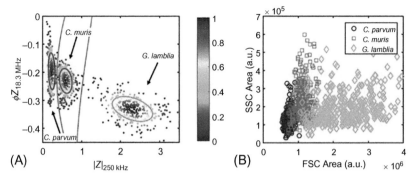

FIG. 6.9 Impedance vs optical detection of parasites. (A) Scatter plot of phase at 18.3 MHz vs impedance magnitude at 250 kHz for all parasite species measured together in a mixed sample. Annotated confidence ellipses contain ~50% of each population. *Green lines* indicate boundaries defining equal probability deviation between adjacent populations. The *colour* of each data point represents the normalised proximity of the event to the respective population mean. (B) Conventional flow cytometry data (SSC vs FSC) for all parasite species in PBS. Individual optical scatter data for each population plotted together. A total of 1000 events are plotted in both (A) and (B). *(Reproduced under the Creative Commons Attribution 4.0 International Licence (https://creativecommons.org/licenses/by/4.0/) from J.S. McGrath, et al., Analysis of parasitic protozoa at the single-cell level using microfluidic impedance cytometry, Sci. Rep. 7(1) (2017) 2601.)*

the second part of the twentieth century. Since 20 years, the numerous electro-chemical biosensors developed have demonstrated obvious interesting features for pathogen detection and analysis. They can be classified into potentiometric, voltammetric, coulometric, impedimetric, and DEP, based on the observed parameter such as electrical potential, current, charge, and dielectric properties. These different categories can be further refined depending if the system is in steady state or not, if the measured quantity is observed in time or frequency domain, and if the analysis uses small or large amplitude signals as described in Section 6.1. Potentiometric based detection have been presented in Section 6.2. With this technique, the bio-recognition process is converted into a potential signal. The technique has a very low limit of detection because the response is a logarithmic value of the analyte's concentration. However, not so many potentiometric biosensors exist for the detection of pathogens. Voltammetry has been introduced in Section 6.3. In this technique, a time-dependent potential or current is applied to the system, and the resulting current or potential characterising the biorecognition event is measured. Its amperometric form is the most common electrochemical method which has been used for pathogen detection and shows very high sensitivity. The major advantage of coulometry is the absence of calibration procedure since it is an absolute method based on Faraday's law, as presented in Section 6.4. This technique is mainly used for titration of solutions and can be implemented for the kinetic study of the respiratory process of a living organism. The ability of modern electronics to accurately measure and integrate current-time functions permits coulometry to be applied even with relatively small amounts of sample material. It requires a fine knowledge of the underlying physicochemical mechanisms only known by experts and has not so much been employed for waterborne pathogen detection. The integration of impedance with bio-recognition elements for the detection of the pathogen (Section 6.5) has led to the development of many devices in recent years. In impedance measurement, a controlled alternative potential of a few milli-volts is applied to the system over a wide range of frequencies. It induces the flow of an electrical current which depends on the biological properties of the system. The detection limits of impedancemetry are still inferior compared to the others techniques. However, there is a growing interest in this technique for biosensing and in particular for waterborne pathogen detection because it is contrary to the previous techniques a label-free method. Dielectrophoresis (Section 6.6) has been mainly used for manipulation (i.e. concentration, separation, displacement) of micro and nanoparticles, including waterborne pathogens. Its use as a characterisation technique is for time being not fully exploited because it requires the implementation of complex microsystems. We believe that the current development of total analysis system technologies will allow its broader use. The effects of scaling on the performances of these devices have been reviewed in Section 6.6. The small size is advantageous for portable or high throughput applications. However, we show that depending on the measured quantity, miniaturisation can not necessarily improve the signal over noise ratio. We review the scientific literature about the use of electrochemical biosensors for the detection of pathogens (Section 6.7). We did not restrict our analysis to studies dedicated to waterborne pathogens since they are not so much studied, though more examples are emerging. Since the interest in detection and analysis of waterborne pathogen is growing it has been noted that electrochemical techniques have a huge potential for water analysis [182]. We mention different approaches developed by various research groups. Voltammetry and impedancemetry are the most cited in recent years. The reported sensors have different complexities, analysis times, sample preparation procedures, detection limits depending on the target pathogen. Improvements are still required to plan large commercialisation of this technique. Indeed, national and international legislations require more and more information about the presence of pathogens in our environment, and in particular in water. This information has to be reliable, accurate, and low cost and should be available in a very short time. To enhance the emergence of electrochemical biosensors for label-free waterborne pathogen detection and analysis, we point out some improvements that are primordial. The variability of electrochemical biosensors needs to be systematically and thoroughly quantified [183]. If some reported systems comprise multiple integrated biosensors to carry out intrinsically this statistical analysis, most works do not report on it. This lack of solid statistical analyses may be an important challenge for commercialisation, dissuading industrialists, and venture capitalists [184]. The inclusion of redundant sensor elements for self-verification, the availability of back-up sensors for enhanced longevity, and the use of recognised and rigorous statistical analysis are highly desirable [185, 186]. The costs of electrochemical biosensors need to be rapidly decreased to be competitive with labelling or the other label-free techniques. For this, biosensors' lifetime can be improved and/or the cost of materials and manufacturing processes can be decreased by using organic electrodes, flexible substrates, printing, and roll-to-roll processes. Analysis time has to be shorter than gold standard techniques. Analysis time includes sample pretreatment, cultivation, extraction, etc., and detection. In order to become more attractive, electrochemical biosensors need to show that they are capable of reaching at least the same detection levels as traditional techniques (10–100 CFU/mL), and there has been recent progress towards these levels, with some systems able to analyse at the single pathogen level. If significant advances have already been accomplished with pure samples, we can regret that only a few devices have been applied to real samples or handle in real conditions. Several obstacles still have to be overcome such as the interfacing of the micro-size biosensors with the macro-size real world [187]. These include their association with different miniaturised elements for automatic multisteps operations (e.g. microfiltration, ultraconcentration, microlysis, micromixer, etc.) in analytical microsystems (i.e. lab-on-a-chip platform) and their integration within wireless networks is probably the route of choice to ensure commercial success. The issue of calibration and conditioning of the bio-recognition species are also challenging questions. The rapid development of new recognition elements and nano-scale transduction electrodes is now playing an important role to fill this gap. We are convinced that electrochemical devices will play a central role in the future deployment of autonomous micro-systems for water quality management. Although an 'old' technique, electro-chemistry is still a field of intense fundamental researches at the intersection between micro and nanotechnologies, chemistry, biology, and physics. New experimental approaches (i.e. electrochemical scanning probe microscopy) and theoretical understanding (i.e. mechanisms governing bioreceptor-nanomaterial interactions and induced electron transfer) are constantly reported in the literature. This co-development and progress of experiment and theory is the perfect methodology to design viable commercial products in the future [188].

References

[1] J.A.V. Butler, Studies in heterogeneous equilibria—part II: the kinetic interpretation of the Nernst theory of electromotive force, Trans. Faraday Soc. 19 (1924) 729–733.

[2] T. Erdey-Gruz, M. Volmer, Zur theorie der wasserstoffüberspannung, Z. Phys. Chem. 150 (1930) 203–213.

[3] R.A. Marcus, Electron transfer reactions in chemistry: theory and experiment, Rev. Mod. Phys. 65 (1993) 599–610.

[4] M.Z. Bazant, M.S. Kilic, B. Storey, A. Ajdari, Towards an understanding of nonlinear electrokinetics at large voltages in concentrated solutions, Adv. Colloid Interface Sci. 152 (2009) 48–88.

[5] I. Giaever, C.R. Keese, A morphological biosensor for mammalian cells, Nature 366 (1993) 591–592.

[6] G.A. Zelada-Guillen, S.V. Bhosale, J. Riu, F.X. Rius, Real-time potentiometric detection of bacteria in complex samples, Anal. Chem. 82 (2010) 9254–9260.

[7] P.T. Kissinger, A.W. Bott, Electrochemistry for the non-electrochemist, Curr. Sep. 20 (2) (2002) 51–53.

[8] J. Heyrovsky, Electrolysis with a dropping mercury cathode, Philos. Mag. 45 (1923) 303–315.

[9] Z.S. Bi, P. Salaun, C.M.G. van den Berg, Study of bare and mercury-coated vibrated carbon, gold and silver microwire electrodes for the determination of lead and cadmium in seawater by anodic stripping voltammetry, Electroanalysis 25 (2013) 357–366.

[10] J. Barek, J. Zima, Eighty years of polarography: history and future, Electroanalysis 15 (2003) 467–472.

[11] D. Macdonald, Reflections on the history of electrochemical impedance spectroscopy, Electrochim. Acta 51 (2006) 1376–1388.

[12] L.R. Faulkner, Understanding electrochemistry: some distinctive concepts, J. Chem. Educ. 60 (1983) 262–264.

[13] C. Margo, J. Katrib, M. Nadi, A. Rouane, A four-electrode low frequency impedance spectroscopy measurement system using the AD5933 measurement chip, Physiol. Meas. 34 (2013) 391–405.

[14] E. Sarró, M. Lecina, A. Fontova, C. Solà, F. Gòdia, J.J. Cairó, R. Bragós, Electrical impedance spectroscopy measurements using a four-electrode configuration improve on-line monitoring of cell concentration in adherent animal cell cultures, Biosens. Bioelectron. 31 (2012) 257–263.

[15] G.S. Popkirov, R.N. Schindler, A new impedance spectrometer for the investigation of electrochemical systems, Rev. Sci. Instrum. 63 (1992) 5366–5372.

[16] A.S. Baranski, T. Krogulec, L.J. Nelson, P. Norouzi, High-frequency impedance spectroscopy of platinum ultramicroelectrodes in flowing solutions, Anal. Chem. 70 (1998) 2895–2901.

[17] C.M. Pettit, P.C. Goonetilleke, C.M. Sulyma, D. Roy, Combining impedance spectroscopy with cyclic voltammetry: measurement and analysis of kinetic parameters for faradaic and nonfaradaic reactions on thin-film gold, Anal. Chem. 78 (2006) 3723–3729.

[18] S.A. Mozaffari, T. Chang, M. Park, Diffusional electrochemistry of cytochrome c on mixed captopril/3-mercapto-1-propanol self-assembled monolayer modified gold electrode, J. Phys. Chem. C 113 (2009) 12434–12442.

[19] H.A. Pohl, J.S. Crane, Dielectrophoresis of cells, Biophys. J. 11 (1971) 711–727.

[20] Y. Zhao, U.-C. Yi, S.K. Cho, Microparticle concentration and separation by traveling-wave dielectrophoresis (twdep) for digital microfluidics, J. Microelectromech. Syst. 16 (2007) 1472–1481.

[21] T. Kakutani, S. Shibatani, M. Sugai, Electrorotation of non-spherical cells: theory for ellipsoidal cells with an arbitrary number of cells, Bioelectrochem. Bioenerg. 31 (1993) 131–145.

[22] J. Lyklema, Fundamentals of Interface and Colloid Science: Solid-Liquid Interface, Academic Press, 1995.

[23] N.G. Green, H. Morgan, Dielectrophoresis of submicrometer latex spheres: 1. Experimental results, J. Phys. Chem. B 103 (1999) 41–50.

[24] D. Morganti, H. Morgan, Characterization of non-spherical polymer particles by combined electrorotation and electroorientation, Colloids Surf. A Physicochem. Eng. Asp. 376 (2011) 67–71.

[25] T. Honegger, K. Berton, E. Picard, D. Peyrade, Determination of clausius–mossotti factors and surface capacitances for colloidal particles, Appl. Phys. Lett. 98 (2011) 181906.

[26] R. Pethig, Dielectrophoresis: status of the theory, technology, and applications, Biomicrofluidics 4 (2010) 039901.

[27]   D. Diamond, S. Coyle, S. Scarmagnani, J. Hayes, Wireless sensor networks and chemo−/ biosensing, Chem. Rev. 108 (2008) 652–679.

[28]   A. Hassibi, R. Navid, R.W. Dutton, T.H. Lee, Comprehensive study of noise processes in electrode electrolyte interfaces, J. Appl. Phys. 96 (2004) 1074–1082.

[29]   R. Hintsche, M. Paeschke, U. Wollenberger, U. Schnakenberg, B. Wagner, T. Lisec, Microelectrode arrays and application to biosensing devices, Biosens. Bioelectron. 9 (1994) 697–705.

[30]   Q.S. Guo, T. Kong, R.G. Su, Q. Zhang, G.S. Cheng, Noise spectroscopy as an equilibrium analysis tool for highly sensitive electrical biosensing, Appl. Phys. Lett. 101 (2012) 093704.

[31]   M.J. Madou, R. Cubicciotti, Scaling issues in chemical and biological sensors, Proc. IEEE 91 (2003) 830–838.

[32]   W. Olthuis, W. Streekstra, P. Bergveld, Theoretical and experimental determination of cell constants of planar-interdigitated electrolyte conductivity sensors, Sens. Actuators B 24 (1995) 252–256.

[33]   D. Landheer, W.R. McKinnon, G.C. Aers, W.H. Jiang, J.J. Deen, M.W. Shinwari, Calculation of the response of field-effect transistors to charged biological molecules, IEEE Sensors J. 7 (2007) 1233–1242.

[34]   H. Anan, M. Kamahori, Y. Ishige, K. Nakazato, Redox-potential sensor array based on extended-gate field-effect transistors with ω-ferrocenylalkanethiol modified gold electrodes, Sens. Actuators B 187 (2013) 254–261.

[35]   Y.F. Jia, X.B. Yin, J. Zhang, S. Zhou, M. Song, K.L. Xing, Graphene oxide modified light addressable potentiometric sensor and its application for ssDNA monitoring, Analyst 137 (24) (2012) 5866–5873.

[36]   J.-H. Ahn, S.-J. Choi, J.-W. Han, T.J. Park, S.Y. Lee, Y.-K. Choi, Double-gate nanowire field effect transistor for a biosensor, Nano Lett. 10 (2010) 2934–2938.

[37]   D.F. Schaffhauser, M. Patti, T. Goda, Y. Miyahara, I.C. Forster, P.S. Dittrich, An integrated field-effect microdevice for monitoring membrane transport in *Xenopus laevis* oocytes via lateral proton diffusion, PLoS One 7 (2012) e39238.

[38]   C. Zuliani, D. Diamond, Opportunities and challenges of using ion-selective electrodes in environmental monitoring and wearable sensors, Electrochim. Acta 84 (2012) 29–34.

[39]   S.V. Dzyadevych, V.N. Arkhypova, A.P. Soldatkin, A.V. El'skaya, C. Martelet, N. Jaffrezic-Renault, Amperometric enzyme biosensors: past, present and future, IRBM 29 (2008) 171–180.

[40]   N.J. Ronkainen, H.B. Halsall, W.R. Heineman, Electrochemical biosensors, Chem. Soc. Rev. 39 (2010) 1747–1763.

[41]   X.J. Huang, L. Aldous, A.M. O'Mahony, F.J. del Campo, R.G. Compton, Toward membrane-free amperometric gas sensors: a microelectrode array approach, Anal. Chem. 82 (2010) 5238–5245.

[42]   F.J. Del Campo, O. Ordeig, F.J. Munoz, Improved free chlorine amperometric sensor chip for drinking water applications, Anal. Chim. Acta 554 (2005) 98–104.

[43]   K.H. Lee, T. Ishikawa, S. Sasaki, Y. Arikawa, I. Karube, Chemical oxygen demand (COD) sensor using a stopped-flow thin layer electrochemical cell, Electroanalysis 11 (1999) 1172–1179.

[44]   J. Fukuda, S. Tsujimura, K. Kano, Coulometric bioelectrocatalytic reactions based on NAD-dependent dehydrogenases in tricarboxylic acid cycle, Electrochim. Acta 54 (2008) 328–333.

[45]   E. Grygolowicz-Pawlak, E. Bakker, Background current elimination in thin layer ion-selective membrane coulometry, Electrochem. Commun. 12 (2010) 1195–1198.

[46] S. Carroll, M.M. Marei, T.J. Roussel, R.S. Keynton, R.P. Baldwin, Microfabricated electrochemical sensors for exhaustive coulometry applications, Sens. Actuators B 160 (2011) 318–326.

[47] K. Cheung, S. Gawad, P. Renaud, Impedance spectroscopy flow cytometry: on-chip label-free cell differentiation, Cytometry A 65 (2005) 124–132.

[48] L.L. Sohm, O.A. Saleh, G.R. Facer, A.J. Beavis, R.S. Allan, D.A. Notterman, Capacitance cytometry: measuring biological cells one by one, Proc. Natl. Acad. Sci. 97 (2000) 10687–10690.

[49] H.E. Ayliffe, A.B. Frazier, R.D. Rabbitt, Electric impedance spectroscopy using microchannels with integrated metal electrodes, J. Microelectromech. Syst. 8 (1999) 50–57.

[50] H. Thielecke, Capillary chip based characterization of small tissue samples, Med. Device Technol. 14 (2003) 18–20.

[51] J. Wegener, C.R. Keese, I. Giaver, Electric cell-substrate impedance sensing (ECIS) as non-invasive means to monitor the kinetics of cell spreading to artificial surfaces, Exp. Cell Res. 259 (2000) 158–166.

[52] R. Ehret, W. Baumann, M. Brischwein, A. Schwinde, B. Wolf, Online control of cellular adhesion with impedance measurements using interdigitated electrode structures, Med. Biol. Eng. Comput. 36 (1998) 365–370.

[53] S. Arndt, J. Seebach, K. Psathaki, H.J. Galla, J. Wegener, Bioelectrical impedance assay to monitor changes in cell shape apoptosis, Biosens. Bioelectron. 19 (2004) 583–594.

[54] Q. Liu, Y. Jinjiang, X. Lidan, C.O.T. Johnny, Z. Yu, W. Ping, Y. Mo, Impedance studies of bio-behavior and chemosensitivity of cancer cells by micro-electrode arrays, Biosens. Bioelectron. 24 (2009) 1305–1310.

[55] L. Ceriotti, A. Kob, S. Drechsler, J. Ponti, E. Thedinga, P. Colpo, R. Ehret, F. Rossi, Online monitoring of BALB/3T3 metabolism and adhesion with multiparametric chip-based system, Anal. Biochem. 371 (2007) 92–104.

[56] K. Khoshmanesh, S. Nahavandi, S. Baratchi, A. Mitchell, K. Kalantar-zadeh, Dielectrophoretic platforms for bio-microfluidic systems, Biosens. Bioelectron. 26 (2011) 1800–1814.

[57] J.K. Valley, S. Neale, H.-Y. Hsu, A.T. Ohta, A. Jamshidi, M.C. Wu, Parallel single-cell light-induced electroporation and dielectrophoretic manipulation, Lab Chip 9 (2009) 1714–1720.

[58] B.H. Lapizco-Encinas, R.V. Davalos, B.A. Simmons, E.B. Cummings, Y. Fintschenko, An insulator-based (electrodeless) dielectrophoretic concentrator for microbes in water, J. Microbiol. Methods 62 (2005) 317–326.

[59] A.D. Goater, J.P.H. Burt, R. Pethig, A combined travelling wave dielectrophoresis and electrorotation device: applied to the concentration and viability determination of cryptosporidium, J. Phys. D Appl. Phys. 30 (1997) L65–L69.

[60] H. Li, R. Bashir, Dielectrophoretic separation and manipulation of live and heat-treated cells of Listeria on microfabricated devices with interdigitated electrodes, Sens. Actuators B 86 (2002) 215–221.

[61] H. Morgan, N.G. Green, Dielectrophoretic manipulation of rod-shaped viral particles, Electrostatics 42 (1997) 279–293.

[62] K.E. Sung, M.A. Burns, Optimization of DNA stretching in microfabricated devices, Anal. Chem. 78 (2006) 2939–2947.

[63] R.W. Clarke, S.S. White, D.-J. Zhou, L.-M. Ying, D. Klenerman, Trapping of proteins under physiological conditions in a nanopipette, Angew. Chem. 117 (2005) 3813–3816.

[64] A.K. Trilling, J. Beekwilder, H. Zuilhof, Antibody orientation on biosensor surfaces: a mini-review, Analyst 138 (2013) 1619–1627.

[65] L. Lu, G. Chee, K. Yamada, S. Jun, Electrochemical impedance spectroscopic technique with a functionalized microwire sensor for rapid detection of foodborne pathogens, Biosens. Bioelectron. 42 (2013) 492–495.

[66] B.K. Jena, C.R. Raj, Electrochemical biosensor based on integrated assembly of dehydrogenase enzymes and gold nanoparticles, Anal. Chem. 78 (2006) 6332–6339.

[67] B. Liang, L. Fang, G. Yang, Y. Hu, X. Guo, X. Ye, Direct electron transfer glucose biosensor based on glucose oxidase self-assembled on electrochemically reduced carboxyl graphene, Biosens. Bioelectron. 43 (2013) 131–136.

[68] S.Y. Lee, R. Matsuno, K. Ishihara, M. Takai, Direct electron transfer with enzymes on nanofiliform titanium oxide films with electron-transport ability, Biosens. Bioelectron. 41 (2013) 289–293.

[69] K. Kerman, M. Kobayashi, E. Tamiya, Recent trends in electrochemical DNA biosensor technology, Meas. Sci. Technol. 15 (2004) R1–R11.

[70] L. Wang, X. Wang, X. Chen, J. Liu, S. Liu, C. Zhao, Development of an electrochemical DNA biosensor with the DNA immobilization based on in situ generation of dithiocarbamate ligands, Bioelectrochemistry 88 (2012) 30–35.

[71] P. Banerjee, A.K. Bhunia, Mammalian cell-based biosensors for pathogens and toxins, Trends Biotechnol. 27 (2009) 179–188.

[72] M. Qu, B.M. Boruah, W. Zhang, Y. Li, W. Liu, Y. Bi, G.F. Gao, R. Yang, D. Liu, B. Gao, A rat basophilic leukaemia cell sensor for the detection of pathogenic viruses, Biosens. Bioelectron. 43 (2012) 412–418.

[73] B. Wicklein, M.A.M. del Burgo, M. Yuste, E. Carregal-Romero, A. Llobera, M. Darder, P. Aranda, J. Ortin, G. del Real, C. Fernandez-Sanchez, E. Ruiz-Hitzky, Biomimetic architectures for the impedimetric discrimination of influenza virus phenotypes, Adv. Funct. Mater. 23 (2013) 254–262.

[74] I. Willner, M. Zayats, Electronic aptamer-based sensors, Angew. Chem. Int. Ed. 46 (2007) 6408–6418.

[75] M. Labib, A.S. Zamay, O.S. Koloyskaya, I.T. Reshetneva, G.S. Zamay, R.J. Kibbee, S.A. Sattar, T.N. Zamay, M.V. Berezovski, Aptamer-based viability impedimetric sensor for bacteria, Anal. Chem. 84 (2012) 8966–8969.

[76] C. Malitesta, E. Mazzotta, R.A. Picca, A. Poma, I. Chianella, S.A. Piletsky, MIP sensors—the electrochemical approach, Anal. Bioanal. Chem. 402 (2012) 1827–1846.

[77] P. Qi, Y. Wan, D. Zhang, Impedimetric biosensor based on cell-mediated bioimprinted films for bacterial detection, Biosens. Bioelectron. 39 (2013) 282–288.

[78] A. Singh, D. Arutyunov, C.M. Szymanski, S. Evoy, Bacteriophage based probes for pathogen detection, Analyst 137 (2012) 3405–3421.

[79] K. Kiilerich-Pedersen, C.R. Poulsen, T. Jain, N. Rozlosnik, Polymer based biosensor for rapid electrochemical detection of virus infection of human cells, Biosens. Bioelectron. 28 (2011) 386–392.

[80] D. Khodagholy, T. Doublet, M. Gurfinkel, P. Quilichini, E. Ismailova, P. Leleux, T. Herve, S. Sanaur, C. Bernard, G.G. Malliaras, Highly conformable conducting polymer electrodes for in vivo recordings, Adv. Mater. 23 (2011) H268–H272.

[81] S. Tria, L.H. Jimison, A. Hama, M. Bongo, R.M. Owens, Sensing of EGTA mediated barrier tissue disruption with an organic transistor, Biosensors 3 (2013) 44–57.

[82] R.D. Henderson, R.M. Guijt, P.R. Haddad, E.F. Hilder, T.W. Lewis, M.C. Breadmore, Manufacturing and application of a fully polymeric electrophoresis chip with integrated polyaniline electrodes, Lab Chip 10 (2010) 1869.

[83] R.D. Henderson, R.M. Guijt, L. Andrewartha, T.W. Lewis, T. Rodemann, A. Henderson, E.F. Hilder, P.R. Haddad, M.C. Breadmore, Lab-on-a-Chip device with laser-patterned polymer electrodes for high voltage application and contactless conductivity detection, Chem. Commun. 48 (2012) 9287–9289.

[84] M. Brun, J.-F. Chateaux, A.-L. Deman, P. Pittet, R. Ferrigno, Nanocomposite carbon-PDMS material for chip-based electrochemical detection, Electroanalysis 23 (2011) 321–324.

[85] Y. Sameenoi, M.M. Mensack, K. Boonsong, R. Ewing, W. Dungchai, O. Chailapakul, D.M. Cropek, C.S. Henry, Poly(dimethylsiloxane) cross-linked carbon paste electrodes for microfluidic electrochemical sensing, Analyst 136 (2011) 3177–3184.

[86] F. Liu, J. Zhang, Y. Deng, D. Wang, Y. Lu, X. Yu, Detection of EGFR on living human gastric cancer BGC823 cells using surface plasmon resonance phase sensing, Sens. Actuators B 153 (2011) 398–403.

[87] T. Cohen-Karni, Q. Qing, Q. Li, Y. Fang, C.M. Lieber, Graphene and nanowire transistors for cellular interfaces and electrical recording, Nano Lett. 10 (2010) 1098–1102.

[88] N. Mohanty, V. Berry, Graphene-based single-bacterium resolution biodevice and DNA transistor: interfacing graphene derivatives with nanoscale and microscale biocomponents, Nano Lett. 8 (2008) 4469–4476.

[89] L. Feng, Y. Chen, J. Ren, X. Qu, A graphene functionalized electrochemical aptasensor for selective label-free detection of cancer cells, Biomaterials 32 (2011) 2930–2937.

[90] C.X. Guo, X.T. Zheng, Z.S. Lu, X.W. Lou, C.M. Li, Biointerface by cell growth on layered graphene–artificial peroxidase–protein nanostructure for in situ quantitative molecular detection, Adv. Mater. 22 (2010) 5164–5167.

[91] Q. He, H.G. Sudibya, Z. Yin, S. Wu, H. Li, F. Boey, W. Huang, P. Chen, H. Zhang, Centimeter-long and large-scale micropatterns of reduced graphene oxide films: fabrication and sensing applications, ACS Nano 4 (2010) 3201–3208.

[92] M.S. Mannoor, H. Tao, J.D. Clayton, A. Sengupta, D.L. Kaplan, R.R. Naik, N. Verma, F.G. Omenetto, M.C. McAlpine, Graphene-based wireless bacteria detection on tooth enamel, Nat. Commun. 3 (2012) 763.

[93] R. Kempaiah, A. Chung, V. Maheshwari, Graphene as cellular interface: electromechanical coupling with cells, ACS Nano 5 (2011) 6025–6031.

[94] T.A. Webster, E.D. Goluch, Electrochemical detection of pyocyanin in nanochannels with integrated palladium hydride reference electrodes, Lab Chip 12 (2012) 5195–5201.

[95] V.M. Tolosa, K.M. Wassum, N.T. Maidment, H.G. Monbouquette, Electrochemically deposited iridium oxide reference electrode integrated with an electroenzymatic glutamate sensor on a multi-electrode array microprobe, Biosens. Bioelectron. 42 (2013) 256–260.

[96] Q. Shen, S.-K. You, S.-G. Park, H. Jiang, D. Guo, B. Chen, X. Wang, Electrochemical biosensing for cancer cells based on TiO$_2$/CNT nanocomposites modified electrodes, Electroanalysis 20 (2008) 2526–2530.

[97] F. Xiao, J. Song, H. Gao, X. Zan, R. Xu, H. Duan, Coating graphene paper with 2D-assembly of electrocatalytic nanoparticles: a modular approach toward high-performance flexible electrodes, ACS Nano 6 (2011) 100–110.

[98] J. Wan, X. Yan, J. Ding, R. Ren, A simple method for preparing biocompatible composite of cellulose and carbon nanotubes for the cell sensor, Sens. Actuators B 146 (2010) 221–225.

[99] R. Mukhopadhyay, When PDMS isn't the best, Anal. Chem. 79 (2007) 3248–3253.

[100] A. Kurian, S. Prasad, A. Dhinojwala, Unusual surface aging of poly(dimethylsiloxane) elastomers, Macromolecules 43 (2010) 2438–2443.

[101] K. Cheung, Implantable microscale neural interface, Biomed. Microdevices 9 (2007) 923–938.

[102] S. Myllymaa, K. Myllymaa, H. Korhonen, J. Töyräs, J.E. Jääskeläinen, K. Djupsund, H. Tanila, R. Lappalainen, Fabrication and testing of polyimide-based microelectrode arrays for cortical mapping of evoked potentials, Biosens. Bioelectron. 24 (2009) 3067–3072.

[103] C. González, M. Rodríguez, A flexible perforated microelectrode array probe for action potential recording in nerve and muscle tissues, J. Neurosci. Methods 72 (1997) 189–195.

[104] L. Giovangrandi, K.H. Gilchrist, R.H. Whittington, G.T.A. Kovacs, Low-cost microelectrode array with integrated heater for extracellular recording of cardiomyocyte cultures using commercial flexible printed circuit technology, Sens. Actuators B 113 (2006) 545–554.

[105] H. Charkhkar, G.L. Knaack, B.E. Gnade, E.W. Keefer, J.J. Pancrazio, Development and demonstration of a disposable low-cost microelectrode array for cultured neuronal network recording, Sens. Actuators B 161 (2012) 655–660.

[106] J. Seymour, N. Langhals, D. Anderson, D. Kipke, Novel multi-sided, microelectrode arrays for implantable neural applications, Biomed. Microdevices 13 (2011) 441–451.

[107] D. Lee, T. Cui, Low-cost, transparent, and flexible single-walled carbon nanotube nanocomposite based ion-sensitive field-effect transistors for pH/glucose sensing, Biosens. Bioelectron. 25 (2010) 2259–2264.

[108] M. Péter, T. Schüler, F. Furthner, P.A. Rensing, G.T. van Heck, H.F.M. Schoo, R. Möller, W. Fritzsche, A.J.J.M. van Breemen, E.R. Meinders, Flexible biochips for detection of biomolecules, Langmuir 25 (2009) 5384–5390.

[109] B.L. Hassler, T.J. Amundsen, J.G. Zeikus, I. Lee, R.M. Worden, Versatile bioelectronic interfaces on flexible non-conductive substrates, Biosens. Bioelectron. 23 (2008) 1481–1487.

[110] G. Priano, G. González, M. Günther, F. Battaglini, Disposable gold electrode array for simultaneous electrochemical studies, Electroanalysis 20 (2008) 91–97.

[111] L. Setti, A. Fraleoni-Morgera, A. Ballarin, A. Filippini, D. Frascaro, C. Piana, An amperometric glucose biosensor prototype fabricated by thermal inkjet printing, Biosens. Bioelectron. 20 (2005) 2019–2026.

[112] S. Gamerith, A. Klug, H. Scheiber, U. Scherf, E. Moderegger, E.J.W. List, Direct ink-jet printing of Ag–Cu nanoparticle and Ag-precursor based electrodes for OFET applications, Adv. Funct. Mater. 17 (2007) 3111–3118.

[113] J. Wang, M. Musameh, Carbon nanotube screen-printed electrochemical sensors, Analyst 129 (2004) 1.

[114] T. Schuler, T. Asmus, W. Fritzsche, R. Möller, Screen printing as cost-efficient fabrication method for DNA-chips with electrical readout for detection of viral DNA, Biosens. Bioelectron. 24 (2009) 2077–2084.

[115] J.D. Newman, A.P.F. Turner, Home blood glucose biosensors: a commercial perspective, Biosens. Bioelectron. 20 (2005) 2435–2453.

[116] D.J. Adlam, D.E. Woolley, A multiwell electrochemical biosensor for real-time monitoring of the behavioural changes of cells in vitro, Sensors 10 (2010) 3732–3740.

[117] M. Brischwein, S. Herrmann, W. Vonau, F. Berthold, H. Grothe, E.R. Motrescu, B. Wolf, Electric cell-substrate impedance sensing with screen printed electrode structures, Lab Chip 6 (2006) 819–822.

[118] R.M. Pemberton, J. Xu, R. Pittson, G.A. Drago, J. Griffiths, S.K. Jackson, J.P. Hart, A screen-printed microband glucose bioensor system for real-time monitoring of toxicity in cell culture, Biosens. Bioelectron. 26 (2011) 2448–2453.

[119] G.C. Jensen, C.E. Krause, G.A. Sotzing, J.F. Rusling, Inkjet-printed gold nanoparticle electrochemical arrays on plastic: application to immunodetection of a cancer biomarker protein, Phys. Chem. Chem. Phys. 13 (2011) 4888–4894.

[120] X.-J. Huang, A.M. O'Mahony, R.G. Compton, Microelectrode arrays for electrochemistry: approaches to fabrication, Small 5 (2009) 776–788.

[121] S.C. Lim, S.H. Kim, Y.S. Yang, M.Y. Lee, S.Y. Nam, J.B. Ko, Organic thin-film transistor using high-resolution screen-printed electrodes, Jpn. J. Appl. Phys. 48 (2009) 081503.

[122] E. Gili, M. Caironi, H. Sirringhaus, Organic integrated complementary inverters with inkjet printed source/drain electrodes and sub-micron channels, Appl. Phys. Lett. 100 (2012) 123303.

[123] L. Zhang, H. Liu, Y. Zhao, X. Sun, Y. Wen, Y. Guo, X. Gao, C. Di, G. Yu, Y. Liu, Inkjet printing high-resolution, large-area graphene patterns by coffee-ring lithography, Adv. Mater. 24 (2012) 436–440.

[124] S. Bae, H. Kim, Y. Lee, X. Xu, J.-S. Park, Y. Zheng, J. Balakrishnan, T. Lei, H.R. Kim, Y.I. Song, Y.-J. Kim, K.S. Kim, B. Ozyilmaz, J.-H. Ahn, B.H. Hong, S. Iijima, Roll-to-roll production of 30-inch graphene films for transparent electrodes, Nat. Nanotechnol. 5 (2010) 574–578.

[125] J. Kafka, N.B. Larsen, S. Skaarup, O. Geschke, Fabrication of an all-polymer electrochemical sensor by using a one-step hot embossing procedure, Microelectron. Eng. 87 (2010) 1239–1241.

[126] Y.T. Wang, Z.Q. Zhang, V. Jain, J. Yi, S. Mueller, J. Sokolov, Z.X. Liu, K. Levon, B. Rigas, M.H. Rafailovich, Potentiometric sensors based on surface molecular imprinting: detection of cancer biomarkers and viruses, Sens. Actuators B 146 (2010) 381–387.

[127] J. Albers, T. Grunwald, E. Nebling, G. Piechotta, R. Hintsche, Electrical biochip technology: a tool for microarrays and continuous monitoring, Anal. Bioanal. Chem. 377 (2003) 521–527.

[128] E. Nebling, T. Grunwald, J. Albers, P. Schafer, R. Hintsche, Electrical detection of viral DNA using ultramicroelectrode arrays, Anal. Chem. 76 (2004) 689–696.

[129] C.F. Ding, F. Zhao, M.L. Zhang, S.S. Zhang, Hybridization biosensor using 2,9-dimethyl-1,10-phenantroline cobalt as electrochemical indicator for detection of hepatitis B virus DNA, Bioelectrochemistry 72 (2008) 28–33.

[130] S. Zhang, Q. Tan, F. Li, X. Zhang, Hybridization biosensor using diaquabis[N-(2-pyridinylmethyl)benzamide-κ2N,O]-cadmium(II) dinitrate as a new electroactive indicator for detection of human hepatitis B virus DNA, Sens. Actuators B 124 (2007) 290–296.

[131] O. Laczka, R.M. Ferraz, N. Ferrer-Miralles, A. Villaverde, F.X. Munoz, F.J. del Campo, Fast electrochemical detection of anti-HIV antibodies: coupling allosteric enzymes and disk microelectrode arrays, Anal. Chim. Acta 641 (2009) 1–6.

[132] H.S. Mandal, Z.D. Su, A. Ward, X.W. Tang, Carbon nanotube thin film biosensors for sensitive and reproducible whole virus detection, Theranostics 2 (2012) 251–257.

[133] U. Jarocka, H. Radecka, T. Malinowski, L. Michalczuk, J. Radecki, Detection of prunus necrotic ringspot virus in plant extracts with impedimetric immunosensor based on glassy carbon electrode, Electroanalysis 25 (2013) 433–438.

[134] I. Ermolina, J. Milner, H. Morgan, Dielectrophoretic investigation of plant virus particles: cow pea mosaic virus and tobacco mosaic virus, Electrophoresis 27 (2006) 3939–3948.

[135] I. Ermolina, H. Morgan, N.G. Green, J.J. Milner, Y. Feldman, Dielectric spectroscopy of tobacco mosaic virus, Biochim. Biophys. Acta 1622 (2003) 57–63.

[136] A.K. Balasubramanian, K.A. Soni, A. Beskok, S.D. Pillai, A microfluidic device for continuous capture and concentration of microorganisms from potable water, Lab Chip 7 (2007) 1315–1321.

[137] F. Grom, J. Kentsch, T. Müller, T. Schnelle, M. Stelzle, Accumulation and trapping of hepatitis A virus particles by electrohydrodynamic flow and dielectrophoresis, Electrophoresis 27 (2006) 1386–1393.

[138] J. Kirkegaard, Aptasensor Development for Detection of Virus in Water, DTU Nanotech, 2016.

[139] A.M. Gall, et al., Waterborne viruses: a barrier to safe drinking water, PLoS Pathog. 11 (6) (2015) e1004867.

[140] L. Liu, D.M. Moore, A survey of analytical techniques for noroviruses, Foods 9 (3) (2020) 318–355.

[141] H. Suzuki, E. Tamiya, I. Karube, Disposable amperometric CO_2 sensor employing bacteria and a miniature oxygen electrode, Electroanalysis 3 (1991) 53–57.

[142] C.M. Ruan, L.J. Yang, Y.B. Li, Immunobiosensor chips for detection of *Escherichia coli* O157: H7 using electrochemical impedance spectroscopy, Anal. Chem. 74 (2002) 4814–4820.

[143] F.G. Perez, I. Tryland, M. Mascini, L. Fiksdal, Rapid detection of *Escherichia coli* in water by a culture-based amperometric method, Anal. Chim. Acta 427 (2001) 149–154.

[144] A. Bajwa, et al., Rapid detection of viable microorganisms based on a plate count technique using arrayed microelectrodes, Sensors 13 (7) (2013) 8188–8198.

[145] C.M. Pandey, et al., Highly sensitive electrochemical immunosensor based on graphene-wrapped copper oxide-cysteine hierarchical structure for detection of pathogenic bacteria, Sens. Actuators B 238 (2017) 1060–1069.

[146] J.D. Brewster, R.S. Mazenko, Filtration capture and immunoelectrochemical detection for rapid assay of *Escherichia coli* O157:H7, J. Immunol. Methods 211 (1998) 1–18.

[147] A.G. Gehring, C.G. Crawford, R.S. Mazenko, L.J. Van Houten, J.D. Brewster, Enzyme-linked immunomagnetic electrochemical detection of *Salmonella typhimurium*, J. Immunol. Methods 195 (1996) 15–25.

[148] A.G. Gehring, D.L. Patterson, S.I. Tu, Use of a light-addressable potentiometric sensor for the detection of *Escherichia coli* O157:H7, Anal. Biochem. 258 (1998) 293–298.

[149] Y. Che, Y. Li, M. Slavik, Detection of *Campylobacter jejuni* in poultry samples using an enzyme-linked immunoassay coupled with an enzyme electrode, Biosens. Bioelectron. 16 (2001) 791–797.

[150] Y. Li, P. Cheng, J.H. Gong, L.C. Fang, J. Deng, W.B. Liang, J.S. Zheng, Amperometric immunosensor for the detection of *Escherichia coli* O157:H7 in food specimens, Anal. Biochem. 421 (2012) 227–233.

[151] R. Miranda-Castro, N. de-los-Santos-Alvarez, M.J. Lobo-Castanon, A.J. Miranda-Ordieres, P. Tunon-Blanco, PCR-coupled electrochemical sensing of *Legionella pneumophila*, Biosens. Bioelectron. 24 (2009) 2390–2396.

[152] C.Y. Yu, G.Y. Ang, C.Y. Yean, Multiplex electrochemical genosensor for identifying toxigenic *Vibrio cholerae* serogroups O1 and O139, Chem. Commun. 49 (2013) 2019–2021.

[153] J. Wang, Electrochemical nucleic acid biosensors, Anal. Chim. Acta 469 (2002) 63–71.

[154] A.J. Baeumner, R.N. Cohen, V. Miksic, J. Min, RNA biosensor for the rapid detection of viable *Escherichia coli* in drinking water, Biosens. Bioelectron. 18 (2003) 405–413.

[155] N. Jaiswal, et al., Electrochemical genosensor based on carboxylated graphene for detection of water-borne pathogen, Sens. Actuators B 275 (2018) 312–321.

[156] S.M. Radke, E.C. Alocilja, A high density microelectrode array biosensor for detection of *E. coli* O157:H7, Biosens. Bioelectron. 20 (2005) 1662–1667.

[157] R. Maalouf, C. Fournier-Wirth, J. Coste, H. Chebib, Y. Saïkali, O. Vittori, A. Errachid, J.-P. Cloarec, C. Martelet, N. Jaffrezic-Renault, Label-free detection of bacteria by electrochemical impedance spectroscopy: comparison to surface plasmon resonance, Anal. Chem. 79 (2007) 4879–4886.

[158] A. Cobb, Development of Low-Cost Impedimetric Biosensors for Clinical Diagnostics and Water Testing, Clemson University, 2016.

[159] N. Li, A. Brahmendra, A.J. Veloso, A. Prashar, X.R. Cheng, V.W.S. Hung, C. Guyard, M. Terebiznik, K. Kerman, Disposable immunochips for the detection of *Legionella pneumophila* using electrochemical impedance spectroscopy, Anal. Chem. 84 (2012) 3485–3488.

[160] J. Monzó, et al., Fundamentals, achievements and challenges in the electrochemical sensing of pathogens, Analyst 140 (21) (2015) 7116–7128.

[161] A.L. Furst, M.B. Francis, Impedance-based detection of bacteria, Chem. Rev. 119 (1) (2019) 700–726.

[162] K. Jiang, et al., Rapid label-free detection of *E. coli* using antimicrobial peptide assisted impedance spectroscopy, Anal. Methods 7 (23) (2015) 9744–9748.

[163] J. Jiang, et al., Smartphone based portable bacteria pre-concentrating microfluidic sensor and impedance sensing system, Sens. Actuators B 193 (2014) 653–659.

[164] M.T. Guler, I. Bilican, Capacitive detection of single bacterium from drinking water with a detailed investigation of electrical flow cytometry, Sensors Actuators A Phys. 269 (2018) 454–463.

[165] R.E. Fernandez, et al., Review: microbial analysis in dielectrophoretic microfluidic systems, Anal. Chim. Acta 966 (2017) 11–33.

[166] M. Kim, et al., A microfluidic device for label-free detection of *Escherichia coli* in drinking water using positive dielectrophoretic focusing, capturing, and impedance measurement, Biosens. Bioelectron. 74 (2015) 1011–1015.

[167] J. Suehiro, et al., Improvement of electric pulse shape for electropermeabilisation assisted dielectrophoretic impedance measurement for high sensitive bacteria detection, Sens. Actuators B 2 (2005) 209–215.

[168] S.A. Muhsin, et al., Smart biosensor for rapid and simultaneous detection of waterborne pathogens in tap water, in: 2019 20th International Conference on Solid-State Sensors, Actuators and Microsystems & Eurosensors XXXIII (Transducers & Eurosensors XXXIII), 2019.

[169] J. Liu, et al., An integrated impedance biosensor platform for detection of pathogens in poultry products, Sci. Rep. 8 (2018) 16109.

[170] K. Kikkeri, et al., A monolithic dielectrophoresis chip with impedimetric sensing for assessment of pathogen viability, J. Microelectromech. Syst. 27 (5) (2018) 810–817.

[171] C. Dalton, A.D. Goater, J. Drysdale, R. Pethig, Parasite viability by electrorotation, Colloids Surf. A Physicochem. Eng. Asp. 195 (2001) 263–268.

[172] B.A. Simmons, V.R. Hill, V. Fintschenko, E.B. Cummings, Concentration and Separation of Bioloigcal Organisms by Ultrafiltration and Dielectrophoresis, Sandia Corporation, USA, 2010.

[173] F. Tokumasu, G.R. Ostera, C. Amaratunga, R.M. Fairhurst, Modifications in erythrocyte membrane zeta potential by *Plasmodium falciparum* infection, Exp. Parasitol. 131 (2012) 245–251.

[174] G. Luka, et al., Label-free capacitive biosensor for detection of cryptosporidium, Sensors 19 (2) (2019) 258–267.

[175] K. Reybier, C. Ribaut, A. Coste, J. Launay, P.L. Fabre, F. Nepveu, Characterization of oxidative stress in Leishmaniasis-infected or LPS-stimulated macrophages using electrochemical impedance spectroscopy, Biosens. Bioelectron. 25 (2010) 2566–2572.

[176] T. Houssin, J. Follet, A. Follet, E. Dei-Cas, V. Senez, Label-free analysis of water-polluting parasite by electrochemical impedance spectroscopy, Biosens. Bioelectron. 25 (2010) 1122–1129.

[177] A. Dibao-Dina, et al., Electrical impedance sensor for quantitative monitoring of infection processes on HCT-8 cells by the waterborne parasite Cryptosporidium, Biosens. Bioelectron. 66 (2015) 69–76.

[178] R. Lejard-Malki, et al., Selective electrohydrodynamic concentration of waterborne parasites on a chip, Lab Chip 18 (21) (2018) 3310–3322.

[179] J.S. McGrath, et al., Analysis of parasitic protozoa at the single-cell level using microfluidic impedance cytometry, Sci. Rep. 7 (1) (2017) 2601.

[180] H. Choi, et al., A flow cytometry-based submicron-sized bacterial detection system using a movable virtual wall, Lab Chip 14 (13) (2014) 2327–2333.

[181] V. Narcisi, M. Mascini, G. Perez, M. Del Carlo, P.G. Tiscar, H. Yamanaka, D. Compagnone, Electrochemical genosensors for the detection of Bonamia parasite: selection of single strand-DNA (ssDNA) probes by simulation of the secondary structure folding, Talanta 85 (2011) 1927–1932.

[182] M.A. Shannon, P.W. Bohn, M. Elimelech, J.G. Georgiadis, B.J. Marinãs, A.M. Mayes, Science and technology for water purification in the coming decades, Nature 452 (2008) 301–310.

[183] D.R. Thévenot, K. Toth, R.A. Durst, G.S. Wilson, Electrochemical biosensors: recommended definitions and classification, Biosens. Bioelectron. 16 (2001) 121–131.

[184] H. Becker, Mind the gap! Lab Chip 10 (2010) 271.

[185] J. Comley, Progress in the implementation of label-free detection—part 1: cell-based assays, Drug Discov. World (2008) 77–88.

[186] J. Comley, Progress in the implementation of label-free detection—part 2: binding analysis assays, Drug Discov. World (2008) 28–49.

[187] B. Brehm-Stecher, C. Young, L.A. Jaykus, M.L. Tortorello, Sample preparation: the forgotten beginning, J. Food Prot. 72 (2009) 1774–1789.

[188] C. Batchelor-McAuley, E.J.F. Dickinson, N.V. Rees, K.E. Toghill, R.G. Compton, New electrochemical methods, Anal. Chem. 84 (2012) 669–684.

Chapter 7

Biosensors for the detection of waterborne pathogens

Helen Bridle[a] and Marc Desmulliez[b]

[a]*Institute of Biological Chemistry, Biophysics and Bioengineering, Heriot-Watt University, Edinburgh, Scotland,* [b]*Institute of Signals, Sensors and Systems, Heriot-Watt University, Edinburgh, Scotland*

Biosensors are devices that comprise a target recognition element, which is biological, coupled with a mechanism of signal transduction. The function of a biosensor is to convert a biological recognition event into a detectable signal by the action of the transducer and signal conditioning circuitry, thereby providing selective quantitative or semiquantitative analytical information (Fig. 7.1). Common recognition elements include enzymes, antibodies, DNA, or even whole cells. These elements bind either cells and cell fragments or nucleic acids, or amplification products from molecular methods present in the solution. This chapter will focus on biosensors, which capture whole pathogens, rather than, for example the amplification products of molecular methods, which will be covered in Chapter 8.

Signal transduction can be performed using electrochemical, optical, mass-sensitive, or thermal means. The first three methods are by far the most popular and will be the focus of this chapter. Various combinations of these transduction mechanisms give rise to many different biosensing technologies. Automation and miniaturisation of biological analytical techniques combined with the development of online and remote measurement equipment can be achieved through biosensor technology. Other, oft-quoted advantages of biosensors include short times of analysis, low cost of assays, portability, and real-time measurements.

Since Clark and Lyon developed the first biosensor for glucose detection in 1962, biosensors have been intensively studied and extensively utilised in various applications, ranging from public health and environmental monitoring to homeland security and food and water safety. In terms of pathogen detection biosensors, Lazcka and colleagues found from an analysis of publication outputs that biosensors ranked 4th in 2007 in a comparison of different technologies, and reported that biosensors are the fastest growing technology for this application (Fig. 7.2) [1], a trend which is continuing today [2].

Waterborne Pathogens: Detection Methods and Applications. https://doi.org/10.1016/B978-0-444-64319-3.00007-1
189

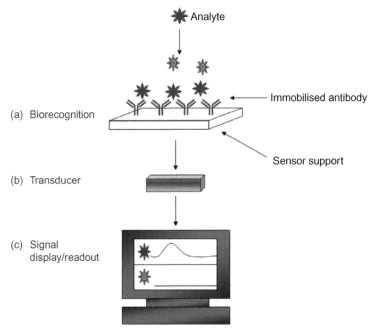

FIG. 7.1 Biosensor components. (A) Analyte interaction with the biorecognition element: this is facilitated by the specificity of the immobilised antibody for its cognate antigen (purple). Other biorecognition elements include enzymes, lectins, receptors, and microbial cells. (B) Signal transduction: converts the interaction of the analyte molecule and analyte into a quantifiable signal. (C) Readout or display: shows the specific signal generated by interaction with the analyte of interest. Yellow: nonspecific analyte. *(Fig. 1 from P.J. Conroy, et al., Antibody production, design and use for biosensor-based applications, Semin. Cell Dev. Biol. 20 (1) (2009) 10–26. Reproduced with permission.)*

This chapter aims to provide an introduction to biosensing technology and review the applications for waterborne pathogens. The first sections of the chapter present a background to biosensors classified under various taxonomies. Biosensors can be divided either by recognition element, signal transduction approach, direct, or indirect detection techniques. Direct detection biosensors are designed such that that the biospecific reaction is directly determined in real-time. In indirect detection, the biosensor detects either the presence of a secondary 'label' added after the initial biorecognition step or the products of that preliminary biochemical reaction. Finally, biosensors can also be categorised into sensors operating in batch (intermittent) and continuous (monitoring) mode. The next sections of the chapter will focus on reviewing the types of biosensors applied to the different types of waterborne pathogens. The final section of the chapter will discuss the advantages and disadvantages of using biosensors for waterborne pathogens.

Biosensor research and development is an enormous field, as evidenced by the large, and growing, number of research publications. Many excellent review articles and books, focussing on different aspects of biosensor technologies, have been published in recent years and the reader is referred to these for more

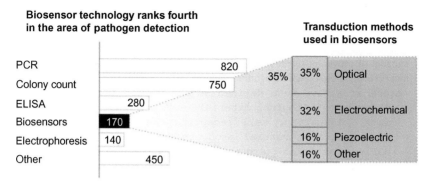

Source: ISI Web of Science. *ca.* 2500 articles found on pathogen detection over the last 20 years.

(A)

Biosensors is the fastest growing technology for pathogen detection

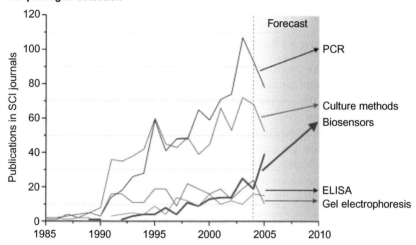

Source: ISI Web of Science. *ca.* 2500 articles found on pathogen detection over the last 20 years.

(B)

FIG. 7.2 (A) An approximate number of articles using different techniques to detect and/or identify pathogenic bacteria. Articles using more than one technique have been excluded to avoid overlap between categories. (B) Time series of the number of works published on the detection of pathogen bacteria over the last 20 years. The fact that certain articles used more than one technique has been accounted for to make this graph. *(Source: ISI Web of Science ca 2500 articles found on pathogen detection over the last 20 years. Fig. 2 from O. Lazcka, F.J.D. Campo, F.X. Muñoz, Pathogen detection: a perspective of traditional methods and biosensors, Biosens. Bioelectron. 22 (7) (2007) 1205–1217. Reproduced with permission.)*

detailed background information [1, 3–7]. This chapter will, therefore, focus on technologies already applied to waterborne pathogens, or on techniques, which offer significant potential. An increasing body of research has been undertaken on food pathogens or in medical diagnostics and, where appropriate, some of the literature on these topics, is included here.

Box: Biosensor definitions

'A biosensor can be defined as an integrated receptor-transducer device, which is capable of providing selective quantitative or semi-quantitative analytical information using a biological recognition element'.

'Biosensors have been defined as analytical devices incorporating a biological material (e.g., tissue, microorganisms, organelles, cell receptors, enzymes, antibodies, nucleic acids, natural products, etc.), a biologically derived material (e.g., recombinant antibodies, engineered proteins, aptamers, etc.) or a biomimic material (e.g., synthetic catalysts, combinatorial ligands and imprinted polymers) intimately associated with or integrated within a physicochemical transducer or transducing microsystem, which may be optical, electrochemical, thermometric, piezoelectric, magnetic or micromechanical'
. [1].

7.1. Performance characteristics

The performance of biosensors can be characterised in terms of sensitivity, specificity, detection limit, assay time, reproducibility, and precision of the results (Table 7.1). The sample volume processed and the performance in complex sample matrices are important factors, especially for waterborne pathogens. Additional considerations are the ease of operation, which could incorporate the degree of automation, the number of stages in the assay protocol or the need for secondary or amplification steps, and the size/portability of the system. These latter factors will influence the potential of the system to be adopted as a field-based biosensor. Furthermore, the cost of any sensor system is crucial to industry adoption. The potential of any biosensor technology for multiplexing is also of interest to create simple tools capable of analysing multiple pathogens in one sample.

In the selection and optimisation of biosensor trade-offs between various performance characteristics have to be made, depending upon the exact application. For example, a field-based early warning system will need to be cheap, robust, fully automated, capable of long-term operation with minimal maintenance, and deliver fast results. In this case, high sensitivity and specificity will be less important and will most likely be sacrificed to achieve the characteristics mentioned previously. Highly sensitive and specific measurements could be performed back at an analytical lab, using larger, more expensive equipment, which may require operation by skilled technicians. The typical size of the biosensor surface is in square millimetres. The sample volume should be as low as possible given the size of the biosensor while providing a meaningful measurement of the number of pathogens present.

Biosensor systems are inherently complex and to fully understand the interplay of convection, diffusion, and reaction involved in the capture of the target analyte on the surface is challenging. The reader is referred to an excellent ar-

TABLE 7.1 Definition of important factors to be considered when evaluating biosensor performance.

Characterisation term	Definition
Limit of detection	Lowest quantity (number or concentration) of pathogens that can be detected by a given method or technique
Sensitivity	Most commonly, this is defined as the ability to detect very few organisms in a sample (rather than the degree of discrimination between various concentration levels). Sensitivity is influenced by both the affinity of the recognition element for the target analyte and the transduction process
Specificity/ selectivity	The ability of the sensors to discern one closely related species or strain from another or even viable organisms from dead ones
Detection range	The concentration range across which the sensor operates. The lower limit is determined by the sensitivity. Ideally, the sensor response will be linear between the limits, allowing easy quantification, and further electronic processing
Reproducibility	The ability for a sensor to provide the same results for the number of pathogens
Processing/sample volume	Total volume necessary for one analysis
Processing/assay time	The total time necessary to prepare the sample, analyse it, and read out the results
Background/ sample matrix	Type of water and water content or other solution in which detection takes place

ticle by Todd Squires, where the physics of analyte transport and reaction dynamics are discussed concerning sensor size and sample delivery, which is very helpful in the design and optimisation of biosensors [8].

7.2. Recognition elements

In the early stages of biosensor research, the recognition elements employed were biological materials, such as antibodies and enzymes. In recent years, novel materials have emerged, either derived from biological materials, for example antibody fragments, or designed to mimic biological materials (biomimetics). These new recognition elements have been developed to offer advantages in terms of cost, stability, long-term storage, sensitivity, or selectivity. However, antibody and nucleic acid-based sensors remain the most popular. Perhaps this will change in the future as the new materials become more established and readily available for waterborne pathogens.

The choice of recognition element depends upon the exact application and involves a trade-off between various performance characteristics. The immobilisation of this material on the sensor surface is also of critical importance for biosensor performance. This section concentrates on antibodies as the most popular recognition, describes a range of recognition elements, followed by a brief discussion of other recognition elements employed as well as consideration of surface immobilisation techniques.

7.2.1 Antibodies

Antibodies (immunoglobulins, Ig) are molecules produced by biological systems in response to a contaminating agent (antigen). Antibodies (Ab) were first applied to detection in the 1950s [9], and are the most widely used biorecognition elements, thanks to their proven sensitivity and specificity [10]. Abs can be used to detect the whole pathogen (their surface proteins) or some of the pathogen components (lysate, enzymes, toxin, spore, pili). Antibody-based methods have been used extensively to detect bacteria, viruses, toxins, and spores, alike [7]. Highly selective and sensitive antibodies are readily available for many pathogens, and there are several well established methods to conjugate antibodies and perform surface immobilisation.

Two categories of antibodies are used in biosensors: monoclonal and polyclonal, which are produced, or 'raised against' the pathogen of interest, in different ways. Monoclonal antibody (mAbs) solutions are produced in vitro from hybridoma cell lines and consist of an identical, well defined population of antibodies that bind to a single epitope. Polyclonal antibodies (pAbs) are produced in vivo and consist of a suite of antibodies that bind to several epitopes on the antigen, which is time-consuming and costly. Although mAb production is quicker, the costs are considerably higher. Since mAbs are raised against a single epitope, mAbs tend to be homogeneous reagents with better-defined specificities and are less likely to undergo nonspecific binding (i.e. binding of the antibody to nonantigenic materials [11]). In contrast, pAbs are nonhomogenous, which can result in increased levels of nonspecific binding. pAbs recognise different epitopes on the same pathogen and some of these antigens may be present in other closely related but nonpathogenic organisms, which may lead to false-positive results. mAbs offer enhanced specificity and can, for example differentiate *B. anthracis* spores from vegetative cells or spores of other *Bacillus* spp. [12]. However, compared to mAbs, pAbs possess greater potential for antibody attachment to the antigenic surface and, importantly for waterborne pathogen detection, have a higher resistance to pH, and salt concentration changes. Besides, mAbs can be too specific, in which case they may not detect species variants that lack a particular epitope. The higher production costs and susceptibility to unfavourable environmental conditions limit the broad use of monoclonal antibodies in field-ready sensors [13].

The biggest advantage of antibody-based probes is the specificity and affinity of these polypeptides to target analytes. Antibodies form tight noncovalent bonds with specific target molecules with apparent K_d values of 10^{-7}–10^{-11} M [14]. Thus, antibodies can interact strongly with the target analyte even in a complex mixture, resulting in a highly specific biosensor. There are, however, several disadvantages with the use of antibodies on biosensors, which limit their use in the field, including the potential for nonspecific binding, the need for physiological pH, and temperature monitoring during immunoassay procedures, varied antibody performance from batch to batch, their sensitivity to chemicals in drinking water, their inability to distinguish viable from nonviable organisms and the high cost of mAbs production [15]. One of the major disadvantages of antibodies is their relative instability to environmental fluctuations, especially high temperature, compared to other peptide-based probes. This limitation may require antibody-based biosensors to be stored in refrigerated containers and can reduce long-term storage and field applicability. Besides, antibodies are only available for organisms that elicit an antigenic response. Some pathogenic organisms are nonimmunogenic but produce toxic metabolites in vivo that produce an immune response. Nonetheless, despite the long list of drawbacks associated with the use of antibodies, they continue to be the excellent option for the selective detection of a broad range of microorganisms.

The antibody isotope, which is most abundant in the blood serum (around 75% of total serum immunoglobilin), is known as IgG ($\sim$150 kDa) (Fig. 7.3). There are five primary antibody isotopes (IgG, IgM, IgD, IgA, and IgE), which vary in their specificity and avidity for pathogens. For example, an IgG type of antibody was proven to be more specific to *Cryptosporidium* and showed higher avidity than an IgM type of antibody [16]. IgG can be divided into F(ab')2 and Fc fragments, of which the fragments F(ab) contain the antigen-binding sites.

Engineered antibody fragments, such as F(ab), have been employed for immunodetection as they retain the specificity of antibodies but offer improvements in production cost and substrate coverage densities, and they can accommodate systems that require small-sized receptors [16]. Initial approaches to the generation of antibody fragments utilised chemical agents for proteolytic cleavage of certain bonds. This method was used to create antibody fragments for a *Cryptosporidium* biosensor [16]. Alternatively, using genetic modifications, recombinant antibodies can be generated. Conroy provides an excellent introduction to recombinant antibody technology and the application to biosensors [7]. At present, there have only been a few biosensors reported using recombinant antibodies but this area is predicted to grow significantly due to the advantages offered by these antibodies [7] (Fig. 7.4).

An alternative approach to antibody fragments followed the discovery of camelid and shark antibodies, composed only of single heavy chains with very small antigen-binding domains [13]. This discovery facilitated the development of thermostable antibodies that retain specificity. Regions from these antibodies have been cloned and expressed as 12–15 kDa single-domain antibodies (sdAbs) that are stable to temperatures as high as 90°C. In 2007, Sherwood et al. developed an unoptimised chemiluminescent assay, the most specific clone could detect 0.1–1 pfu/well of Ebola virus antigens within 30 min [17]. These

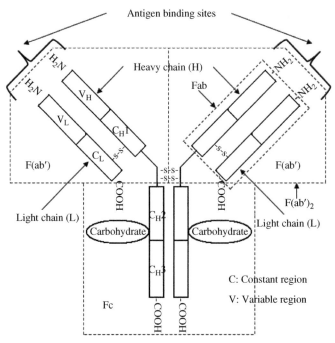

FIG. 7.3 IgG is made up of two pairs of polypeptide chains linked together by a disulphide bond. Each pair of polypeptides comprises one heavy chain (gamma, γ) and one light chain (Kappa, κ, or lambda, λ), also connected by disulphide bonds. Each light chain is composed of two domains: constant domain (CL) and a variable domain (VL). The heavy chain has three constant domains (CH1, CH2, and CH3) and one variable domain (VH). The variable regions of the heavy chain and the light chain are the regions of antigen interaction. The Fc (fragment crystallisation) fragment does not have binding properties. It contains the antibody effector functions, such as complement activation, cell membrane receptor interaction, and transplacental transfer. F(ab')2 consists of two identical F(ab') [fragment antigen binding] held together by disulphide bond at the hinge region. Each Fab' fragment has one light chain (VL and CL) and two domains of the heavy chain (VH and CH1). The carbohydrate moieties are located at the constant heavy chain domain CH2. Fab fragment is the fragment that does not have any hinge region with the thiol group but does have other domains like Fab' [16].

highly sensitive and selective sdAb probes could be used in any antibody-based biosensor designed to detect infectious agents [13].

Phage antibody technology has also been developed offering advantages over traditional antibodies in terms of specificity, sensitivity, and robustness [7]. In this approach, a fragment of the antibody is produced and displayed on the surface of a bacteriophage.

7.2.2 Alternative recognition elements

This section will include a short introduction to other types of recognition elements, including peptides, enzymes, carbohydrates, DNA/RNA, aptamers, whole cells, and biomimetics.

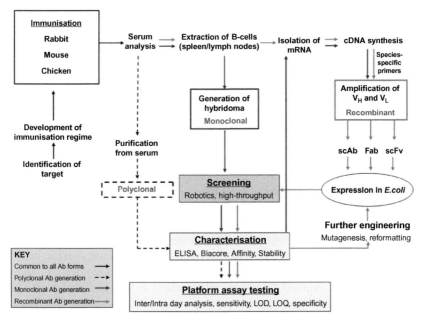

FIG. 7.4 Overview of antibody generation, screening, and characterisation. This flow diagram illustrates the overall steps in the generation of polyclonal, monoclonal, and recombinant antibodies. *LOD*, the limit of detection; *LOQ*, the limit of quantitation. *(Fig. 4 from P.J. Conroy, et al., Antibody production, design and use for biosensor-based applications, Semin. Cell Dev. Biol., 20 (1) (2009) 10–26. Reproduced with permission.)*

Peptides are short chains of amino acids and could, therefore, be considered to include antibody fragments. Other types of peptides of interest as recognition elements in biosensors include antimicrobial peptides (AMPs), which play an important role in the immune system response to pathogen infection. AMPS consists of 15–45 amino acids and semiselectively bind to microbial surfaces. However, there are two main problems encountered with this type of recognition element. Firstly, there is the antimicrobial activity which can destroy the pathogens of interest. Secondly, since AMP target interaction is primarily electrostatic these probes are particularly sensitive to changes in solution ionic strength which is likely to be especially disadvantageous for environmental applications.

Enzymes have been commonly used in biosensors as they are highly specific [18]. However, disadvantages include the expense and instability of purified enzymes. Recently, genetically engineered enzymes have been applied to enhance the sensitivity and selectivity of enzyme-based biosensors, through a variety of strategies, for example introducing modifications to increase the accessibility of the active site to improve surface immobilisation or to enhance electron transfer capabilities [19].

DNA/RNA biosensors are based on the interaction between a nucleic acid target and its complementary probe and widely used for the detection of the products of molecular methods, for example amplification products from polymerase chain reaction [20]. As this chapter focuses on biosensors for the detection of whole pathogens, further discussion of this type of recognition element is left to Chapter 8 which covers molecular methods. Peptide nucleic acids are artificially synthesised polymer similar to DNA or RNA and these have been applied to the detection of antimicrobial resistance using biosensors [21].

Aptamers are small (2–25 kDa) artificial nucleic acid sequences that exhibit high affinity and specificity to target probes. This type of recognition element is produced through a series of selection and amplification steps known as Systematic Evolution of Ligands by EXponetial enrichment (SELEX) to select a high-affinity probe. The advantages of aptamers are that following the SELEX identification they can be synthesised easily in large quantities and they can be produced against any target, even those that do not elicit an immunogenic response. However, one main drawback is that they are sensitive to nuclease attack, although this can potentially be mitigated via chemical modifications.

Carbohydrates are a large class of biomolecules that are often involved in pathogen interactions, for example many pathogens recognise specific carbohydrate sequences in the human gut in the process of initiating infection. These interactions have inspired the use of carbohydrates as recognition elements for pathogen detection. Advantages compared to antibodies is that carbohydrates do not denature on exposure to temperature or pH alterations and hence are more stable, as well as the fact that carbohydrates are smaller, thus allowing for higher surface density on the biosensor surface, which can improve sensitivity. However, disadvantages are the low affinity of the carbohydrate-protein interactions (10^2–10^3 lower K_d values than antibody recognition) and the obtainable specificity.

The specificity of carbohydrates is relatively broad, which can be a challenge for accurate detection of the intended target pathogen. However, in some cases, such broad specificity may prove advantageous. For example, some antibodies may be so highly specific to the extent that mutants that differ slightly from the original target are not detected; carbohydrates do not suffer this limitation. One route to improve carbohydrate specificity has been to combine carbohydrates with lectins. Lectins are a group of proteins that strongly bind to specific carbohydrate moieties. Concanavalin A, a widely used mannose- and glucose-binding lectin, has been employed in biosensors for pathogens.

Whole-cell biosensors employ living cells as the biological recognition element. The main advantage of this approach is that toxicity can be assessed; this approach is being considered for the detection of low-concentration micropollutants in water. However, widespread adoption of these sensors could be challenged by safety permissions, especially when the cells have been genetically modified, although some whole cell-based sensors are commercially available.

An early example of the use of living systems for the detection of harmful agents is that of canaries in coal mines. The canaries were employed as a means of monitoring the presence of carbon monoxide in the mine. In recent years, many different cell types have been employed in biosensors, with bacteria being particularly popular. Mammalian cells have also been utilised as recognition elements. This was reviewed by Banerjee and Bhunia, with a particular focus on pathogens relevant to food, environment, medical, and biosecurity applications [22].

The unique aspect of this type of sensor is its ability to simulate physiological responses in vivo. Another advantage of the whole-cell approach is that these systems offer sensitivity to a group of pathogens and therefore, could offer a greater ability to detect new and emerging pathogens. Furthermore, this approach can detect the toxicity of pathogens as opposed to quantitative concentration determination. As has been particularly interesting for micropollutants whole-cell biosensors could detect synergistic effects. However, in water applications, the failure of some whole-cell biosensor systems to detect all pathogens can be a drawback. The presence of nonviable pathogens in water supplies can be a drawback. The presence of nonviable pathogens in water supplies can indicate problems with treatment systems and is therefore valuable information. Other disadvantages of whole-cell biosensors are specificity, reliability, robustness; long-term storage, and shelf-life; as well as price.

Biomimetics is chemicals designed to mimic biological recognition with one example being molecularly imprinted polymers. These polymers are synthesised with the intended target, or more commonly an analogue of the target present. Once the polymer is stably synthesised these 'target' molecules are washed out, leaving recognition cavities corresponding to the target. This approach has been widely used for small molecules, though less so for the detection of whole cells; it has also been combined with biosensors. The advantages of this approach are that these polymers are more stable and robust than antibodies, although a major drawback can be the challenges of design of an appropriate synthesis route as well as postprocessing of the polymer.

7.2.3 Immobilisation strategies

Once a particular recognition element has been selected, the development of biosensors requires the functionalisation of the biosensor surface to incorporate the selected biorecognition elements. This surface immobilisation is one of the most critical steps in biosensor development because biosensor performance (sensitivity, dynamic range, reproducibility, and response time) depends upon the recognition element retaining its original properties after immobilisation. For example, some methods of antibody immobilisation are nondirectional, and therefore, some of these antibodies may thus end up immobilised in orientations such that target recognition is the difficulty. The stability of the immobilisation, as well as the extent of surface coverage, are also important factors to consider in the selection of a strategy.

Some examples of existing immobilisation strategies include biomolecule physisorption, entrapment, and encapsulation into polymers or membranes, silanisation, and self-assembled monolayer (SAM) formation coupled to biomolecule cross-binding or covalent bonding using linkages such as protein A or protein G, biotin-streptavidin or *N*-hydroxysulfosuccinimide and 1-ethyl-3-[3-dimethylaminopropyl] carboiimide hydrochloride. Details of these methods can be found in the excellent book Bioconjugate Techniques [23] as well as in other reviews [24, 25]. Besides, the modified surface has to be inert and biocompatible so that it does not affect the sample composition or integrity in any way; it should also guarantee a constant signal baseline. Blocking of the unmodified surface is also required, using, for example bovine serum albumin (BSA) to prevent nonspecific binding.

The easiest way to immobilise antibodies is to unspecifically deposit or adsorb them on the sensing surface. Nonetheless, most authors maintain that direct immobilisation by chemical conjugation or crosslinking of precise functional groups in the antibody (i.e. amines, carboxylates, carbohydrates) better preserves their integrity and functionality while promoting more organised structures. This was found, for example, in a study of the best immobilisation strategies for *Cryptosporidium* [26].

7.3. Transduction methods

7.3.1 Optical

Many optical biosensors rely on fluorescence which was covered in Chapter 5. Lateral flow assays (LFAs) are a common type of sensor, well known for the home pregnancy test example application. In this type of assay, the test sample floes along a solid substrate past a functionalised area that would capture the target of interest leading to a colour change. Most examples of the earlier mentioned types of detection approaches were covered in Chapter 5. This chapter concentrates on those methods in which the binding of the analyte causes a change in the optical properties of the surface, which can then be utilised for detection. One example of this is surface plasmon resonance (SPR), where binding of the target changes the refractive index at the biosensor surface. Other examples include the use of waveguides, fibre optics, or ellipsometry [7].

SPR measures changes in the refractive index at the interface between a planar metal surface and a dielectric material [27]. Analyte-binding events are detected by coupling photons from a light source to surface plasmons and then measuring a change in the properties of the reflected light (Fig. 7.5). Detectors that measure intensity, incident angle, wavelength, or phase of the reflected light have been designed. SPR sensors have been used to detect a range of analytes such as antibiotics, vitamins, hormones, pesticides as well as bacteria and protozoa. A comprehensive review of the use of SPR has been published by Homola [28]. Alternative techniques operating on a similar principle of detecting changes in refractive index or resonant reflection are the use of silicon microring resonators and photonic crystal biosensors.

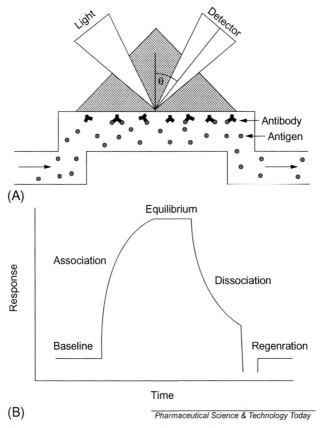

(A)

(B)

FIG. 7.5   Schematic diagram depicting a typical surface plasmon resonance (SPR) experiment. (A) The ligand (shown as an antibody) is immobilised on the biosensor chip surface. The analyte (which represents the antigen) passes through a microfluidic flow cell and SPR is used to monitor the change in the refractive index caused as analyte accumulates on the biosensor surface. (B) Salient features of a typical sensorgram. Before injecting an analyte, the baseline response should be stable. An increase in response during the association phase represents complex formation in real-time. Equilibrium is achieved when an equal number of analyte molecules associate with and dissociate from the surface at the same time. The surface can be washed and the decay rate of the complex obtained during the dissociation phase. Following regeneration, the binding response should return to the starting baseline position. *(Source: Fig. 1 from reference D.G. Myszka, R.L. Rich, Implementing surface plasmon resonance biosensors in drug discovery, Pharm. Sci. Technol. Today 3 (9) (2000) 310–317. Reproduced with permission.)*

Another method of optical biosensing is those techniques facilitated by waveguides [7]. Waveguides are typically made of glass, quartz, or polymer films with a high refractive index embedded within a lower refractive index material. Incident linear laser beams are, therefore, constrained within the waveguide by total internal reflection (TIR), resulting in an evanescent wave. A proportion of the light penetrates the bilayer and is reflected into the waveguide after undergoing a phase shift that interferes with the transmitted light. Changes in this interference pattern can thus be interpreted as changes in the biolayer allowing for the detection of analyte binding events.

Interferometry is based on changes in the refractive index profile with the evanescent field volume of a waveguide due to analyte binding. Optical waveguide lightmode spectroscopy utilises the change in resonance angle of polarised light diffracted by a grating and incoupled into a thin waveguide layer upon binding. TIR fluorescence is a method where the incident light from TIR excites molecules in the evanescent field, which produces a measurable fluorescent evanescent wave.

7.3.2 Electrochemical

Electrochemical techniques study interfacial phenomena by looking at the relationship between current and potential. A perturbation in either the current or the potential of the working electrode is imposed, and the response of the system to those perturbations is observed. In terms of electrochemical biosensors, various approaches can be employed for the detection of the target including amperometry, potentiometry, conductometry, and voltammetry. Other electrochemical methods such as impedance were covered in Chapter 6.

Amperometric biosensors operate with a constant potential applied between a working electrode and a reference electrode (Fig 7.6A) [29]. On the interaction of the target analyte with the working electrode surface, a redox reaction occurs, generating a current that is proportional to the concentration of the electroactive target. Since picoampere current measurements are possible, ultrasensitive amperometric biosensors can be obtained [30]. This is the most widely used technique, particularly for microbial detection [30].

Potentiometric biosensors monitor the potential at a working electrode for a reference electrode, and this potential demonstrates concentration-dependent behaviour. A species-selective working electrode gives outstanding sensitivity and selectivity, although a challenge for this type of sensor is that a highly stable and accurate reference electrode is required [31]. Light-addressable potentiometric sensors have also been developed (Fig 7.6B) [32], based on semiconductor activation by a light-emitting diode, and have been applied to the detection of *Escherichia coli* [33].

Conductometric biosensors measure the conductivity change in the solution due to the production or consumption of ionic species, for example by the metabolic activity of microorganisms. Conductance measurements are very fast and sensitive, though selectivity is relatively poor [34].

Voltammetric biosensors measure both current and potential. The peak current position is used for identification, while the peak current density is proportional to the concentration of the corresponding species. The advantages of this type of electrochemical biosensor are that the low noise allows for highly sensitive measurements and that simultaneous detection of multiple analytes is easily achieved.

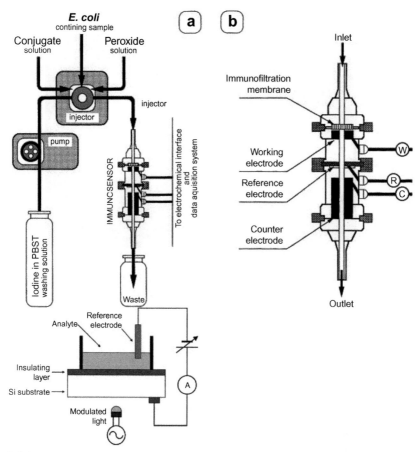

FIG. 7.6 (A) Schematic of the enzyme immunofiltration assay system (A) and the flow-injection immunosensor (B). (C) Schematic of the LAPS set-up. *(Source: Panel (A) Fig. 1 from reference I. Abdel-Hamid, et al., Highly sensitive flow-injection immunoassay system for rapid detection of bacteria, Anal. Chim. Acta 399 (1) (1999) 99–108; Panel (B) Fig. 1 from reference T. Yoshinobu, et al., Portable light-addressable potentiometric sensor (LAPS) for multisensor applications, Sens. Actuators B 95 (1) (2003) 352–356. Reproduced with permission.)*

7.3.3 Mass-sensitive

Quartz crystal microbalance (QCM) biosensors have been employed for the detection of a range of analytes from proteins such as lysozyme and BSA [35], DNA sequences from pathogens like *E. coli*, as well as the detection of intact pathogens, including *E. coli* [36] and *Cryptosporidium parvum* [37]. In QCM a quartz disc is sandwiched between two electrodes, normally made of gold. As quartz is piezoelectric, the crystal can be excited by applying an AC voltage across the electrodes and will exhibit a resonance frequency (Fig. 7.7). A mass change on the sensor surface results in a shift of the resonance frequency.

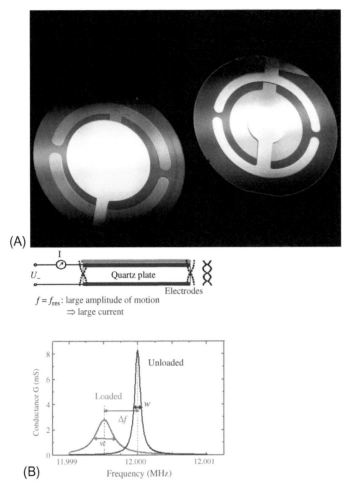

FIG. 7.7 QCM schematic (A) image of a typical QCM electrode/sensing area. (B) schematic of operation showing the shear movement. *(Wikicommons Source: (A) http://commons.wikimedia. org/wiki/File:Quartz_resonators_with_front_and_back_electrodses.jpg?uselang=en-gb. (B) http:// commons.wikimedia.org/wiki/File:QCM_principle.gif?uselang=en-gb.)*

QCM sensors are well known for their high sensitivity $(0.1\,\mathrm{Hz}(\mathrm{ng\,cm}^{-2})^{-1}$ at $5\,\mathrm{MHz}$ [37]), and high specificity. Furthermore, this type of biosensor presents a large tolerance to high temperatures, is label-free and is relatively inexpensive.

QCM devices typically operate at frequencies between 1 and $10\,\mathrm{MHz}$ and the entire piezoelectric substrate is used for wave propagation. In contrast surface acoustic wave (SAW) sensors operate at higher frequencies ($50\,\mathrm{MHz}$ to low GHz) and the acoustic energy is confined to a thin surface layer on the substrate. Given that the sensitivity is proportional to the square of the frequency for this type of sensors the principle advantage of SAW over QCM is higher sensitivity.

Piezoelectric-excited millimetre-sized cantilever (PEMC) sensors have been applied for the detection of several toxins, proteins, biomarkers, and pathogenic microorganisms such as *B. anthracis, E. coli,* and *C. parvum* [38]. This type of mass-sensitive sensor is a two-layered system with different functions for each layer (Fig. 7.8) [39]. The piezoelectric layer, usually made of lead zirconate titanate (PZT) acts as an actuator and sensor. An alternating current is passed through the PZT layer. The nth resonant mode is obtained at the frequency, Fn, depending upon the effective spring constant (a function of the beam thickness, width, length, and Young's modulus) and the effective mass of the cantilever in air. The other layer which is usually made of silica or glass is functionalised with recognition elements to bind the target microorganism. As in the QCM sensor, mass binding to the cantilever reduces the resonant frequency. PEMC sensors are extremely sensitive to mass changes and were determined to be 0.3–2 fg/Hz using paraffin additions [40].

Magnetoelastic (ME) sensors work due to the application of a magnetic field that causes them to oscillate. These sensors are fabricated from the amorphous ferromagnetic ribbon, which is very cheap, and offers the possibility of remote wireless remote query sensing, thus making them ideal for incorporation in automated systems. In a time-varying magnetic field, the ME sensor oscillates with a characteristic frequency that depends upon the properties, physical dimensions, shape, and mass of the material that the sensor is made of. Changes caused to the resonance frequency of the sensor are a result of mass loaded to the surface of the sensor (Fig. 7.9) [41].

As with all of the earlier techniques microcantilever sensors can operate in the resonance frequency mode where a mass change on the sensor leads to a shift in the resonance frequency. Alternatively, these cantilevers can be operated in deflection mode, where binding of the target to the sensor surface leads to bending of the cantilever which can be read-out by piezoelectric or optical means. While the dimensions of these cantilevers are on the orders of micrometres typical deflections are the nanometre range. Microcantilever sensors were

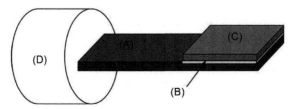

FIG. 7.8 Schematic of a PEMC-b sensor (A) The PZT layer, (B) the adhesive layer, (C) the non-piezoelectric glass layer, and (D) the nonconductive epoxy used to protect the electrodes on the encapsulated end of the PZT. For a schematic of PEMC-a, the reader is referred to Ref. [39]. *(Source: reprinted from Fig. 1 with permission from reference G.A. Campbell, R. Mutharasan, A method of measuring* Escherichia coli *O157:H7 at 1 cell/mL in 1 liter sample using antibody functionalized piezoelectric-excited millimeter-sized cantilever sensor, Environ. Sci. Technol. 41 (5) (2007) 1668–1674. Copyright (2007) American Chemical Society.)*

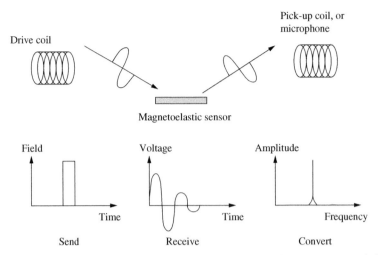

FIG. 7.9 Drawing illustrating the remote query nature of magnetoelastic sensors. A magnetic filed impulse is applied to the sensor; the transient response is captured and then converted into the frequency domain using a fast Fourier transform. The resonant frequency is tracked to provide chemical and environmental information. *(Source: Fig. 1 from reference C. Ruan, et al., Magnetoelastic immunosensors: amplified mass immunosorbent assay for detection of* Escherichia coli *O157:H7, Anal. Chem. 75 (23) (2003) 6494–6498. Reproduced with permission. Copyright (2003) American Chemical Society.)*

developed from atomic force microscopy (AFM); are typically made from silicon, silicon nitrate, or silicon oxide; and have been applied in a wide range of sensing applications [42] (Fig. 7.10).

7.4. Biosensors for waterborne viruses

Very few biosensors have been reported for viruses in general, and waterborne viruses in particular [43]. Sen suggested that the reason for this is that the epitopes on the capsid of enteroviruses are very small rendering immunological methods not successful or sensitive enough for direct detection in water [44]. However, it seems that recent developments in biosensor transduction are beginning to address this challenge. Novel optical methods of detection described in the following section, appear to dominate.

In 2012 Pineda et al. reported direct detection of rotavirus using antibody-based photonic crystal biosensors. The light incident a photonic crystal from a 90 angle is strongly reflected at a single wavelength at which an optical resonant reflection occurs. This resonant wavelength is altered upon binding of materials to the photonic crystal surface, measured in this research using optical fibres. The authors claim this system is cheap enough for disposable sensors. A 30-min assay of a partially processed water sample yielded a detection sensitivity of 36 virus focus forming units without the use of any external reagents [45]. An alternative antibody-based sensor for rotavirus by Jung et al. utilised a graphene

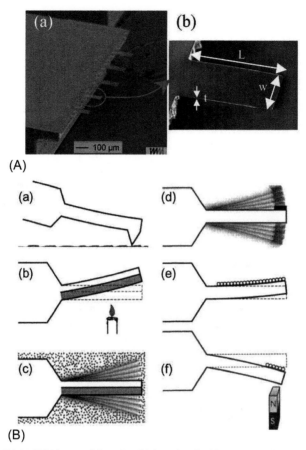

FIG. 7.10 (A) (a) SEM image of five microfabricated (poly)silicon rectangular cantilevers of different lengths. They are all supported by the same silicon chip. (b) Zoomed image of a cantilever with length $L=100\,\mu m$, width $W=40\,\mu m$, and thickness $t=0.5\,\mu m$. (B) schematic of possible uses for a cantilever transducer (side view). (A) AFM force sensors, (B) temperature/heat sensor, (C) viscoelasticity sensor, (D) mass sensor (end load), (E) stress sensor, and (F) sensor to monitor the presence of magnetic beads on its surface. *((A) Images by IMM, Mainz, Germany. Source: Fig. 1 from reference R. Raiteri, et al., Micromechanical cantilever-based biosensors, Sens. Actuators B 79 (2) (2001) 115–126. Reproduced with permission. (B) Source: Fig. 2 from reference R. Raiteri, et al., Micromechanical cantilever-based biosensors, Sens. Actuators B 79 (2) (2001) 115–126. Reproduced with permission.)*

oxide (GO) surface, and a secondary gold nanoparticle label, which binds to captured rotavirus and quenches the GO surface fluorescence through FRET [11]. However, the LOD in this approach was much higher at 10^5 pfu/mL.

Silicon microring resonators have also been employed as virus biosensors (Fig. 7.11). In this set-up, binding of biomolecules to the functionalised microring sensor causes small changes in the effective refractive index, resulting in a detectable shift in resonance wavelength. Shang et al. detected norovirus (NV)

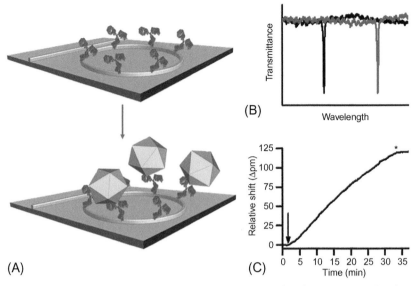

FIG. 7.11 Illustration of viral detection using silicon photonic microring resonators. Microrings functionalised to prevent antibodies (purple) specific for the virus of interest (blue icosahedron) (A) support optical resonances at particular wavelengths that are extremely sensitive to antigen binding-induced changes on the microring, the resonance shifts to longer wavelengths (B), and this shift is monitored in real-time (C) to allow quantification of the target virus. In (C) the arrow indicates the time at which the sample is introduced and the asterisk (*) indicates the return to running buffer. *(Source: Fig. 1 reproduced with permission from M.S. McClellan, L.L. Domier, R.C. Bailey, Label-free virus detection using silicon photonic microring resonators, Biosens. Bioelectron. 31 (2012) 388–392.)*

using carbohydrate functionalised silicon microrings with a LOD of 250 ng/mL [46]. However, the authors stress that the focus was on examining carbohydrate-mediated host–virus interactions and that antibodies would offer an improved LOD. Work by McClellan demonstrates the improved detection limit using antibodies with this transduction approach, reaching a 10 ng/mL LOD with an assay time of 45 min, although not using a waterborne virus [47].

QCM-based virus biosensors have been used in several examples [48, 49], including salmonella detection with a LOD of 10–20 cfu/mL [50]. An antibody-based SAW device for viruses was presented in 2008, capable of detecting on the order of 10^3 viruses/mL [51]. While the sensitivity was reduced in different water matrices, the performance of this sensor was maintained in sewage effluents and river water.

Viral biosensors for waterborne pathogens were reviewed in 2015 with the authors predicting that the advantages of biosensors for viruses could lead to this being the gold standard method in the future [48]. There has been recent work on optical and electrochemical methods for norovirus detection, along with much progress on methods for medical diagnostics [52], which is similar for hepatitis viruses [53]. However, further development and studies are required to further apply biosensors to waterborne virus detection, extending and increasing the work of the last few years and to equal the studies focussing on bacteria [2].

7.5. Biosensors for waterborne bacteria

Of all the pathogen types, bacteria are the most studied about biosensors. The information in this section summarises the biosensor technology applied to a range of different bacteria over the past couple of decades, with particular focus on the testing of water samples and on whole bacteria detection. Where the information is available, the performance characteristics of the biosensors, for example LOD, assay time, specificity are given. *E. coli* has been widely studied and *Salmonella* has also been the subject of much recent research, especially for foodborne applications. A section on some of the recent work on AMR detection is included.

7.5.1 *E. coli*

Comprehensive reviews of biosensors for this pathogen have been published over the years, for example in 2015 [54], 2017 [55], and 2018 [56], and older work concentrating on food [57] and water [43]. General reviews of biosensors for pathogens in food and water have also recently summarised the state-of-the-art in optical [55, 58] and electrochemical systems [54, 59] and online approaches for water [54]. An overview will be given here highlighting some of the developments and most interesting approaches; publications for biosensors and this pathogen have been increasing over the last few years and detection limits lowering [60].

Many different optical, electrochemical, and mass-sensitive biosensors have been developed and optimised. In the previous edition of this book, we found that a comparison of the LOD achieved by optical and electrochemical transduction methods was relatively similar, with both methods achieving a LOD of 10^2 cells/mL but that recent developments in mass-sensitive techniques had allowed reaching the limit of single-cell detection. Electrochemical methods have been growing in popularity through optical remain popular [60], and LODs have improved for all methods to tens of colony-forming units per millilitre or less.

In 2003, a commercially available (Analyte 2000, Research International) fibre optic waveguide biosensor was applied using antibodies to capture the bacteria and fluorescent labels for detection. Positive samples were then subjected to further analysis by traditional culture methods or molecular methods.

An alternative antibody-based indirect sensor with fluorescent labelling was reported by Ho et al. in 2004. In this approach, capture antibodies were immobilised on the interior surface of a microcapillary, through which the test sample subsequently flowed. Next, liposome secondary antibody conjugate was passed through the capillary, followed by a rinse to remove any unbound conjugate.

The final detection step involved the lysis of these liposomes to release the encapsulated fluorescent molecules, giving a LOD of 360 cells/mL in 45 min. Liposomes can encapsulate 10^5–10^6 fluorescent molecules, thus offering a means of signal amplification, which resulted in the low LOD reported.

Another optical biosensor approach taken by Song and Kwon was a photodiode array, using a secondary antibody conjugated to alkaline phosphatase, giving a LOD of 10^4 cells [61]. As signal detection depends upon the action of the enzyme-producing a blue precipitate, the absorbance of which is detected. This sensor is very sensitive to pH.

In 2005, Su and Li compared direct detection with SPR and QCM using the same immunocapture strategy and 1 mL samples, achieving LOD of 10^5 and 10^6, respectively. In 2005, Taylor et al. attempted to improve the SPR detection by pretreatment of the bacteria as well as the use of a secondary antibody for signal amplification. Again 1 mL samples were used, here at a flow rate of 50 μL/min. Bacterial modifications were either heat and ethanol treatment to render cells nonviable or lysis with detergent. Compared to Su and Li, the SPR LOD for untreated *E. coli* was higher at 10^7. The best LOD was obtained was for the detergent lysed cells (10^4 cfu/mL), explained by better capture of small cell fragments and improved delivery of these fragments to the sensor surface since smaller fragments diffuse faster. In 2006, Subramanian et al. trialled a different immobilisation strategy, using SAM to immobilise the capture antibody closer to the SPR surface, and a secondary antibody to increase the mass bound to the surface, improving the LOD from 10^6 to 10^3 cfu/mL. A commercially available Spreeta SPR instrument was applied in 2007 to detect *E. coli* in matrices like milk and apple juice with a LOD between 10^2 and 10^3 cfu/mL. A later SPR study for this pathogen incorporated gold nanorods into a standard sandwich immunoassay, generating a fourfold improvement in sensitivity [62] (Fig. 7.12). Integration with magnetic separation technologies, nanoparticle

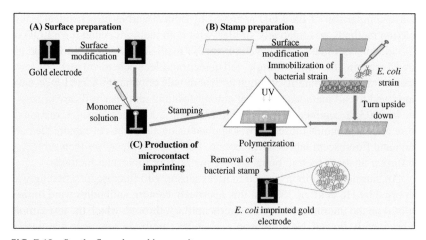

FIG. 7.12   See the figure legend in opposite page.

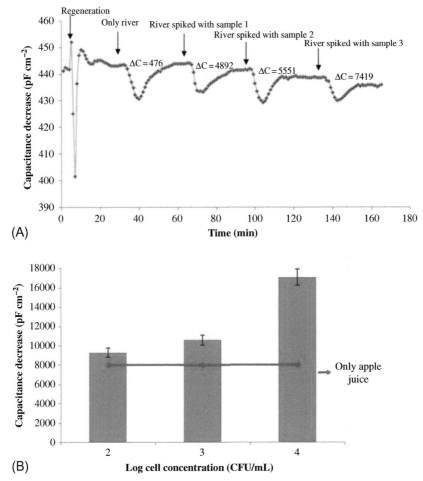

FIG. 7.12 (A) Schematic representation of microcontact imprinting of *E. coli* onto the polymer-modified surfaces. (A) preparation of electrode surface, (B) preparation of bacteria stamps, (C) production of the microcontact imprinting. Fig. 1 from Ref [58]. (B) Real sample experiments (A) real-time responses of *E. coli* imprinted capacitive biosensor indicating the sequential injection of *E. coli* spiked river water sample, sample 1: 1.0×10^2 CFU/mL, sample 2: 1.0×10^3 CFU/mL, sample 3: 1.0×10^4 CFU/mL, (B) real-time responses of *E. coli* imprinted capacitive biosensor with *E. coli* spiked apple juice sample flow rate: $100 \mu L$/min; sample volume: $500 \mu L$; running buffer: 10 mM sodium phosphate, pH: 7.4; regeneration buffer: pure ethyl alcohol, 10 mg/mL lysozyme solution (in 10 mM Tris–HCl buffer, pH 8.0, with 1 mM EDTA) and 50 mM glycine-HCl, pH 2.5; *T* 25°C. *(Fig. 6 from reference N. Idil, et al., Whole cell based microcontact imprinted capacitive biosensor for the detection of Escherichia coli, Biosens. Bioelectron. 87 (2017) 807–815. Reproduced with permission.)*

enhancement, and different operating modalities such as long-range SPR [63] has brought LODs down to between 3 and 50 cfu/mL [64]. The 3 cfu/mL was achieved by comparing SPR sensor preparation strategies, for example different methods of antibody immobilisation or use of nanoparticles, with the latter proving most effective [58].

In 2005, Rijal et al. reported the detection of 70 cells/mL in 30 min based on evanescent field absorption [65], and in the same year, Zhu et al. reported 10^2 cells/mL in 2 h using a fluorescent sandwich immunoassay [66], both using optical fibres. In 2011, U-bent optical fibres were adopted in an attempt to improve the LOD. However, during the 20 min allowed for detection, detection of concentrations lower than 10^2 cfu/mL was not possible due to diffusion-limited transport to the sensor surface [67]. Bacteriophages have also been used on optical fibres, with a LOD similar to the previous reports of 10^3 cfu/mL [68]. In 2016 pH-sensitive hydrogel nanofibre-light addressable potentiometric sensor based on sugar fermentation was shown to achieve a LOD of 20 cfu/mL [69].

A 2016 review of optical biosensors found that the integration of nanomaterials and microfluidics were useful to overcome limitations such as limited LOD and complexity of analysis [64]. Paper-based colorimetric LFAs have been reported to achieve 5 cfu/mL in 30 min when combined with magnetic preconcentration, and similar performance was reported using a nanoparticle aggregation approach.

Techniques using colorimetric sensing and metabolic activity are also utilised, for example those based on β-galactosidase [70] and ATP [71], though the latter is not specific without some additional isolation step. The ColiSense enzyme-based system is a field-portable system delivering results in 75 min [70] (Fig. 7.13).

Electrochemical methods for *E. coli* detection are popular with many different approaches developed; the field has been reviewed for foodborne applications and in 2017 for *E. coli* O157:H7 [60]. A few label based biosensors have been developed with the most successful being an amperometric approach with nanoparticle amplification reaching LODs of 148 cfu/mL in the buffer and 309 cfu/mL in tap water. In terms of label-free detection approaches various methods have been trialled including impedance, potentiometric, conductometric, capacitive, and amperometric with LODs ranging from 1 to 250 cfu/mL; the best systems were a capacitive biosensor at 10 cfu/mL and an impedometric one achieving 1 cfu/mL [72]. More recently a couple of other amperometry-based systems have reported single colony-forming units per millilitre LODs, using nanomaterials in signal amplification [54, 59]. The review concluded that there were some examples of systems with low LODs but that material development would be key to enhancing this performance, which seems to have been borne out by recent work [73]. The review also raised the issue of specificity as an important factor for improvement and highlighted that sample processing for operation in complex matrices was critical [60]. Other more recent studies include

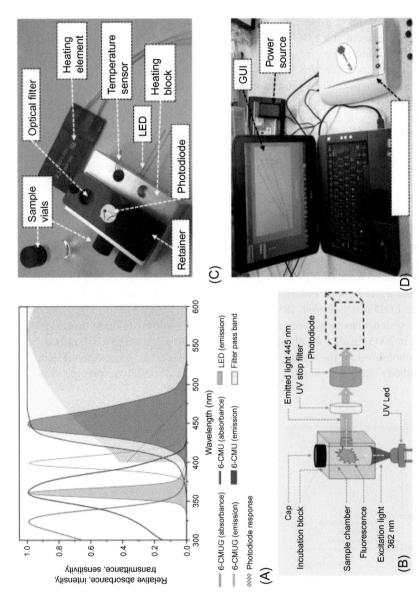

FIG. 7.13 ColiSense system design and construction. (A) Normalised spectra of chemical components of the assay and optical components of the fluorescence detection system. (B) Schematic of the incubation and fluorescence detection system. (C) The physical realisation of key system components. (D) ColiSense instrument with the power source and Graphical user interface (GUI). *(Source: Fig. 2, from reference B. Heery, et al., ColiSense, today's sample today: a rapid on-site detection of β-D-glucuronidase activity in surface water as a surrogate for E. coli, Talanta 148 (2016) 75–83. Reproduced with permission.)*

a microwire sensor with DEP capture [74], molecularly imprinted polymers as recognition elements [75], using a photochemical immobilisation technique for rapid immunosensor preparation [76], though without reaching single figure colony-forming units per millilitre LOD. A fully automated microfluidic impedance biosensor did however achieve 50 cfu/mL with a water sample [77] and a capacitive device had a LOD of 70 cfu/mL and was shown to work with river water [78]. Using magnetic beads and nanoporous membranes to assist concentration in an impedance biosensor 10 cfu/mL was reported [79].

Various mass-sensitive biosensors have been developed including QCM, QCM-D, PEMC, ME, and μC. Poitras and Tufenkji described a QCM-D sensor for the detection of *E. coli* in 2009. While the LOD was similar to the previous study, this work highlighted how the dissipation slope could be a more accurate and rapid parameter to detect the presence of bacterial cells [36] (Fig. 7.14 below). In 2014, another QCM system was reported though the LOD remained similar to previous work [80], and work in 2016 used QCM to monitor the growth of *E. coli* [81]. It is only recently that techniques such as the use of aptamers (achieving a LOD of 34 cfu/mL in 40 min) [82] and nanomaterial-based enhancement [83, 84] have successfully improved performance.

Campbell and Mutharasan have explored the potential of PEMC biosensors with a variety of pathogens. The advantage of the PEMC set-up adopted by this group is the recirculating flow system, enabling the batch processing of up to 1 L of the water sample. For the *E coli* immunosensor, at a flow rate of 1.5 mL/min, the LOD was 1 cell/mL [85] (Fig. 7.15 below).

ME systems have also been developed with different recognition strategies. Firstly, in 2003, an alkaline phosphatase-labelled anti-*E. coli* O157:H7 antibody was immobilised on the sensor surface, with the mass change associated with the antibody–antigen binding reaction amplified by biocatalytic precipitation

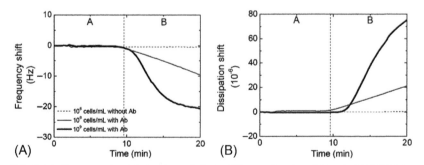

FIG. 7.14 Representative (A) *f* shifts, and (B) *D* shifts as a function of *E. coli* O157:H7 injection time for various cell concentrations (solid lines). The dashed line shows (A) *f* shifts, and (C) *D* shifts as a function of *E. coli* O157:H7 (10^8 cells/mL) injection time for a sensor crystal prepared without antibody. *(Fig. 3 from reference C. Poitras, N. Tufenkji, A QCM-D-based biosensor for* E. coli *O157:H7 highlighting the relevance of the dissipation slope as a transduction signal, Biosens. Bioelectron. 24 (7) (2009) 2137–2142. Reproduced with permission.)*

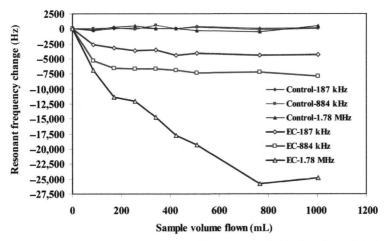

FIG. 7.15 Resonant frequency response of PEMC-b sensor to the binding of *E. coli* O157:H7 (EC) at a concentration of 1 bacteria/mL (total sample volume was 1 L) for the dominant resonant modes investigated. The control experiments were an anti-EC PEMC-b sensor exposed only to PBS solution in the same fashion as the EC sample. *(Source: Fig. 5 from reference G.A. Campbell, R. Mutharasan, A method of measuring* Escherichia coli *O157:H7 at 1 cell/mL in 1 liter sample using antibody functionalized piezoelectric-excited millimeter-sized cantilever sensor, Environ. Sci. Technol. 41 (5) (2007) 1668–1674. Reproduced with permission. Copyright (2007) American Chemical Society.)*

of 5-bromo-4-chloro-3-indolyl phosphate [86]. Secondly, in 2009, the sensor surface was functionalised with mannose and con A was employed to mediate the bacteria mannose reaction [87]. This latter system reported the detection of as little as 60 cells/mL with a sample volume of 2.5 mL, incubated for 2.5 h (Fig. 7.16).

Finally, a μC system operated in bending mode was capable of detection of 10^6 CFU/mL within 2 h [88]. Cantilever's progress was made achieving a LOD of 100 cells/mL when operating in dynamic mode overcoming liquid damping by utilising a micron-sized gap at the cantilever end, where cells are electrokinetically collected [89].

7.5.2 Salmonella

For this bacterium, all three transduction methods have been trialled, and there have been several recent reviews of different aspects of biosensor detection, particularly in the areas of food safety [90] and electrochemical detection [91, 92] as well as a lot of interest in using nanomaterials [93, 94], which is covered more in Chapter 9. A review on the application of biosensors to food applications concluded none of the technologies were yet ready for use in this market but noted that low detection limits have been achieved with progress required in analysis time, validation, portability and autonomy of systems, with electrochemical approaches being promising [91].

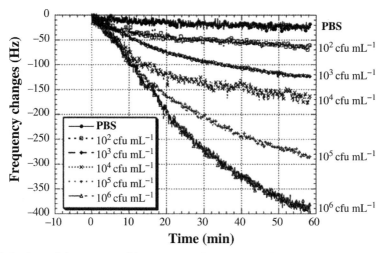

FIG. 7.16   Real-time response of the magnetoelastic *E. coli* O157:H7 sensor resonance frequency to concentration in pH10.0 PBS-containing BCIP. *(Fig. 6 from C. Ruan, et al., Magnetoelastic immunosensors: amplified mass immunosorbent assay for detection of* Escherichia coli *O157:H7, Anal. Chem. 75 (23) (2003) 6494–6498. Reproduced with permission. Copyright (2003) American Chemical Society.)*

In terms of optical sensors, FRET-based optical fibres yielded a LOD of 10^4 cfu/mL within 5 min [15]. An optical grating coupler, where APTES was found to be an optimal surface immobilisation strategy for the capture antibody, had an identical LOD [95] as did an LFA [34]. In this LFA, liposomes were employed for signal amplification improving on previous work by several orders of magnitude. Other LFAs have quantum dot nanobeads at a LOD of 10^2 cfu/mL [96] while another system used SERS detection to achieve a LOD of 27 cfu/mL [93]. An alternative quantum dot nanobead microfluidic system achieved a LOD of 43 cfu/mL using [97].

Using smartphone imaging and an immunomagnetic separation with immunofluorescent detection strategy on a microfluidic device a LOD of 58 cfu/mL was achieved [98]. A similar process with a magnetic nanoparticle column for concentration/separation with Prussian blue-based visual detection on a smartphone reported 14 cfu/mL detection [99]. Integrating sample processing with detection is becoming more common and another optical biosensor used immunomagnetic beads and a 3D microfluidic system for concentration (from up to ~200 mL of the sample) before absorbance measurements for a LOD of 17 cfu/ mL [100]. Optical approaches have also been used with the products of molecular methods for *Salmonella* [101].

Oh et al. reported a LOD for a SAM protein G based SPR immunosensor in 2004 of 10^2 cfu/mL [102]. Waswa et al. improved this technique in 2006 with a protein A immobilised immunosensor which achieved 23 cfu/mL [103]. Recent work has not improved on these limits [92, 104].

As for *E. coli*, amperometric biosensors were developed in the 1990s [43] whereas the use of EIS is more recent. Novel statistical signal processing was combined with impedance spectroscopy to create an immunosensor capable of detecting 500 cfu/mL in 6 min [105], and this LOD is similar to a lot of work with different techniques in the last few years. Recently, there has been a lot of development of different types of electrochemical biosensors. In 2018, bacteriophage was utilised to capture *Salmonella* in a capacitive biosensor with a LOD of 200 cfu/mL from a 300 μL sample; one strength of the system was the ability to regenerate the surface and thus allow reuse of the biosensor up to forty times [31]. In 2019 an impedimetric detection approach was integrated with a microfluidic device for rapid food-based detection with a LOD of 300 cfu/mL [106]. In 2020 an impedance aptasensor reached 80 cfu/mL, using nickel nanowire bridges to amplify the signal [107]. Other impedimetric biosensors achieved 7/8 cell/mL, with the use of DEP for cell focussing before detection, and was also able to distinguish serotypes and viability [90, 99].

In 2020 a voltammetric detection system based on enzymatic production of catalysate, linking the urease to the bacteria via an antibody and gold nanoparticles, and measurement of the resistance generated was created, reporting a LOD of 10 cfu/mL [108]. The same group also integrated a similar strategy with a colorimetric detection protocol with the same level of LOD but also reported the magnetic grid separation step of sample preconcentration allowed the use of 50 mL samples [109].

Finally, in terms of mass-sensitive biosensors, QCM, μC, and ME Salmonella biosensors have all been developed. In 2002, Wong et al. were able to distinguish between three serotypes (A, B, and D) of salmonella and reported a LOD of 10^4 cfu/mL [110]. This LOD was improved by 2 orders of magnitude by Su and Li in 2005 by the use of both a different mode of detection (impedance measurements of motional resistance) and IMS beads for amplification [111]. In 2018 incorporating a preenrichment step (for 2 h) enabled a QCM immunosensor to achieve 10 cfu/mL [112].

A bending mode μC achieved a LOD of 10^6 cfu/mL whereas SEM observations indicated that signal was obtained with just 25 bacterial cells bound on the surface [113]. This highlights a key problem with biosensing technologies. While the systems may offer sufficient sensitivity to detect low numbers of pathogens, the capture efficiency is often low thus raising the LOD. A resonant mode μC achieved a LOD of 5×10^3 cfu/mL [114] and a recent paper integrating a resonant mode μC with either microfluidic delivery or taking a dip and dry approach reported LODs of 10^5 and 10^3 cfu/mL, respectively [115]. The advantage of this latter work was that the μC signal was read-out using piezoelectric means rather than laser deflection, thus potentially enabling portable devices.

In 2016 a ME biosensor was developed for salmonella detection in soil using phages as a recognition element and the further suggested work to optimise this proof-of-concept has not yet been published [116].

One new system was the use of an NMR biosensor, which represents a different approach and while the LOD was currently high there are many potential improvements to be made, including optimising the probes and creating more portable detection devices [107]. Detection is performed by immunocoupling of the bacteria to a magnetic contrast agent, in this case, gadolinium.

There is some progress towards commercialisation through systems such as the RBD 3000 Micro PRO and the Rapid B are not simplified for example field use [91]. There seems to have been a great deal of recent focus on biosensors for the food industry, and there is great potential for these techniques to transfer to water testing with appropriate sample processing technology.

7.5.3 *Campylobacter*

Campylobacter biosensor detection was reviewed in 2013 describing a range of different approaches including some DNA based sensors though mostly antibody-based whole bacteria detecting systems [117]. An issue with immunosensors for *Campylobacter* is species cross-reactivity [118]. One exception was the two reported QCM set-ups in which lectins were used as the recognition element, with a LOD of 10^3 cfu/mL. In 2016, a sandwich assay was implemented on a QCM platform with gold nanoparticles and this approach lowered the LoD to 150 cfu/mL [119]. In 2017 another QCM set-up with gold nanoparticles and magnetic immunoseparation reported a LoD of 20–30 cfu/mL [120].

A further review in 2019 looked at optical and electrochemical biosensors concluding the work was promising, especially as lower LODs were now being achieved but that more was needed to device long-term stability and that aptamers were thought to offer greater sensitivity than antibodies [75]. There are several examples of SPR being used with the best LODs being in the range of 10^2–10^3 cfu/mL before the subtractive inhibition assay approach utilised in 2019 enabling 131 cfu/mL detection [121]. In 2004, a TIR fluorescence sandwich assay was developed by Sapsford et al. [122]. The LOD was reported to be 10^2–10^3 cfu/mL depending on the strain and species of bacteria; the difference was accounted for by variation in the antibody affinity for the various species and strains. In 2009, Bruno and colleagues reported an aptamer-based quantum dot sandwich assay capable of rapidly delivering results (15–20 min) using a portable handheld fluorometre. The LOD achieved was as low as 2.5 cfu/mL rising to 10–250 cfu/mL in food matrices. A FRET assay in 2020 achieved 10 cfu/mL in PBS and 100 cfu/mL in poultry liver [121].

More examples of electrochemical sensors, including amperometric, impedimetric, and conductometric, are reported though the LoDs mostly remained in the same range [122], even in more recent work with screen printed electrodes that incorporated enzyme-linked immune magnetic set-up. However, the advantages of this potentiostatic device are portability and the ability to detect different species [123]. One amperometric biosensor for which the authors claimed single bacteria detection would be possible within 10 min using signal amplification based on high-ion flux through a membrane. The device utilised a planar lipid bilayer into which the capture antibodies were embedded, coated onto a stainless steel working electrode. Binding of the pathogens alters the membrane allowing current to pass.

In 2019, the volatile organic compounds emission of *C. jejuni* were analysed to develop metal oxide gas sensors and proof-of-principle demonstrated, with further work required to lower detection limit and evaluate applicability in complex samples [124]. In 2020 the technique was utilised to monitor milk samples at 10^4–10^5 cfu/mL levels [125]. A system of culturing on-chip, combined with optical detection, was utilised to screen for antimicrobial resistance in *Campylobacter* samples [126].

7.5.4 Cholerae

Both traditional techniques and emerging biosensor technologies for cholera were reviewed in 2016 [127]. The majority of approaches detect either DNA [128–132], toxin [127, 133–135], or antibody [136] (these latter two not applying to waterborne testing), with only a few studies detecting the bacteria themselves.

The Ravichandran group developed a dry-reagent dipstick LFA in 2011 for cholera with a LOD of 10^2 cfu/mL. A paper-based test with an optical readout was recently tested in Haiti and shown to be a reliable field-test system [137]. Another rapid diagnostic strip-based test designed novel antibodies for stool testing, though the LOD was high at 1×10^4 cfu/mL [138]. A lower LoD of 10 cfu was obtained with a dipstick system though overnight incubation was required; the system was demonstrated with 200 mL samples of environmental water though [139].

In 2006, Jyoung and colleagues utilised SPR for the detection of cholera with a detection range between 10^5 and 10^9 cells/mL and improvements in the implementation of the SPR approach lowered the LoD to 50 cells/mL in a 2017 study [140]. In 2010, Sungkanak and colleagues utilised a μC for the ultrasensitive detection of *Vibrio cholerae* O1. The authors immbolised antibody using a SAM method and achieved a LOD of 10^3 cfu/mL which they suggest could be further improved by the redesign of the cantilever.

In 2016 an electrochemical antibody-based sensor was applied to cholera with a LoD of 1 cfu/mL with performance tested in different conditions, for example pH and temperature, from 5 mL samples in a couple of hours [131]. In 2017 another electrochemical biosensor reported a LoD of 5 cfu/mL and correlated performance against traditional techniques with a water sample [139].

Cell-based biosensors have been reported to be very fast at generating a response and a hybridoma based bioluminescent system for cholera detection was reported in 2016 [141]. Interestingly the system performed well in a variety of matrices, including freshwater, wastewater, and seawater samples, and a LoD of 50 cfu/mL was reported (12 cfu per assay). Another recent approach, from 2019, which performed well in pond water samples and required no preenrichment is a sensor based on particle diffusometry and the products of an isothermal amplification step [142].

More work is required to develop field-ready diagnostics for waterborne cholera [133], and biosensors are thought to be highly promising especially those combined with nanotechnology [143].

7.5.5   *Shigella*

The TIR fluorescent sandwich assay reported by Sapsford et al. in 2004 for *Campylobacter* was also applied for the detection of *Shigella*, with LoDs varying based on the species tested, due to antibody specificity [122]. The sensitivity could be improved by increasing the incubation time, though this effect was reversed in complex samples and cross-reactivity of the antibody with *E. coli* was also observed. In 2017, an aptamer search was performed for use in a fluorescent biosensor, with the advantage of species-level discrimination and lack of cross-reactivity with other bacteria [144]. Aptamer use is a recent topic of interest for *Shigella* biosensors and in 2018 an electrochemical aptasensor was used in spiked water samples, at 10^2 cfu/mL [145]. In 2019, a colorimetric aptasensor, using gold nanoparticles, was reported with a LoD of 80 cfu/mL and a 20 min detection time, aimed at food testing, though it could be adapted for water testing [146]. DNA detecting *Shigella* biosensors have also been reported [73], for example a low-cost, portable evanescent wave fibre biosensor [147] and a nanoparticle incorporating fluorescent system [148].

7.6.   Biosensors for waterborne protozoa

The field of biosensors for protozoan detection was recently reviewed in 2019, with the authors concluding that there were several promising studies, typically utilising antibodies and highlighting the critical importance of integration with appropriate sample processing and the potential of nanomaterials to enhance signals [149]. Another important area of future study to enable improved biosensor performance was increased understanding and knowledge relating to oocyst surface chemistries. Examples in the following section are given for the state-of-the-art in protozoan biosensor detection. No reports have been found of systems offering any level of viability determination, despite this being an important factor [150].

Poitras et al. detected *C. parvum* oocysts in clean water using a QCM biosensor with dissipation monitoring (QCM-D), with a detection range of 3×10^5 to 10^7 oocysts/mL, using a flow rate of 50 μL/min [37]. The flow was repeatedly stopped to allow time for the reagents to adsorb and react (60 min for the oocysts). Furthermore, the influence of the background matrix on detection was tested in solutions containing either biological interferents such as bacteria, particularly *E. coli* O157:H7 and *Enterococcus faecalis*, or nonbiological ones such as latex microspheres or humic and fulvic acids, commonly found in natural waters. A decline in the performance of up to 64% was measured depending on the interferent [37]. Poitras et al also demonstrated that the initial slopes in f and D (Fig. 7.17) could be used as a rapid means to detect oocysts, requiring just 5 min for *C. parvum* quantification, when utilising the initial slopes methodology [37]. The volume of solution held in the flow cell allowing 60 min for oocyst binding was 40 μL.

C. parvum has been used as the target analyte in only two SPR experiments to date, carried out by Kang et al. (Fig. 7.18) [151, 152]. The LoD was highly dependent on the biological recognition strategy employed. Using strepavidin-biotin for immobilisation of antibody on the surface followed by continuous oocyst flow gave a LOD of 1×10^6 oocysts/mL. This high number is due to the low capture efficiency of the surface immobilised antibody, which is a common problem for biosensors [153]. Decrease of the LoD to 100 oocysts/mL was possible by labelling the oocysts with biotin. This recognition strategy thus takes advantage of the high affinity, rapid reaction between the surface immobilised strepavidin and biotin. The disadvantage of this method is that centrifugation is required in the sample processing making integration of this detection method into a continuous flow system very difficult, without the use of a different mechanism to label and wash the cells. Interestingly, although the LoD was 100 oocysts/mL the total volume injected into the sensor was only 20 μL (2 μL/min for 10 min). Assuming an even distribution of oocysts through the buffer, this would suggest that the number of bound oocysts was around 2. This technology could, therefore, detect clinically relevant levels of oocysts, by placing a greater burden on the sample processing stage that is currently undertaken [154]. At such low flow rates, the sample must be concentrated by at least 500,000 times, which, without enrichment, would lead to a very sparsely

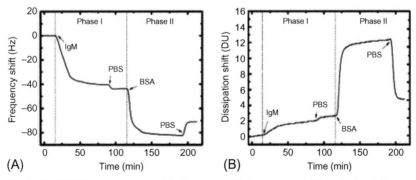

FIG. 7.17 QCM-D measurements of the 1 overtone (A) frequency and (B) dissipation shifts during the physisorption of anti-*C. parvum* antibodies onto gold-coated quartz crystals (phase I) followed by BSA adsorption (phase II). *(Reproduced with permission from C. Poitras, J. Fatisson, N. Tufenkji, Real-time microgravimetric quantification of* Cryptosporidium parvum *in the presence of potential interferents, Water Res. 43 (10) (2009) 2631–2638.)*

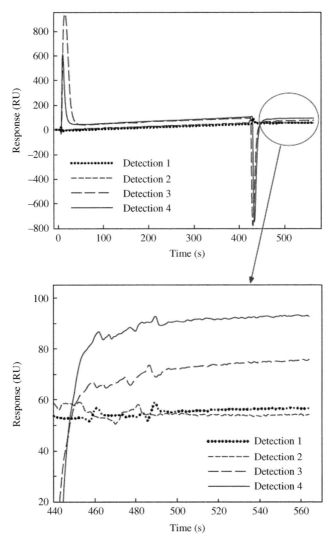

FIG. 7.18 Performance tests of the *Cryptosporidium* sensor chip during SPR-based inhibition assay. Detection 1: *C. parvum* oocyst in HBS-EP buffer; Detection 2: a mixture of *Bacillus stearothermophilus* spore, *Chlamydomonas reinhardtii*, *E. coli*, and *C. parvum* oocyst in HBS-EP buffer; Detection 3: *C. parvum* oocyst in tap water; and Detection 4: *C. parvum* oocyst in reservoir water. The concentrations of the used microorganisms were 1×10^5 cells mL^{-1} in all cases. *(Fig. 5 from reference C.D. Kang, et al., Surface plasmon resonance-based inhibition assay for real-time detection of* Cryptosporidium parvum *oocyst, Water Res. 42 (6–7) (2008) 1693–1699. Reproduced with permission.)*

populated matrix as other contaminants would also be concentrated. However, this problem of requiring extensive sample preparation is not exclusive to SPR biosensing technologies.

Cantilever sensors have also been applied for *Cryptosporidium*, including both static mode cantilevers with an incorporated microfluidic channel for sample delivery [155] as well as PEMC. The most successful approach is the use of PEMC, offering both the lowest LOD and largest sample processing. *C. parvum* can be detected with PEMC sensors not only in deionised water but also in PBS and other background matrices such as milk. Xu and Mutharasan proved that such oocysts can be detected under a recirculating flow of 1 mL/min both in PBS and in 25% milk in PBS background. Campbell and Mutharasan achieved detection of *Cryptosporidium parvum* oocysts in PBS the range of 100–1000 oocysts/mL in less than 15 min and suggested that detection of 1–10 oocysts/mL could be possible [156]. Both experiments were conducted in flow cell systems with sensor cell volumes of 120 µL and 90 µL, respectively. However, sensitivity is reduced in the presence of interferents, with a reported decrease in detection of around 45% in milk, and no testing has yet been performed in finished drinking water [157]. The same authors have utilised the same biosensor set-up for detection of *Giardia lamblia* with a LOD of 1–10 cysts/mL in 15 min in a variety of water matrices (buffer, tap, and river water) [158] (Fig. 7.19).

In 2019, a capacitive biosensor was developed reporting a LoD of 40 oocysts per mm^2, though it still needs to be tested with real drinking water samples [159]. In the same year, another electrochemical sensor was developed for the detection of *Cryptosporidium* DNA [160], and molecular methods are discussed in more detail in Chapter 8.

Recent work has utilised aptamers for *Cryptosporidium* detection, combining the identified aptamer with electrochemical detection and obtaining a LOD of 100 oocysts [161]. Further investigations will establish whether the identified aptamers, here for *C. parvum*, are capable of genus-specific detection and/or confer any species specificity. The method was coupled with magnetic beads and trialled in drinking water in 2019 and the LOD reduced to 50 oocysts [162]. Another alternative recognition element that has been trialled is lectins, though further work is needed to confirm specificity [163].

There are very few reports of other protozoan biosensors. In 2004 Wang et al. performed serological toxoplasma detection, based on the specific agglutination of antigen-coated gold nanoparticles, averaging 10 nm in diameter, in the presence of the corresponding antibody causes a frequency change that is monitored by a piezoelectric device [164]. The year after, the same group utilised enzymatic catalysis-induced precipitation of an insoluble product monitored by EIS and CV as well as QCM for serological *Toxoplasma* detection [165]. More recently *Toxoplasma* biosensor development has been focussed on the diagnosis of infection, rather than detection of the protozoan itself; although there are several examples of systems exploiting techniques including SPR, piezoelectric, electrochemical, and laser-induced fluorescence, the approaches do not apply to water testing [166].

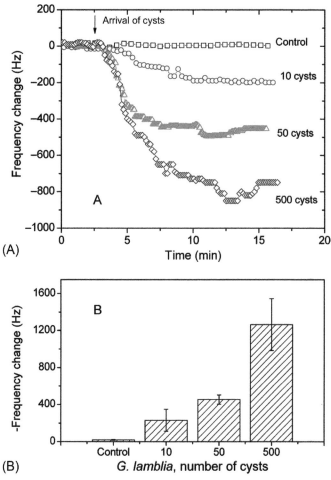

(A)

(B)

FIG. 7.19 (A) Response time of the sensor to addition of 1 mL of sample containing *G. lamblia* samples to the flow loop set in recirculation mode at 2.4 mL/min. (B) Average frequency responses ($n = 4$) to various concentrations. *Fig. 2 from S. Xu, R. Mutharasan, Rapid and sensitive detection of* Giardia lamblia *using a piezoelectric cantilever biosensor in finished and source waters, Environ. Sci. Technol. 44 (5) (2010) 1736–1741. Reproduced with permission. Copyright (2010) American chemical society.*

7.7. Biosensors for the detection of waterborne parasites

In terms of waterborne parasites, the biosensor community has mainly studied schistosoma, utilising the presence of antibodies within the host serum as the detection marker. Various techniques have been proposed including an amperometric immunosensor with a LOD of 0.36 µg/mL [167], piezoelectric immunosensors [168, 169], and SPR [170]. While this approach is useful for diagnostics, it would be difficult to adapt for waterborne detection.

7.8. Biosensors for antimicrobial resistance

Biosensors are also applied for the detection of antimicrobial resistance (AMR), adopting different approaches including screening for growth or lysis in the presence of antibiotics or detection of the relevant genes. This is an area with a lot of recent research with numerous articles published in 2020, and several recent reviews [5, 171–174]. There are several commercially available tools: some are based on automation of the standard broth microdilution method, using colorimetric or fluorogenic read-out, for example BD Phoenix; biosensor wise LifeScale offer a cantilever based system [5].

Electrochemical impedance spectroscopy and differential pulse voltammetry were used to track the growth of bacteria encapsulated in growth-media containing hydrogels on screen-printed electrodes, and whether the presence of antibiotics impacted growth curves [175]. Another system utilising growth was a low-cost paper-based device combining growth and colorimetric readout of AMR susceptibility for point-of-care urine infection testing [4]. QCM has also been applied in this way [176] as have cantilevers [64].

A localised SPR based optical fibre biosensor operated by surface capture of the bacteria followed by exposure to antibiotics with susceptibility noted by the measurement of antibiotic-induced lysis [177]. Electrochemical sensors have been used to detect AMR genes [21, 178].

A system, with DEP to enhance capture and synthetic peptides probes to discriminate AMR strains, using graphene FET were applied to achieve detection of AMR bacteria in just 5 min but the LOD was relatively high (10^4 cfu/mL) and the sample volume low [179].

Further improvements are required, exploring further technologies and then increasing automation and parallelisation of systems to simplify processes and lower cost, as well as sample processing [165]. Besides, the majority of work is focussed on medical applications and a greater focus on environmental applications is required.

7.9. Summary and future outlook

In this concluding section, we will summarise the state-of-the-art concerning biosensors for waterborne pathogens, discuss the advantages and disadvantages of this approach to detection, and consider key areas of research and development critical for biosensors to meet the demands of waterborne pathogen detection.

The vision expounded by Rose and Grimes in 2001, who summarised the view of a colloquium panel of water experts, of a distributed network of sensors continuously monitoring in real-time is far from a reality. However, biosensors for waterborne pathogens are a thriving and growing area of research, with much recent progress. Indeed, since the first edition of this book, there are has been significant progress in lowering the limits of detection. Then we concluded that LODs were generally on the order of 10^2 cells/mL for bacteria whereas now there are a variety of systems reporting LODs less than 10 cells/mL and single-cell level detection has been demonstrated. This holds for all types of transduction technology, with literature indicating that optical methods are the most widely studied, followed by electrochemical and then mass-sensitive. All types of biosensors have also been applied to protozoa though the most successful technology here is that of PEMC. For viruses, optical methods are the most popular, and successful, with a photonic crystal biosensor developed by Pineda et al. achieving a LOD of 36 virus-forming units. Other virus biosensors have been developed including electrochemical and mass-sensitive approaches, though the technologies have not been widely applied to waterborne pathogens. A variety of techniques have also been applied to the challenge of AMR detection and this is an important area of future development, particularly to translate finding from medical applications into systems useful for environmental applications.

In general, there is a far greater focus on foodborne pathogen biosensor development and it would be useful to translate the technologies developed here to waterborne pathogen studies. While there are many similar pathogens there might be differences in the sample volumes and types of interferents which will impact implementation. Indeed sample processing is a clear challenge that needs to be addressed so that the colony-forming units per millilitre LODs reported can be relevant to the testing of larger volumes of water. Again, this is another area where there has been positive development over the last few years as we previously concluded that the biosensor literature often overlooked the application to larger real-world samples but there are a lot of recent reports of work in milk and juice and food samples as well as integrated systems incorporating sample processing steps along with biosensor detection. This is critical as a recent review highlighted the need for systems to be user-friendly and simple to facilitate wider uptake; automated set-ups with portable pretreatment is the goal.

There is also a lot of outstanding work to address questions of stability, reproducibility, robustness, and suitability for long-term operation for field deployment. Further studies characterising performance in complex matrices, particularly different water types, should be done and validated against traditional techniques.

The final areas for improvement include specificity and the generation of additional information. For example, any advances enabling species and viability discrimination would be extremely beneficial for real-world applications and this has been recognised in a couple of systems offering some level of viability assessment as well as in various reviews of the field.

In summary, the advantages of biosensors include sensitive detection with rapid results as well as the potential ease of portability of some detection schemes. However, disadvantages include issues of specificity; this is important to avoid false positives and indeed for some waterborne pathogens availability of recognition elements is a challenge in itself. Another important issue is discrimination between species as well as between viable and nonviable pathogens.

Future work should concentrate on the challenges mentioned previously, and electrochemical schemes may have particular advantages here in offering information regarding species and viability. Materials development is another important area and the recent positive impact of the inclusion of a variety of nanomaterials and nanotechnology approaches has contributed to the lowering of the LOD; more details are given in Chapter 9. Multiplexed detection is another development that would be extremely useful and miniaturised detection platforms, such as those described in Chapter 10, represent one solution. Further integration with sample processing should also be undertaken.

In conclusion, there has been significant recent progress in biosensor development, single pathogen level detection has been demonstrated and there is great potential in the further adoption of nanotechnology and microfluidics as well as the requirement for a greater number of investigations to consider waterborne applications and integrate sample processing methods to create more relevant systems.

References

[1] O. Lazcka, F.J.D. Campo, F.X. Muñoz, Pathogen detection: a perspective of traditional methods and biosensors, Biosens. Bioelectron. 22 (7) (2007) 1205–1217.

[2] K.R. Srivastava, et al., Biosensors/molecular tools for detection of waterborne pathogens (Chapter 13), in: M.N.V. Prasad, A. Grobelak (Eds.), Waterborne Pathogens, Butterworth-Heinemann, 2020, pp. 237–277.

[3] A. Ahmed, et al., Biosensors for whole-cell bacterial detection, Clin. Microbiol. Rev. 27 (3) (2014) 631.

[4] P.J.W. He, et al., Laser-patterned paper-based sensors for rapid point-of-care detection and antibiotic-resistance testing of bacterial infections, Biosens. Bioelectron. 152 (2020) 112008.

[5] J. Dietvorst, et al., Current and near-future technologies for antibiotic susceptibility testing and resistant bacteria detection, TrAC Trends Anal. Chem. 127 (2020) 115891.

[6] F.Y. Ramírez-Castillo, et al., Waterborne pathogens: detection methods and challenges, Pathogens (Basel, Switzerland) 4 (2) (2015) 307–334.

[7] P.J. Conroy, et al., Antibody production, design and use for biosensor-based applications, Semin. Cell Dev. Biol. 20 (1) (2009) 10–26.

[8] T.M. Squires, R.J. Messinger, S.R. Manalis, Making it stick: convection, reaction and diffusion in surface-based biosensors, Nat. Biotechnol. 26 (4) (2008) 417–426.

[9] R.S. Yalow, S.A. Berson, Assay of plasma insulin in human subjects by immunological methods, Nature 184 (1959) 1648–1649.

[10] P.B. Luppa, L.J. Sokoll, D.W. Chan, Immunosensors—principles and applications to clinical chemistry, Clin. Chim. Acta 314 (2001) 1–26.

[11] J.H. Jung, et al., A graphene oxide based immuno-biosensor for pathogen detection, Angew. Chem. Int. Ed. 49 (2010) 5708–5711.

[12] M.K. Swiecki, et al., Monoclonal antibodies for Bacillus anthracis spore detection and functional analyses of spore germination and outgrowth, J. Immunol. Methods 176 (2006) 6076–6084.

[13] E.R. Goldman, et al., Facile generation of heat-stable antiviral and antitoxin single domain antibodies from a semisynthetic llama library, Anal. Chem. 78 (2006) 8245–8255.

[14] A.K. Abbas, H. Lichtman, Cellular and Molecular Immunology, Elsevier, Philadelphia, 2005.

[15] S. Ko, S.A. Grant, A novel FRET-based optical fiber biosensor for rapid detection of Salmonella typhimurium, Biosens. Bioelectron. 21 (7) (2006) 1283–1290.

[16] R. Das, Cryptosporidium Detection Through Antibody Immobilization on a Solid Surface, Bangladesh University of Engineering and Technology, 2007.

[17] L.J. Sherwood, et al., Rapid assembly of sensitive antigen-capture assays for Marburg virus, using in vitro selection of llama single-domain antibodies, at biosafety level 4, J Infect Dis 196 (2007) S213–S219.

[18] S.F. D'Souza, Microbial biosensors, Biosens. Bioelectron. 16 (6) (2001) 337–353.

[19] M. Campàs, B. Prieto-Simón, J.-L. Marty, A review of the use of genetically engineered enzymes in electrochemical biosensors, Semin. Cell Dev. Biol. 20 (1) (2009) 3–9.

[20] A. Sassolas, B.D. Leca-Bouvier, L.J. Blum, DNA biosensors and microarrays, Chem. Rev. 108 (1) (2008) 109–139.

[21] E.A. Obaje, et al., Carbon screen-printed electrodes on ceramic substrates for label-free molecular detection of antibiotic resistance, J. Interdiscip. Nanomed. 1 (3) (2016) 93–109.

[22] P. Banerjee, A.K. Bhunia, Mammalian cell-based biosensors for pathogens and toxins, Trends Biotechnol. 27 (3) (2009) 179–188.

[23] G.T. Hermanson, Bioconjugate Techniques, Academic Press, 2008.

[24] X. Munoz-Berbel, et al., Impedance based biosensors for pathogen detection, in: M. Zourob (Ed.), Principles of Bacterial Detection: Biosensors, Recognition Receptors and Microsystems, Springer, 2008.

[25] A. Sassolas, L.J. Blum, B.D. Leca-Bouvier, Immobilization strategies to develop enzymatic biosensors, Biotechnol. Adv. 30 (3) (2012) 489–511.

[26] D. Gavriilidou, H. Bridle, Comparison of immobilization strategies for *Cryptosporidium parvum* immunosensors, Biochem. Eng. J. 68 (2012) 231–235.

[27] D.G. Myszka, R.L. Rich, Implementing surface plasmon resonance biosensors in drug discovery, Pharm. Sci. Technol. Today 3 (9) (2000) 310–317.

[28] J. Homola, Surface plasmon resonance sensors for detection of chemical and biological species, Chem. Rev. 108 (2) (2008) 462–493.

[29] I. Abdel-Hamid, et al., Highly sensitive flow-injection immunoassay system for rapid detection of bacteria, Anal. Chim. Acta 399 (1) (1999) 99–108.

[30] M. Pohanka, P. Skladal, Electrochemical biosensors—principles and applications, J. Appl. Biomed. 6 (2008) 57–64.

[31] S. Niyomdecha, et al., Phage-based capacitive biosensor for Salmonella detection, Talanta 188 (2018) 658–664.

[32] T. Yoshinobu, et al., Portable light-addressable potentiometric sensor (LAPS) for multisensor applications, Sens. Actuators B 95 (1) (2003) 352–356.

[33] A.G. Gehring, D.L. Patterson, S.-I. Tu, Use of a light-addressable potentiometric sensor for the detection of *Escherichia coli* O157:H7, Anal. Biochem. 258 (2) (1998) 293–298.

[34] J.A. Ho, et al., Lipsome-based immunostrip for the rapid detection of *Salmonella*, Anal. Bioanal. Chem. 391 (2) (2008) 479–485.

[35] G. Olanya, et al., Protein interactions with bottle-brush polymer layers: effect of side chain and charge density ratio probed by QCM-D and AFM, J. Colloid Interface Sci. 349 (1) (2010) 265–274.

[36] C. Poitras, N. Tufenkji, A QCM-D-based biosensor for *E. coli* O157:H7 highlighting the relevance of the dissipation slope as a transduction signal, Biosens. Bioelectron. 24 (7) (2009) 2137–2142.

[37] C. Poitras, J. Fatisson, N. Tufenkji, Real-time microgravimetric quantification of *Cryptosporidium parvum* in the presence of potential interferents, Water Res. 43 (10) (2009) 2631–2638.

[38] H. Bridle, et al., Detection of *Cryptosporidium* in miniaturised fluidic devices, Water Res. 46 (6) (2012) 1641–1661.

[39] G.A. Campbell, R. Mutharasan, A method of measuring *Escherichia coli* O157:H7 at 1 cell/mL in 1 liter sample using antibody functionalized piezoelectric-excited millimeter-sized cantilever sensor, Environ. Sci. Technol. 41 (5) (2007) 1668–1674.

[40] D. Maraldo, et al., Method for label-free detection of femtogram quantities of biologics in flowing liquid samples, Anal. Chem. 79 (7) (2007) 2762–2770.

[41] C. Ruan, et al., Magnetoelastic immunosensors: amplified mass immunosorbent assay for detection of *Escherichia coli* O157:H7, Anal. Chem. 75 (23) (2003) 6494–6498.

[42] R. Raiteri, et al., Micromechanical cantilever-based biosensors, Sens. Actuators B 79 (2) (2001) 115–126.

[43] S. Xu, R. Mutharasan, The coming together of the sciences: biosensors for the detection of waterborne pathogens using antibodies and gene-based recognition chemistries, in: K. Sen, N.J. Ashbolt (Eds.), Environmental Microbiology: Current Technology and Water Applications, Caister Academic Press, Norfolk, 2011.

[44] K. Sen, The needle in a haystack: detection of microbes in source and drinking water using molecular methods, in: K. Sen, N.J. Ashbolt (Eds.), Environmental Microbiology: Current Technology and Water Applications, Caister Academic Press, Norfolk, 2011.

[45] M.F. Pineda, et al., Rapid specific and label-free detection of porcine rotavirus using photonic crystal biosensors, IEEE Sens. J. 9 (4) (2009) 470–477.

[46] J. Shang, et al., An organophosphonate strategy for functionalizing silicon photonic biosensors, Langmuir 28 (2012) 3338–3344.

[47] M.S. McClellan, L.L. Domier, R.C. Bailey, Label-free virus detection using silicon photonic microring resonators, Biosens. Bioelectron. 31 (2012) 388–392.

[48] Z. Altintas, et al., Biosensors for waterborne viruses: detection and removal, Biochimie 115 (2015) 144–154.

[49] A. Afzal, et al., Gravimetric viral diagnostics: QCM based biosensors for early detection of viruses, Chem. Aust. 5 (1) (2017) 7.

[50] F. Salam, Y. Uludag, I.E. Tothill, Real-time and sensitive detection of *Salmonella typhimurium* using an automated quartz crystal microbalance (QCM) instrument with nanoparticles amplification, Talanta 115 (2013) 761–767.

[51] M. Bisoffi, et al., Detection of viral bioagents using a shear horizontal surface acoustic wave biosensor, Biosens. Bioelectron. 23 (9) (2008) 1397–1403.

[52] Y. Saylan, et al., An alternative medical diagnosis method: biosensors for virus detection, Biosensors 9 (2) (2019) 65.

[53] S. Hassanpour, et al., Diagnosis of hepatitis via nanomaterial-based electrochemical, optical or piezoelectrical biosensors: a review on recent advancements, Microchim. Acta 185 (12) (2018) 568.

[54] J. Riu, B. Giussani, Electrochemical biosensors for the detection of pathogenic bacteria in food, TrAC Trends Anal. Chem. 126 (2020) 115863.

[55] R.E. Ionescu, Biosensor platforms for rapid detection of *E. coli* bacteria, in: A. Samie (Ed.), Recent Advances on Physiology, Pathogenesis and Biotechnological Applications, IntechOpen, 2017, https://doi.org/10.5772/67392.

[56] M.R. Nurliyana, et al., The detection method of *Escherichia coli* in water resources: a review, J. Phys. Conf. Ser. 995 (2018) 012065.

[57] O. Tokarskyy, D.L. Marshall, Immunosensors for rapid detection of *Escherichia coli* O157:H7—perspectives for use in the meat processing industry, Food Microbiol. 25 (1) (2008) 1–12.

[58] Ö. Torun, et al., Comparison of sensing strategies in SPR biosensor for rapid and sensitive enumeration of bacteria, Biosens. Bioelectron. 37 (1) (2012) 53–60.

[59] Z. Zhang, J. Zhou, X. Du, Electrochemical biosensors for detection of foodborne pathogens, Micromachines 10 (4) (2019) 222.

[60] M. Xu, R. Wang, Y. Li, Electrochemical biosensors for rapid detection of *Escherichia coli* O157:H7, Talanta 162 (2017) 511–522.

[61] J.M. Song, H.T. Kwon, Photodiode array on-cip biosensor for the detection of *E. coli* O157:H7 pathogenic bacteria, Methods Mol. Biol. 503 (2009) 325–355.

[62] N.-S. Eum, et al., Enhancement of sensitivity using gold nanorods antibody conjugator for detection of *E. coli* O157:H7, Sens. Actuators B 143 (2) (2010) 784–788.

[63] C.-J. Huang, et al., Long-range surface plasmon-enhanced fluorescence spectroscopy biosensor for ultrasensitive detection of *E. coli* O157:H7, Anal. Chem. 83 (3) (2011) 674–677.

[64] S.M. Yoo, S.Y. Lee, Optical biosensors for the detection of pathogenic microorganisms, Trends Biotechnol. 34 (1) (2016) 7–25.

[65] K. Rijal, et al., Detection of pathogen *Escherichia coli* O157:H7 AT 70 cells/mL using antibody-immobilized biconical tapered fiber sensors, Biosens. Bioelectron. 21 (6) (2005) 871–880.

[66] P. Zhu, et al., Detection of water-borne *E. coli* O157 using the integrating waveguide biosensor, Biosens. Bioelectron. 21 (4) (2005) 678–683.

[67] R. Bharadwaj, et al., Evanescent wave absorbance based fiber optic biosensor for label-free detection of *E. coli* at 280 nm wavelength, Biosens. Bioelectron. 26 (7) (2011) 3367–3370.

[68] S.M. Tripathi, et al., Long period grating based biosensor for the detection of *Escherichia coli* bacteria, Biosens. Bioelectron. **35** (1) (2012) 308–312.

[69] P.M. Shaibani, et al., The detection of *Escherichia coli* (*E. coli*) with the pH sensitive hydrogel nanofiber-light addressable potentiometric sensor (NF-LAPS), Sens. Actuators B 226 (2016) 176–183.

[70] B. Heery, et al., ColiSense, today's sample today: a rapid on-site detection of β-D-glucuronidase activity in surface water as a surrogate for *E. coli*, Talanta 148 (2016) 75–83.

[71] C.B. Hansen, et al., Monitoring of drinking water quality using automated ATP quantification, J. Microbiol. Methods 165 (2019) 105713.

[72] M. Barreiros dos Santos, et al., Label-free ITO-based immunosensor for the detection of very low concentrations of pathogenic bacteria, Bioelectrochemistry 101 (2015) 146–152.

[73] L. Zhang, et al., Detection of shigella in milk and clinical samples by magnetic immunocaptured-loop-mediated isothermal amplification assay, Front. Microbiol. 9 (2018) 94.

[74] W. Choi, et al., Dielectrophoresis-based microwire biosensor for rapid detection of *Escherichia coli* K-12 in ground beef, LWT—Food Sci. Technol. (2020) 109230, https://doi.org/10.1016/j.lwt.2020.109230.

[75] P. Vizzini, et al., Electrochemical and optical biosensors for the detection of *Campylobacter* and *Listeria*: an update look, Micromachines 10 (8) (2019) 500.

[76] M. Cimafonte, et al., Screen printed based impedimetric immunosensor for rapid detection of *Escherichia coli* in drinking water, Sensors 20 (1) (2020) 274.

[77] Z. Altintas, et al., A fully automated microfluidic-based electrochemical sensor for real-time bacteria detection, Biosens. Bioelectron. 100 (2018) 541–548.

[78] N. Idil, et al., Whole cell based microcontact imprinted capacitive biosensor for the detection of *Escherichia coli*, Biosens. Bioelectron. 87 (2017) 807–815.

[79] K.Y. Chan, et al., Ultrasensitive detection of *E. coli* O157:H7 with biofunctional magnetic bead concentration via nanoporous membrane based electrochemical immunosensor, Biosens. Bioelectron. 41 (2013) 532–537.

[80] V.K. Thanh Ngo, et al., Quartz crystal microbalance (QCM) as biosensor for the detecting of *Escherichia coli* O157:H7, Adv. Nat. Sci. Nanosci. Nanotechnol. 5 (4) (2014) 045004.

[81] F. Tong, Y. Lian, J. Han, On-line monitoring the growth of *E. coli* or HeLa cells using an annular microelectrode piezoelectric biosensor, Int. J. Environ. Res. Public Health **13** (12) (2016) 1254.

[82] X. Yu, Rapid and Sensitive Detection of *Escherichia coli* O157:H7 Using a QCM Sensor Based on Aptamers Selected by Whole-Bacterium SELEX and a Multivalent Aptamer System, 2018, Theses and Dissertations Retrieved from https://scholarworks.uark.edu/etd/2776.

[83] B.D. Ventura, et al., Quartz crystal microbalance sensors: new tools for the assessment of organic threats to the quality of water, in: The Handbook of Environmental Chemistry, Springer, 2019, pp. 1–28.

[84] A. Nehra, et al., Determination of *E. coli* by a graphene oxide-modified quartz crystal microbalance, Anal. Lett. 50 (12) (2017) 1897–1911.

[85] G.A. Campbell, R. Mutharasan, A method of measuring *Escherichia coli* O157:H7 at 1 cell/mL in 1 liter sample using antibody functionalised piezoelectric-excited millimeter-sized cantilever sensor, Environ. Sci. Technol. 41 (2007) 1669–1674.

[86] C. Ruan, et al., Magnetoelastic immunosensors: amplified mass immunosorbent assay for detection of *Escherichia coli* O157:H7, Anal. Chem. 75 (23) (2003) 6494–6498.

[87] Q. Lu, et al., Wireless, remote-query, and high sensitivity *Escherichia coli* O157:H7 biosensor based on the recognition action of concanavalin A, Anal. Chem. 81 (14) (2009) 5846–5850.

[88] J. Zhang, H.F. Ji, An anti *E. coli* O157:H7 antibody-immobilized microcantilever for the detection of *Escherichia coli* (*E. coli*), Anal. Sci. 20 (4) (2004) 585–587.

[89] S. Leahy, Y. Lai, A cantilever biosensor based on a gap method for detecting *Escherichia coli* in real time, Sens. Actuators B 246 (2017) 1011–1016.

[90] J. Liu, et al., An integrated impedance biosensor platform for detection of pathogens in poultry products, Sci. Rep. 8 (1) (2018) 16109.

[91] N.F.D. Silva, et al., Electrochemical biosensors for *Salmonella*: state of the art and challenges in food safety assessment, Biosens. Bioelectron. 99 (2018) 667–682.

[92] S. Cinti, et al., Electrochemical biosensors for rapid detection of foodborne salmonella: a critical overview, Sensors 17 (8) (2017) 1910.

[93] P. Pashazadeh, et al., Nano-materials for use in sensing of *Salmonella* infections: recent advances, Biosens. Bioelectron. 87 (2017) 1050–1064.

[94] S. Savas, et al., Nanoparticle enhanced antibody and DNA biosensors for sensitive detection of *Salmonella*, Materials 11 (9) (2018) 1541.

[95] N. Kim, I.-S. Park, W.-Y. Kim, *Salmonella* detection with a direct-binding optical grating coupler immunosensor, Sens. Actuators B 121 (2) (2007) 606–615.

[96] J. Hu, et al., Rapid screening and quantitative detection of *Salmonella* using a quantum dot nanobead-based biosensor, Analyst 145 (6) (2020) 2184–2190.

[97] L. Hao, et al., A microfluidic biosensor based on magnetic nanoparticle separation, quantum dots labeling and mno$_2$ nanoflower amplification for rapid and sensitive detection of *Salmonella typhimurium*, Micromachines 11 (3) (2020) 281.

[98] L. Wang, et al., Rapid and sensitive detection of *Salmonella typhimurium* using nickel nanowire bridge for electrochemical impedance amplification, Talanta 211 (2020) 120715.

[99] I. Jasim, et al., An impedance biosensor for simultaneous detection of low concentration of *Salmonella* serogroups in poultry and fresh produce samples, Biosens. Bioelectron. 126 (2019) 292–300.

[100] L. Zheng, et al., Optical biosensor for rapid detection of *Salmonella typhimurium* based on porous gold@platinum nanocatalysts and a 3D fluidic chip, ACS Sens. 5 (1) (2020) 65–72.

[101] Y. Zhang, et al., Label-free visual biosensor based on cascade amplification for the detection of *Salmonella*, Anal. Chim. Acta 1075 (2019) 144–151.

[102] B.-K. Oh, et al., Surface plasmon resonance immunosensor using self-assembled protein G for the detection of *Salmonella paratyphi*, J. Biotechnol. 111 (1) (2004) 1–8.

[103] J.W. Waswa, C. Debroy, J. Irudayaraj, Rapid detection of *Salmonella enteritidis* and *Escherichi coli* using surface plasmon resonance biosensor, J. Food Process Eng. 29 (4) (2006) 373–385.

[104] S.H. Hyeon, W.K. Lim, H.J. Shin, Novel surface plasmon resonance biosensor that uses full-length Det7 phage tail protein for rapid and selective detection of *Salmonella enterica* serovar *Typhimurium*, Biotechnol. Appl. Biochem. (2020).

[105] V. Nandakumar, et al., A methodology for rapid detection of *Salmonella typhimurium* using label-free electrochemical impedance spectroscopy, Biosens. Bioelectron. 24 (4) (2008) 1039–1042.

[106] J. Liu, et al., A microfluidic based biosensor for rapid detection of Salmonella in food products, PLoS One 14 (5) (2019) e0216873.

[107] B. Wu, et al., Nuclear magnetic resonance biosensor based on streptavidin–biotin system and poly-L-lysine macromolecular targeted gadolinium probe for rapid detection of *Salmonella* in milk, Int. Dairy J. 102 (2020) 104594.

[108] Y. Hou, et al., An ultrasensitive biosensor for fast detection of *Salmonella* using 3D magnetic grid separation and urease catalysis, Biosens. Bioelectron. 157 (2020) 112160.

[109] L. Wang, et al., An ultrasensitive biosensor for colorimetric detection of *Salmonella* in large-volume sample using magnetic grid separation and platinum loaded zeolitic imidazolate Framework-8 nanocatalysts, Biosens. Bioelectron. 150 (2020) 111862.

[110] Y.Y. Wong, et al., Immunosensor for the differentiation and detection of *Salmonella* species based on a quartz crystal microbalance, Biosens. Bioelectron. 17 (8) (2002) 676–684.

[111] X.-L. Su, Y. Li, A QCM immunosensor for *Salmonella* detection with simultaneous measurements of resonant frequency and motional resistance, Biosens. Bioelectron. 21 (6) (2005) 840–848.

[112] A. Fulgione, et al., QCM-based immunosensor for rapid detection of *Salmonella typhimurium* in food, Sci. Rep. 8 (1) (2018) 16137.

[113] B.L. Weeks, et al., A microcantilever-based pathogen detector, Scanning 25 (6) (2003) 297–299.

[114] Q. Zhu, W.Y. Shih, W.-H. Shih, In situ, in-liquid, all-electrical detection of *Salmonella typhimurium* using lead titanate zirconate/gold-coated glass cantilevers at any dipping depth, Biosens. Bioelectron. 22 (12) (2007) 3132–3138.

[115] C. Ricciardi, et al., Online portable microcantilever biosensors for Salmonella enterica serotype enteritidis detection, Food Bioproc. Tech. 3 (2010) 956–960.

[116] M.K. Park, B.A. Chin, Novel approach of a phage-based magnetoelastic biosensor for the detection of *Salmonella enterica* serovar *Typhimurium* in soil, J. Microbiol. Biotechnol. 26 (2016) 12.

[117] X. Yang, J. Kirsch, A. Simonian, Campylobacter spp. detection in the 21st century: a review of the recent achievements in biosensor development, J. Microbiol. Methods 95 (1) (2013) 48–56.

[118] S.C. Ricke, et al., Developments in rapid detection methods for the detection of foodborne *Campylobacter* in the United States, Front. Microbiol. 9 (2019) 3280.

[119] N.A. Masdor, Z. Altintas, I.E. Tothill, Sensitive detection of *Campylobacter jejuni* using nanoparticles enhanced QCM sensor, Biosens. Bioelectron. 78 (2016) 328–336.

[120] H. Wang, et al., Rapid and sensitive detection of *Campylobacter jejuni* in poultry products using a nanoparticle-based piezoelectric immunosensor integrated with magnetic immunoseparation, J. Food Prot. 81 (8) (2018) 1321–1330.

[121] N.A. Masdor, et al., Subtractive inhibition assay for the detection of *Campylobacter jejuni* in chicken samples using surface plasmon resonance, Sci. Rep. 9 (1) (2019) 13642.

[122] K.E. Sapsford, et al., Detection of *Campylobacter* and *Shigella* species in food samples using an array biosensor, Anal. Chem. 76 (2) (2004) 433–440.

[123] L. Fabiani, et al., Development of a sandwich ELIME assay exploiting different antibody combinations as sensing strategy for an early detection of *Campylobacter*, Sens. Actuators B 290 (2019) 318–325.

[124] E. Núñez-Carmona, M. Abbatangelo, V. Sberveglieri, Innovative sensor approach to follow *Campylobacter jejuni* development, Biosensors 9 (1) (2019) 8.

[125] E. Núñez-Carmona, et al., Nanostructured MOS sensor for the detection, follow up, and threshold pursuing of *Campylobacter jejuni* development in milk samples, Sensors 20 (7) (2020) 2009.

[126] L. Ma, M. Petersen, X. Lu, Identification and antimicrobial susceptibility testing of *Campylobacter* using a microfluidic lab-on-a-chip device, Appl. Environ. Microbiol. 86 (2020) AEM.00096-20.

[127] F. Cecchini, et al., *Vibrio cholerae* detection: traditional assays, novel diagnostic techniques and biosensors, TrAC Trends Anal. Chem. 79 (2016) 199–209.

[128] A. Narmani, et al., Highly sensitive and accurate detection of *Vibrio cholera* O1 OmpW gene by fluorescence DNA biosensor based on gold and magnetic nanoparticles, Process Biochem. 65 (2018) 46–54.

[129] K.-F. Low, Z.M. Zain, C.Y. Yean, A signal-amplified electrochemical DNA biosensor incorporated with a colorimetric internal control for *Vibrio cholerae* detection using shelf-ready reagents, Biosens. Bioelectron. 87 (2017) 256–263.

[130] Y. Wang, et al., Nanoparticle-based lateral flow biosensor combined with multiple cross displacement amplification for rapid, visual and sensitive detection of *Vibrio cholerae*, FEMS Microbiol. Lett. 364 (23) (2017) fnx234.

[131] M. Rahman, et al., Ultrasensitive biosensor for the detection of *Vibrio cholerae* DNA with polystyrene-co-acrylic acid composite nanospheres, Nanoscale Res. Lett. 12 (1) (2017) 474.

[132] N. Khemthongcharoen, et al., Piezoresistive microcantilever-based DNA sensor for sensitive detection of pathogenic *Vibrio cholerae* O1 in food sample, Biosens. Bioelectron. 63 (2015) 347–353.

[133] T. Ramamurthy, et al., Diagnostic techniques for rapid detection of *Vibrio cholerae* O1/O139, Vaccine 38 (2020) A73–A82.

[134] S. Achtsnicht, et al., Sensitive and rapid detection of cholera toxin subunit B using magnetic frequency mixing detection, PLoS ONE 14 (7) (2019) e0219356.

[135] A.E. Valera, et al., On-chip electrochemical detection of cholera using a polypyrrole-functionalized dendritic gold sensor, ACS Sensors 4 (3) (2019) 654–659.

[136] Q. Palomar, et al., Controlled carbon nanotube layers for impedimetric immunosensors: high performance label free detection and quantification of anti-cholera toxin antibody, Biosens. Bioelectron. 97 (2017) 177–183.

[137] R. Briquaire, et al., Application of a paper based device containing a new culture medium to detect *Vibrio cholerae* in water samples collected in Haiti, J. Microbiol. Methods 133 (2017) 23–31.

[138] W. Chen, et al., Development of an immunochromatographic lateral flow device for rapid diagnosis of *Vibrio cholerae* O1 serotype Ogawa, Clin. Biochem. 47 (6) (2014) 448–454.

[139] Y. Li, et al., An electrochemical strategy using multifunctional nanoconjugates for efficient simultaneous detection of *Escherichia coli* O157:H7 and *Vibrio cholerae* O1, Theranostics 7 (4) (2017) 935–944.

[140] R.A. Taheri, et al., Evaluating the potential of an antibody against recombinant OmpW antigen in detection of *Vibrio cholerae* by surface plasmon resonance (SPR) biosensor, Plasmonics 12 (5) (2017) 1493–1504.

[141] P. Zamani, et al., A luminescent hybridoma-based biosensor for rapid detection of *V. cholerae* upon induction of calcium signaling pathway, Biosens. Bioelectron. 79 (2016) 213–219.

[142] K.N. Clayton, et al., Particle diffusometry: an optical detection method for *Vibrio cholerae* presence in environmental water samples, Sci. Rep. 9 (1) (2019) 1739.

[143] H.R. Al-abodi, et al., Novel gold nanobiosensor platforms for rapid and inexpensive detection of *Vibrio cholerae*, Rev. Med. Microbiol. 31 (2) (2020) 70–74.

[144] M.-S. Song, et al., Detecting and discriminating *Shigella sonnei* using an aptamer-based fluorescent biosensor platform, Molecules (Basel, Switzerland) **22** (5) (2017) 825.

[145] S.S. Zarei, S. Soleimanian-Zad, A.A. Ensafi, An impedimetric aptasensor for *Shigella dysenteriae* using a gold nanoparticle-modified glassy carbon electrode, Microchim. Acta 185 (12) (2018) 538.

[146] J. Feng, et al., Naked-eyes detection of *Shigella flexneri* in food samples based on a novel gold nanoparticle-based colorimetric aptasensor, Food Control 98 (2019) 333–341.

[147] R. Xiao, et al., Portable evanescent wave fiber biosensor for highly sensitive detection of *Shigella*, Spectrochim. Acta A Mol. Biomol. Spectrosc. 132 (2014) 1–5.

[148] N. Elahi, et al., A fluorescence nano-biosensors immobilization on iron (MNPs) and gold (AuNPs) nanoparticles for detection of *Shigella* spp, Mater. Sci. Eng. C 105 (2019) 110113.

[149] S. Jain, et al., Current and emerging tools for detecting protozoan cysts and oocysts in water, TrAC Trends Anal. Chem. 121 (2019) 115695.

[150] A. Rousseau, et al., Assessing viability and infectivity of foodborne and waterborne stages (cysts/oocysts) of *Giardia duodenalis*, *Cryptosporidium* spp., and *Toxoplasma gondii*: a review of methods, Parasite (Paris, France) **25** (2018) 14.

[151] C.D. Kang, et al., Surface plasmon resonance-based inhibition assay for real-time detection of *Cryptosporidium parvum* oocyst, Water Res. 42 (6–7) (2008) 1693–1699.

[152] C.D. Kang, et al., Performance enhancement of real-time detection of protozoan parasite, *Cryptosporidium* oocyst by a modified surface plasmon resonance (SPR) biosensor, Enzyme Microb. Technol. 39 (3) (2006) 387–390.

[153] H. Li, R. Bashir, Dielectrophoretic separation and manipulation of live and heat-treated cells of *Listeria* on microfabricated devices with interdigitated electrodes, Sens. Actuators B 86 (2–3) (2002) 215–221.

[154] H.V. Smith, R.A.B. Nichols, *Cryptosporidium*: detection in water and food, Exp. Parasitol. 124 (1) (2009) 61–79.

[155] H. Bridle, et al., Static mode microfluidic cantilevers for detection of waterborne pathogens, Sensors Actuators A Phys. 247 (2016) 144–149.

[156] G.A. Campbell, R. Mutharasan, Near real-time detection of *Cryptosporidium parvum* oocyst by IgM-functionalized piezoelectric-excited millimeter-sized cantilever biosensor, Biosens. Bioelectron. 23 (7) (2008) 1039–1045.

[157] S. Xu, R. Mutharasan, Detection of Cryptosporidium parvum in buffer and in complex matrix using PEMC sensors at 5 oocysts mL^{-1}, Anal. Chim. Acta 669 (1–2) (2010) 81–86.

[158] S. Xu, R. Mutharasan, Rapid and sensitive detection of *Giardia lamblia* using a piezoelectric cantilever biosensor in finished and source waters, Environ. Sci. Technol. 44 (5) (2010) 1736–1741.

[159] G. Luka, et al., Label-free capacitive biosensor for detection of *Cryptosporidium*, Sensors 19 (2) (2019) 258.

[160] H. Ilkhani, H. Zhang, A. Zhou, A novel three-dimensional microTAS chip for ultra-selective single base mismatched *Cryptosporidium* DNA biosensor, Sens. Actuators B 282 (2019) 675–683.

[161] A. Iqbal, et al., Detection of *Cryptosporidium parvum* oocysts on fresh produce using DNA aptamers, PLoS ONE 10 (9) (2015) e0137455.

[162] A. Iqbal, et al., Development and application of DNA-aptamer-coupled magnetic beads and aptasensors for the detection of *Cryptosporidium parvum* oocysts in drinking and recreational water resources, Can. J. Microbiol. 65 (11) (2019) 851–857.

[163] D. Gavriilidou, H. Bridle, High capture efficiency of lectin surfaces for *Cryptosporidium parvum* biosensors, Biochem. Eng. J. 135 (2018) 79–82.

[164] H. Wang, et al., A piezoelectric immunoagglutination assay for *Toxoplasma gondii* antibodies using gold nanoparticles, Biosens. Bioelectron. 19 (7) (2004) 701–709.

[165] Y. Ding, et al., Enzyme-catalyzed amplified immunoassay for the detection of *Toxoplasma gondii*-specific IgG using Faradaic impedance spectroscopy, CV and QCM, Anal. Bioanal. Chem. 382 (2005) 1491–1499.

[166] A.H. Khan, R. Noordin, Serological and molecular rapid diagnostic tests for *Toxoplasma* infection in humans and animals, Eur. J. Clin. Microbiol. Infect. Dis. 39 (1) (2020) 19–30.

[167] G.D. Liu, et al., Renewable amperometric immunosensor for *Schistosoma japonium* antibody assay, Anal. Chem. 73 (14) (2001) 3219–3226.

[168] Z. Wen, et al., A novel liquid-phase piezoelectric immunosensor for detecting *Schistosoma japonicum* circulating antigen, Parasitol. Int. 60 (3) (2011) 301–306.

[169] Z.Y. Wu, et al., Quartz-crystal microbalance immunosensor for *Schistsoma-japonicum*-infected rabbit serum, Anal. Sci. 19 (3) (2003) 437–440.

[170] A. van Remoortere, et al., Profiles of immunoglobulin M (IgM) and IgG antibodies against defined carbohydrate epitopes in sera of *Schistosoma*-infected individuals determined by surface plasmon resonance, Infect. Immun. **69** (4) (2001) 2396–2401.

[171] V. Gaudin, Advances in biosensor development for the screening of antibiotic residues in food products of animal origin—a comprehensive review, Biosens. Bioelectron. 90 (2017) 363–377.

[172] A. Mehlhorn, P. Rahimi, Y. Joseph, Aptamer-based biosensors for antibiotic detection: a review, Biosensors 8 (2) (2018) 54.

[173] A. Schumacher, et al., In vitro antimicrobial susceptibility testing methods: agar dilution to 3D tissue-engineered models, Eur. J. Clin. Microbiol. Infect. Dis. 37 (2) (2018) 187–208.

[174] A.A. Adegoke, G. Singh, T.A. Stenström, Biosensors for monitoring pharmaceutical nanocontaminants and drug resistant bacteria in surface water, subsurface water and wastewater effluent for reuse (Chapter 16), in: A.M. Grumezescu (Ed.), Nanoparticles in Pharmacotherapy, William Andrew Publishing, 2019, pp. 525–559.

[175] S. Hannah, et al., Rapid antibiotic susceptibility testing using low-cost, commercially available screen-printed electrodes, Biosens. Bioelectron. 145 (2019) 111696.

[176] P.I. Reyes, et al., Magnesium zinc oxide nanostructure-modified quartz crystal microbalance for dynamic monitoring of antibiotic effects and antimicrobial resistance, Procedia Technol. 27 (2017) 46–47.

[177] P. Nag, et al., Beta-lactam antibiotics induced bacteriolysis on LSPR sensors for assessment of antimicrobial resistance and quantification of antibiotics, Sens. Actuators B 311 (2020) 127945.

[178] A. Butterworth, et al., SAM composition and electrode roughness affect performance of a DNA biosensor for antibiotic resistance, Biosensors 9 (1) (2019) 22.

[179] N. Kumar, et al., Dielectrophoresis assisted rapid, selective and single cell detection of antibiotic resistant bacteria with G-FETs, Biosens. Bioelectron. 156 (2020) 112123.

Chapter 8

Molecular methods for the detection of waterborne pathogens

Kimberley Gilbride

Department of Chemistry and Biology, Ryerson University, Toronto, ON, Canada

8.1. Pathogen detection

Many molecular methods have been developed over the last 20 years and they can provide rapid, sensitive, and quantitative tools for the detection of infectious agents in clinical samples and evaluation of the microbiological quality of food and water. In many cases, they do not require the isolation or culturing of the microorganism in order to provide qualitative and quantitative information about the presence and enumeration of the target organism in the samples. Furthermore, they can be designed to specifically detect a specific virulent trait carried by the organism. The varieties of techniques that have been designed can be used to detect a range of target molecules, each with the goal of providing information for the prevention of infectious diseases. Much of the research to date has focused on detection of pathogens in clinical samples where pathogen concentrations are high and pathogen identification is already suspected. Many of these techniques are now being modified to be used in environmental settings where the detection in samples is complicated by mixed communities, interfering contaminants, and unknown pathogenic targets. In addition, the application of molecular methods to drinking and recreational water testing has the added challenge of low levels of pathogens which requires enrichment or concentration of the samples before detection processing.

This chapter will review the molecular methods that have been studied for the detection of viruses, bacteria, protozoa and cyanobacteria in raw and processed water samples. In some cases they facilitate the identification, genotyping, enumeration, viability assessment, and source-tracking of human and animal contamination. Limitations of each method will also be reviewed in the light of the adoption of the method for the use in routine water monitoring.

Waterborne Pathogens: Detection Methods and Applications. https://doi.org/10.1016/B978-0-444-64319-3.00008-3

8.2. Why molecular methods?

Water is a tasteless, odourless, and colourless liquid that is an excellent solvent and found in all organic tissues. Life as we know it would not exist without water since it makes up ~ 75% of the volume of the human body. Therefore, we require a continuous supply of potable water–water that is safe to drink. Ensuring that water is chemically and biologically free from pollutants that can cause illness or disease requires knowledge about the potential contaminant, as well as testing methods that are both specific and sensitive.

On the biological side, microbial water quality monitoring tests requires the detection and identification of the waterborne pathogens. Current culture-dependent microbial analysis is not practical to routinely detect all the various potential pathogens. Aside from the large number of possible pathogens (bacteria, protozoa, viruses, etc.), many do not grow on laboratory media, do not transport well from the water source to the lab, and require tests that are expensive and time-consuming. Therefore, traditionally, water quality monitoring has been based on the presence or absence of indicator organisms (organisms whose presence indicates faecal contamination) and the potential presence of pathogens but not necessarily the presence of a specific pathogen [1, 2]. An ideal indicator organism would possess the following criteria [3]:

1. The organism should be present whenever enteric pathogens are present, and absent when the pathogens are absent.
2. The organism should be useful for all types of water.
3. The organism should have a life span similar to the pathogen.
4. The organism should not reproduce in water.
5. Should be harmless to humans.
6. Testing should have a high level of specificity and sensitivity.
7. Testing should be easy to perform and not too costly.

Coliforms and faecal coliforms have been recognised as good indicators of faecal contamination and possess all the qualities of an 'ideal' indicator most of the time. *Escherichia coli* is the classic microbiological indicator and for the most part, still the most commonly analysed to evaluate the level of faecal contamination [4, 5]. Other indicator organisms that have been used include faecal *Enterococci* [6], *Clostridium perfringens* spores [7], and somatic and male-specific coliphages [8–10].

E. coli and other bacterial indicators have been found to have good correlation with most bacterial pathogens, however, several studies have shown that both *E. coli* and *Enterococci* are not well correlated with *Salmonella* or *Campylobacter* [3, 11]. Recent studies have also questioned the suitability of bacterial indicators for viral or protozoan pathogens [5, 12–15]. Therefore, all presently used indicators fall short on one or more of the criteria as an ideal indicator organism. Furthermore, culture-based methods can be lengthy to perform, labour intensive and unable to detect unculturable species [16]. Their limitations have sparked a renewed quest to develop alternative methods to ensure absolute safety of drinking water. The development and use of molecular methods have recently shown great promise as fast, sensitive, specific, and quantitative ways to detect waterborne pathogens.

8.2.1 Waterborne pathogens

Waterborne pathogens fall into several groups of organisms, including bacteria, cyanobacteria, protozoa, viruses, and helminths. Although helminths are an interesting pathogenic group of parasites, this chapter will concentrate on the microbial pathogens included in the other four groups.

Currently there is no universal method to collect, detect, and identify all pathogenic microbes from a water sample. Along with the use of indicator organisms that may or may not correlate well with pathogen levels, the collection of the sample can also influence the ability to detect the pathogen. For example, detection of some viral pathogens from drinking water may require the concentration of up to 1000 L [17]. In the case of protozoan detection, operator skill is paramount as detection relies on manual scanning of concentrated samples on microscopic slides, and the identification of suspect objects as cysts and oocysts using further specialised microscopy [18].

Molecular methods can provide sensitive, rapid, and quantitative analytical tools for detecting specific pathogens, including emergent strains and indicators. They have been used to evaluate food and water, both raw and treated/processed. The advantages of molecular methods are many. First, they circumvent the need to culture the pathogens, a procedure that is not efficient for some organisms. They can be very specific for particular species and provide further phylogenetic information about the strains. In some cases, strains can be used to track the identification of the source of the contamination in the water source [3, 19]. Owing to an amplification step in most molecular methods, very minute amounts of pathogens can be detected. Lastly, many molecular methods lend themselves to the possibility of automation which can lead to real-time analysis and provide information for microbial risk assessment purposes [4].

However, there still exist several disadvantages of the molecular methods that need to be overcome for the methods to be adopted as viable routine tests. First, although much research has looked at detection of pathogens by molecular methods, there exists no standardisation of protocols. For instance, with PCR based protocols, the primer/probes, conditions, and even the extraction of DNA from the samples vary from lab to lab with varying detection success [5]. Most of the molecular methods require concentration of the sample so suitable volumes can be analysed. In the case of viruses and protozoa, the infective does is very small requiring large volumes to be concentrated to ensure the pathogens are detected. Second, since most molecular methods rely on detection of a single macromolecule and in most of the cases, RNA or DNA, the molecule of choice needs to be extracted from the sample. The variability within the samples themselves can create problems with the initial extraction of the nucleic acid. Contaminants in the samples, for example, humic acids or metals in environmental samples, can inhibit subsequent amplification and produce false negatives. Therefore, comparison of gene numbers from different samples is problematic [20]. As the appearance and concentration of inhibitors are not necessarily known beforehand, it is left to the operator to determine the best extraction method on a sample to sample basis which is time consuming. Lastly, detection of genomic sequences in a sample is not a definitive test for infectivity since DNA can be found in samples outside of cells, in dead cells or in cells in a viable but not culturable state (VTNC) [17, 20].

Albeit the cons to molecular method, much research is still underway to overcome the disadvantages and identify suitable detection targets in pathogens. Molecular methods definitely hold much promise as techniques that can be validated and standardised for use as routine water testing assays of the future. In general, molecular tools have shifted the paradigm from detection of indicator organisms on culture media to direct detection and enumeration of known pathogens, unculturable pathogens, and emerging pathogens.

8.3. Molecular methods

Molecular methods are based on the ability to isolate and detect specific molecules carried by pathogenic organisms. Several techniques for the analysis of microbial communities and identification of bacterial species have been investigated. The development of many taxonomic tools has been based on the composition of fatty acids, peptides, or genomic material contained in the organism. The following sections will review the techniques that have been developed using each of the cellular macromolecules as their basis.

8.3.1   Fatty acids

Qualitatively and quantitatively, fatty acid composition in bacteria is highly conserved and quite stable if culture conditions are highly standardised. The use of fatty acids as a taxonomic tool has been discussed for almost 50 years, and have been successfully applied to various taxa for over 20 years [21]. Differences in fatty acid chain lengths, positions of double bonds, and binding of functional groups potentially make them useful for identification purposes.

Whole cell fatty acid methyl ester (FAME) analysis can be carried out by gas chromatography. Lipski et al. analysed the fatty acid profiles of nitrite-oxidising species and found that each of the genera, *Nitrobacter*, *Nitrococcus*, *Nitrospina*, and *Nitrospira* had a distinct profile. Furthermore, specific fatty acids could be assigned to particular species, for example, 11-methyl-hexadecanoic acid was detected only in *Nitrospira moscoviensis* and can be used as potential lipid marker for detection of this species in environmental and enrichment samples. Cluster analysis of the fatty acid profiles for the organism were in accordance with 16s rRNA sequence-based phylogeny for the nitrite-oxidising bacteria [22].

Fame analysis has also been used to identify and quantify individual fatty acids that are useful to assess the relative differences and similarities among and between microbial communities from ground water [23]. Fig. 8.1 shows the gas chromatograph results obtained from one ground water well on two different occasions. Changes in the profile are attributed to differences in the composition of the communities.

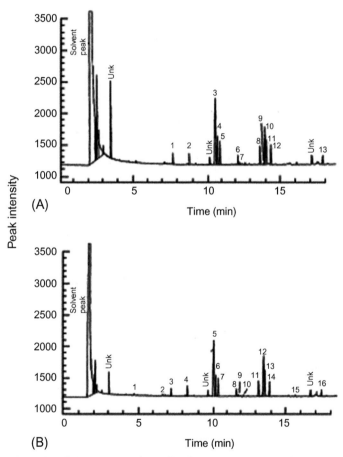

FIG. 8.1 Examples of FAME community profiles from groundwater samples. Community profile from well 63D (A) sampled 6/20/96; (1) 14:0, (2) 15:0 Anteiso, (3) 16:1ω7c/15 Iso 2OH, (4) 16:1ω5c, (5) 16:0, (6) 17:0 Anteiso, (7) 17:1ω8c, (8) 18:3ω6c, (9) 18:2ω6,9c/18:0 Anteiso, (10) 18:1ω9c, (11) 18:1ω7c/ω9t/ω12t, (12) 18:0, (13) 20:0. (B) Sampled 8/15/96; (1) 12:0, (2) 13.566 (unknown), (3) 14:0, (4) 15:0 Anteiso, (5) 16:1ω7c/15 Iso 2OH, (6) 16:1ω5c, (7) 16:0, (8) 17:0 Anteiso, (9) 17:1ω8c, (10) 17:0, (11) 18:3ω6c, (12) 18:1ω9c, (13) 18:1ω7c/ω9t/ω12t, (14) 18:0, (15) 19:0, (16) 20:0.*(Reproduced with permission from A.M. Glucksman, H.D. Skipper, R.L. Brigmon, J.W.S. Domingo, Use of the MIDI-FAME technique to characterize groundwater communities, J. Appl. Microbiol. 88(4) (2000) 711–719.)*

The main advantage of fatty acid methyl ester (FAME) analysis is that it is cheap, easy, and can be automated. However, most of the studies to date have used fatty acid to generate community profiles and monitor changes in community structure over time or space and not to detect individual strains in samples. Parveen et al. [24] found that FAME was not sensitive enough to differentiate between *E. coli* from human and nonhuman sources. Using fatty acid isopropyl ester (FAPEs) analysis which has a simpler protocol than FAMEs, microbial community structure was monitored in an activate sludge secondary waste water treatment system [25]. Werker et al. [25] found that the method allowed for the generation of meaningful and quantitative data for changes in activated sludge quality but could not be directly compared to FAME profiles without a correction factor.

The main disadvantage of FAME analysis for detection and identification of individual strains lies in the power of resolution of the method. Analysis is limited due to the restricted number of strains and FAME profiles available in the BAME@LMG database. Currently, only about 50% of published strains have standard FAME profiles [21]. To circumvent this problem, several recent studies have incorporating machine learning techniques to allow bacterial identification at the species level using FAME data. Machine learning techniques are defined as the design and development of algorithms that allow one to use computational models to predict an outcome based on empirical data, such as from a FAME database. Slabbinck et al. [21]were able to identify species in the three genera, *Bacillus*, *Paenibacillus*, and *Pseudomonas* using a FAME database and a laboratory information management system along with three different machine learning techniques: artificial neural networks (ANN), support vector machines (SVM), and random forests (RF). This approach was superior to the commercially available FAME analysis system, Sherlock microbial identification system [21].

Although the techniques are promising for taxonomic purposes, the detection of faecal coliforms from environmental and water samples is still in the investigative stage and they are a long way from use as routine tests for the detection and identification of pathogens from water samples. Furthermore, more investigation is required to determine the correlation between FAME data and 16s rRNA data at the genus and species level.

8.3.2 Proteins

Proteins are the most abundant macromolecule in the cell comprising over half of the biomass in microbial cells. They represent the structure, composition, and function associated with individual cells. Identification of individual protein molecules that represent unique pathogenic characteristics could be used as identification of microbes in a sample. Several approaches have been taken to develop a fast, low volume, and automated methods based on microbial protein composition, including mass spectrometry and immunological techniques.

8.3.2.1 Mass spectrometry

Mass spectrometry (MS) is an analytical technique that can be used for determining the mass of a particle, for determining the elemental composition of a sample or molecule, and for determining the chemical structures of molecules, such as peptides and other chemical compounds. It is able to do this by measuring the mass-to-charge ratio of charged particles. The machine uses an ion source to generate charged molecules, then it sorts the ions by their mass and with the use of a detector measures the value of an indicator quantity and thus provides data for calculating the abundances of each ion present. The technique has both qualitative and quantitative uses. These include identifying unknown compounds, determining the isotopic composition of elements in a molecule, and determining the structure of a compound by observing its fragmentation. Other uses include quantifying the amount of a compound in a sample. This technique is widely used with several modifications depending on the application.

Analysis of pathogenic and opportunistic free-living amoeba have been successfully differentiated between using matrix-assisted laser desorption/ionisation mass spectrometry (MALDI-TOF MS). Moura et al. [26] were able to generate 20 biomarker fingerprints for *Acanthaamoeba*, *Balamuthia*, and *Naegleria fowleri* with this method in < 15 min. The method was extended by Donohue et al. [27] to describe the detection of waterborne pathogen *Aeromonas* where they found that the mass spectra data contained enough information to distinguish between genera, species, and strains. The limitations of this method are that variations in culture conditions and analytical parameters can significantly influence the mass spectra. To use this method for routine identification of pathogens from unknown samples will require an extensive increase in the number and diversity of references strains catalogued in the mass spectra library [27].

De Bruyne et al. [28] applied MALDI-TOF MS with machine learning techniques (support vector machines (SVMs) and random forests (RFs)) to identify species within the genera *Leuconostoc*, *Fructobacillus*, and *Lactobacillus*. The two machine learning techniques improved the identification accuracy by 4%-10% to 94%, and 98% for *Leuconostoc* and *Fructobacillu*s, respectively. It has also been used to successfully identity *Aeromonas* isolates from environmental water samples [27]. The amount of the sample that can be analysed was found to be critical and identification of isolates can be limited by the availability of the peptide mass fingerprints of type strains in the spectral database. The ability to identify bacteria and many other organisms very rapidly, accurately, and relatively inexpensively makes MALDI-TOF an emerging technology for the future [29].

Furthermore, the recent fabrication of titanium bacterial chips (TBC) for use with MALDI-MS has shown that they can be used as bacterial sensors to directly analyse captured cells [30]. The lowest detectable concentration of bacterial cells was 1×10^4 CFU/mL. The device has yet to be tried on environmental samples for the ability to capture bacterial cells.

8.3.2.2 Immunological techniques

Antibody-based techniques have also been investigated for their ability to detect bacteria in environmental samples [31]. The development of several immunological methods for the detection and enumeration of pathogenic organisms in water sources has become a more recent focus. Most of these studies have targeted either the O or H antigens of *E. coli* O157. Albeit the limitation that *E. coli* O157 may lose its O antigenicity under starvation conditions and still retain its ability to produce toxin causing false negatives [31], many studies have shown successful immunological detection of the pathogen from water samples.

8.3.2.3 ELISA

Enzyme-linked immunosorbent assay (ELISA) can be used to detect and measure either antibody or antigen. The most widely used versions of the ELISA for the detection of bacterial cells in suspensions (water, etc.) is competitive ELISA and the sandwich ELISA [32]. In the latter case, specific antibody is bound to the bottom of microtiter plates and then the sample is added to the wells. If the target antigen is present it will bind to the antibody and can be detected with a conjugate. A conjugate is another antibody attached to an enzyme that converts a supplied substrate to a colormetric, fluorescent, or chemilluminescent product that can be detected.

Park et al. [33, 34] has shown that a sandwich ELIZA assay could be used to detect *E. coli* O157:H7 and *Salmonella typhimurium*. They were able to detect 1.8×10^3 CFU/mL of *E. coli* within 30 min and 9.2×10^3 CFU/mL of *Salmonella* within 20 min.

Combining immunological detection with selective preenrichment can increase the recovery and detection of *E. coli* O157 [35, 36] and is necessary to obtain sensitive detection of contaminant organisms from samples [37]. This aspect of sample processing, along with other sample concentration and isolation methods, is discussed in Chapter 4.

Many immunology-based methods can be coupled with other methods for pathogen detection, for instance, immunomagnetic separation IMS on magnetic beads (Fig. 8.2) has been coupled with MALDI-TOF spectrometry for detection of *Staphylococcal* enterotoxin B, and with flow cytometry for detection of *Listeria monocytogenes* [38]. Setterington et al. [39] isolated *E. coli* O157:H7 cells via IMS and labelled them with biofunctionalised electroactive polyaniline (immuno-PANI). Subsequently, the labelled cell complexes were deposited onto a carbon electrode (SPCE) sensor and pulled to the electrode surface by an external magnetic field, to amplify the electrochemical signal generated by the polyaniline. The developed biosensor could detect as low as 7 CFU of *E. coli* O157:H7 in 70 min (from sampling to detection).

8.3.3 Nucleic acids

Nucleic acids (DNA and RNA) are polymers composed of nucleotides. DNA's role as the hereditary material is supported by its structure which is extremely

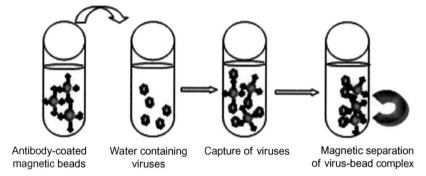

Antibody-coated   Water containing   Capture of viruses   Magnetic separation
magnetic beads   viruses   of virus-bead complex

FIG. 8.2 Immunomagnetic separation (IMS). Magnetic beads coated with antibodies can be used to bind to antigens on the surface of target microbes in water. This figure shows the capture of viral particles and then the concentration of the bead-attached pathogen by a magnet.*(Reproduced with permission from I.A. Hamza, L. Jurzik, K. Überla, M. Wilhelm, Methods to detect infectious human enteric viruses in environmental water samples, Int. J. Hyg. Environ. Health (2011) https://doi. org/10.1016/j.ijheh.2011.07.014.)*

stable and can be reproduced from generation to general with high fidelity. RNA is produced directly from DNA and can represent the metabolic activity of the cell. These molecules make excellent molecular target for detecting organisms and identifying specific phylogenetic sources. The following methods all are based on the detection of nucleic acids and take advantage of the use of this stable genotype characteristic in organisms.

8.3.3.1 FISH

Fluorescence in situ hybridisation (FISH) coupled with confocal scanning laser microscopy is a microscopic based method where samples can be placed on microscope slides and hybridised to fluorescently labelled DNA probes that will bind specifically to nucleic acid sequences in the cells. The presence and location of the probe can be visualised (Fig. 8.3.). If the probes have been designed properly, they can be specific for sequences unique to pathogens found in the samples. It is a powerful tool for detection and identification of specific organisms and can differentiation between dead and alive cells but lacks sensitivity and enrichment is often necessary [40]. Dilute samples may result in false negatives however the concentration of samples to increase sensitivity may increase inhibitor concentrations and the likelihood of inhibiting the hybridisation reaction thereby also causing false negatives.

The application of FISH for the detection of pathogens in water samples has met with varying success. Although detection of *E. coli* O157 has been successful from surface waters [35], wastewater samples [41], sewage [42, 43], clinical samples, and drinking water biofilms [44] it has not been successful with quantifying waterborne *E. coli* O157 from water samples with single-copy gene probes without an enrichment step [37].

Fluorescent in situ hybridisation

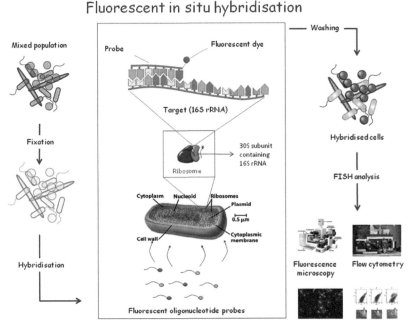

FIG. 8.3 Fluorescent in-situ hybridisation. A probe that will hybridise to a specific target DNA sequence can be labelled with a fluorophore. The labelled probe and the target DNA are then denatured and allowed to re-anneal. The labelled probes can be visualised under a microscope and the location (cell) harbouring the sequence can be detected.(*Image taken from http://www.biovisible. com/indexRD.php?page=fish.*)

8.3.3.2   PCR

The polymerase chain reaction (PCR) is a technique that allows specific sequences of DNA to be synthesised over and over again to produce many copies of the same sequence [45]. While one copy would be hard to detect among so many other DNA sequences, many copies of the same sequence can be easily detected. The amplification process is possible by using the ability of DNA polymerase to synthesise a new strand of DNA complementary to the offered template strand. Because DNA polymerase can add a nucleotide only onto a preexisting 3′-OH group, it requires a primer. This primer can be complementary to a portion of the target sequence or to a portion of the DNA adjacent to the target sequence to which it can add the first nucleotide. If two primers flank the target sequence, it is possible to delineate a specific region of template sequence that the researcher wants to amplify. The sequence can be made over and over again in each round of replication, which at the end of the PCR reaction, will be accumulated in millions of copies (amplicons) (Fig. 8.4).

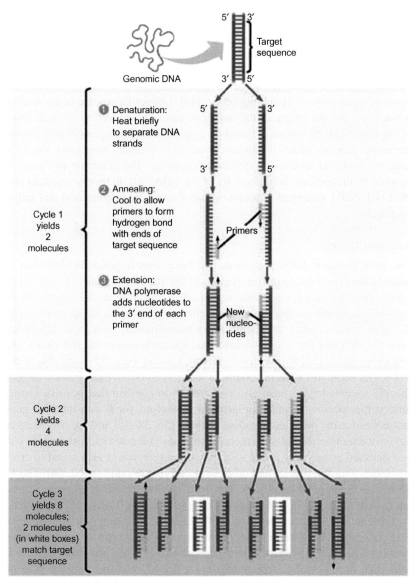

FIG. 8.4 A schematic of the polymerase chain reaction. Amplification of a region of DNA is accomplished by designing two primes that are complementary to the sequence flanking the target region. A heat stable polymerase extends the primers and produces copy of the target region. The dsDNA copy is denatured and synthesised once more. This cycle is repeated many times (often 30 cycles) which produces many copies of the target which can be detected (http://www.foodsafetywatch.com/public/1050.cfm).

Viral targets

For direct surveillance of viral pathogens with PCR, the DNA sequences that have been used for pathogen detect usually target a small portion of the viral genome such as the 145 bp fragment in the 5′ noncoding region (5′NCR) of Enterovirus [46], the vp7 gene of Rotavirus [47], and the hexon and fibre genes of adenovirus [48], among others [49]. Longer targets are being studied in the hope that the length of the amplicon can be correlated to the infectivity of the virus. This technique is based on amplification of large or in some cases the entire genome which if damaged would not produce amplicons since the primers would be attached to different fragments. The detection of damaged genome would indicate noninfectivity of the virus [50]. In reverse transcriptase PCR (RT-PCR), damaged genomes would also not be polymerised and fail to be detected.

Bacterial targets

The most common DNA sequences that have been targeted for detection of bacteria have been the conserved regions of the 16S rRNA gene [51–53]. These genes contain nine hypervariable regions that constitute useful targets for all types of diagnostic testing. As not enough variability exists between strains of the same species, other targets have also been studied, such as the 23srRNA gene [54, 55] and the internal transcribed spacer between the 16S rRNA and 23S rRNA genes [52, 56]. With enrichment, Lin and Tsen [57] were able to detect *E. coli* down to 1 CFU in 100 mL of water. Aside from these general targets, specific species targets have also been tested to confirm that positive testing reflects the specific organism in question. Therefore, for *E. coli* the *uidA* gene that codes for the β-D-glucuronidase enzyme [56, 58–62] and the *tir* gene that targets the translocated intimin receptor [63] have been tested. Although *E. coli* was detected at very low levels with the *uidA* target, it was also found to cross react with *Shigella* species and other *uidA* carrying organisms [62, 64]. The *tir* gene has shown good specificity, however, whether single copy genes targets are sensitive enough for routine monitoring practices still needs to be studied.

Genetic determinants encoding virulence-associated traits such as the shiga toxin (*stx*) and the intimin producing attaching and effacing (*eae*) genes of diarrheagenic *E. coli* O157:H7 and non-O157 strains [62, 65], the invasion associated gene A (*invA*) of *S. typhimurium* [13, 66, 67] and the invasion plasmid antigen H(*ipaH*) gene of *Shigella flexneri* [68, 69] involved in the entry into intestinal epithelial cells have also been examined.

Protozoan targets

Numerous studies have investigated various DNA sequences of protozoa as potential molecular targets to confirm their presence in food and water samples. The use of PCR to confirm *Giardia* and *Cryptosporidium* presence has been employed for over a decade however, they still are used only as confirmatory and not as stand-alone monitoring tests. As with bacteria, the small subunit rRNA gene (18S rRNA gene) has been the focus of several studies with both *Cryptosporidium* and *Giardia* [70–72]. Although the method is the most sensitive for species identification it still needs to be coupled with sample concentration methods because of the low level of parasitic DNA. Other genes have also been studies such as the heat shock protein (*Hsp70*) for *Cryptosporidium parvum* [73] and *Giardia* [70], the elongation factor 1a (*Giardia*) and the triose phosphate isomerise gene (*tpi*) (*Giardia*) [74]. As all species of *Cryptosporidium* oocysts respond to a heat shock by producing the protein hsp70 [75], the mRNA gene coding for hsp70 can also be used as a viability marker. However, in the view of some water companies, this method has not been sufficiently validated for being of practical use [76].

More recently, Wang et al. [77] has developed a microarray containing a large variety of targets to simultaneously detect multiple protozoan parasites in samples. They used the previously tested conserved target genes such as rRNA and *hsp*70 to identify genus or species but also included highly variable genes such as *cp1* (cysteine protease 1) for *Entamoeba*, *c4* (species specific open reading frame) for *Giardia* and *cowp* (*Cryptosporidium* oocyst wall protein), *ptg* (polythreonine-rich glycoprotein), the *TRAP-C2* gene (thrombospondin-related anonymous protein 2) and *p23* (protein 23) for *Cryptosporidium* to be able to identify down to the species, assemblage or genotype level. As few as five trophozoitres of *Giardia* could be accurately detected by this method [77]. It remains to be determined if this method can be incorporated into routine testing of environmental water samples.

8.3.3.3   qPCR, RT-qPCR, and ddPCR

The PCR reaction generates copies of the target sequence exponentially initially. Because of inhibitors of the polymerase reaction found in the sample, reagent limitation, accumulation of pyrophosphate molecules, and self-annealing of the accumulating product, the PCR reaction eventually ceases to amplify target sequence at an exponential rate and a 'plateau effect' occurs, making the end point quantification of PCR products unreliable. Only during the exponential phase of the PCR reaction is it possible to extrapolate back to determine the starting quantity of the target sequence contained in the sample (Fig. 8.5). This attribute makes quantitative PCR (qPCR) and real-time quantitative PCR (RT-qPCR) methods so attractive. Although primers to already identified targeted genes are being used in qPCR and RT-qPCR, there is no consensus on what genes should be targeted or what sets of primers currently available are the best target candidates. The level of detection for waterborne pathogens (*Campylobacter jejuni, Salmonella, Shigella,* and *Yersinia*) from spiked stool samples has been found to range from 10^3 to 10^5 CFU/mL [78]. The main drawback of PCR technologies is its susceptibility to inhibition due to contaminants in the samples. Droplet digital PCR (ddPCR) which has only been commercially available since 2011, has the capability to partition the PCR reaction into thousands of individual reaction droplets

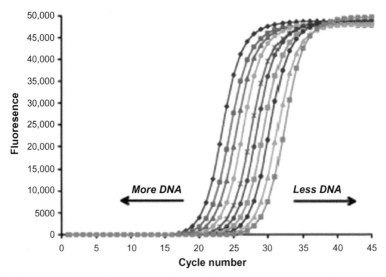

FIG. 8.5 In quantitative PCR, the amplification of the products can be tracked during the reaction by use of dye such as SYBR green that binds to the dsDNA products. The fluorescence can be used to predict the amount of DNA target present at the start of the reaction. The threshold cycle (Ct) value denotes how many cycles of PCR are required for the amount of PCR product (measured by fluorescence) to reach a defined threshold value. The more target DNA present in a sample, the lower the Ct value will be, as the threshold is reached sooner (http://www.langfordvets.co.uk/lab_pcr_fact.htm).

(approx. 20,000) prior to amplification, thereby increasing the ratio of target to nontarget DNA and showing increased tolerance for inhibition. It has been used to shown to improve precision of target numbers for water samples [79, 80].

8.3.3.4   NASBA

Nucleic acid sequence based amplification (NASBA) is a mRNA based technology. Measuring mRNA can confirm the presence and viability of a target organism. As NASBA exclusively amplifies RNA, the presence of DNA in samples will not produce false positives. The method can be carried out isothermally which reduces the need for specialised equipment which makes it ideal as a routine monitoring tool for environmental samples (Fig. 8.6.) [81].

NASBA has been used to monitor and identify *Vibrio cholerae* in ballast water. It required a 6 h enrichment step however 1 CFU/100 mL of *V. cholerae* could be detected in a background of RNA/DNA from other bacterial species present in the ballast water within 9 h [82]. Conventional real-time PCR was found to be more sensitive with the detection of 1 CFU/mL in 7 h presumably because DNA not RNA was the target molecule.

NASBA has also been successfully developed for the detection of a number of RNA viruses from food and water [83, 84]. The advantage of this method is

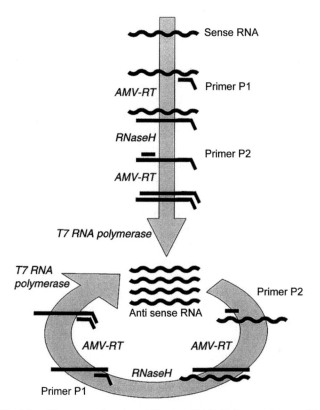

FIG. 8.6 Nucleic acid sequence based amplification (NASBA) is a used to amplify RNA sequences. RNA template is given to the reaction mixture, the first primer attaches to its complementary site at the 3′ end of the template. Reverse transcriptase synthesises the opposite, complementary DNA strand and RNAse H destroys the RNA template The second primer attaches to the 5′ end of the DNA strand and T7 RNA polymerase produces a complementary RNA strand which can be used again, so this reaction is cyclic (http://www.google.ca/search?q=nucleic+acid+sequence+based+amplification+figures&hl=en&prmd=imvns&tbm=isch&tbo=u&source=univ&sa=X&ei=JFSqUNmuE5TOyAHGpYGoCA&sqi=2&ved=0CDIQsAQ&biw=1619&bih=737).

the low detection limits, rapid assay times, and in some cases, with stringent controls, quantifiable results. As many viruses are hard to culture, the use of molecular methods such as NASBA can provide information about the presence and infectivity of the virus.

8.3.3.5 LAMP

Loop-mediated isothermal amplification (LAMP) was designed using the *Bst* polymerase which is a polymerase that has high strand displacement activity to amplify target DNA or RNA once it has been converted to DNA by a reverse transcriptase step. Using specially designed primers the polymerase produces products with loop structures. There is no need to raise or lower the temperature as in conventional PCR since the *Bst* polymerase is able to function at a single temperature with high sensitivity [85].

LAMP assays have been applied to the detection of viral, bacterial and protozoan pathogens. Bakheit et al. [86] developed a LAMP assay to detect and identify *Cryptosporidium* species DNA in faecal samples. Based on targeting the *Hsp-70* gene target, they found that LAMP was able to detect *Cryptosporidium* species DNA in samples that were negative by nested PCR suggesting that the advantage of LAMP is in the detection of target DNA in environmental samples with low concentrations of the pathogenic organism. Lee et al. [87] showed that high specificity and sensitivity of LAMP in the detection of *Mycobacterium tuberculosis* when used in conjunction with an ELISA-hybridisation assay. Suzuki et al. [88] showed that real-time RT-LAMP was better than real-time RT-PCR for the detection of *Noroviruses* from municipal wastewater. Chang et al. [89] used a LAMP based protocol (Fig. 8.7) to simultaneously detect four different pathogens (two viral and two bacterial) from ornamental fishes. When combined with a microfluidic system, NASBA has the potential to be carried out without any manual operation.

To date, however, few studies have been conducted using this assay for the detection of waterborne pathogens from drinking water sources. A recent report [90] has shown that it may be useful for detecting *Toxoplasma* DNA in water resources, an important detecting advancement, since there is little information about the prevalence of *Toxoplasma gondii* in water sources although several cases of waterborne toxoplasmosis have been documented. Gallas-Lindemann et al. [90] were able to detect *T. gondii* by LAMP in almost 10% of their wastewater samples although drinking waters sources were negative. LAMP has the potential to be a sensitive, specific, rapid, and cost effective assay to detect DNA from waterborne pathogens as there is no need for gel electrophoresis and it could be developed as an in situ method for detection of pathogens in contaminated water [42].

8.3.3.6 Microarray

DNA microarrays are small, solid supports onto which sequences from thousands of different genes are immobilised, or attached, at fixed locations. The supports themselves are usually glass microscope slides but can also be silicon chips or nylon membranes. The DNA is printed, spotted, or actually synthesised directly onto the support. A microarray works by exploiting the ability of a given mRNA molecule obtained from the target sample in question to bind specifically to, or hybridise to, the DNA template on the solid support that is complimentary to it. A researcher uses the location of each spot in the array to identify a particular gene sequence. With the aid of a computer, the amount of mRNA bound to the spots on the microarray is precisely measured, generating a profile of gene expression in the cell (Fig. 8.8). Microarrays are a significant advancement in the detection of multiple genes or organisms in samples rapidly because they screen for a very large number of sequences and because of their small compact size.

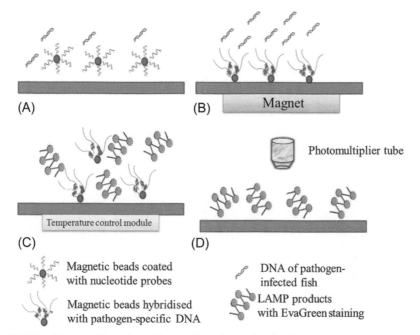

FIG. 8.7 Schematic illustration of the experimental procedure for detection of aquaculture pathogens in the microfluidic loop-mediated isothermal amplification (LAMP) system. (A) Thermal lysis and hybridisation between the magnetic beads and pathogen DNA. (B) Pathogen DNA isolation occurs when an external magnetic field is applied. (C) LAMP reaction of the extracted DNA. (D) Optical detection of the LAMP products.*(Reproduced with permission from W.H. Chang, et al., Rapid isolation and detection of aquaculture pathogens in an integrated microfluidic system using loop-mediated isothermal amplification, Sensors Actuators B Chem. (2013) https://doi. org/10.1016/j.snb.2011.12.054.).*

Many microarrays have been developed that have targeted multiple water-borne pathogens to those specifically aimed at detecting strains of a specific species. Like PCR, the targets chosen have ranged from universal ones like 16SRNA, 18S rRNA, and 23S rRNA [54, 71, 91] to those that can discriminate between different strains of a species [92, 93]. The utility of microarrays is their ability to directly detect specific waterborne pathogens instead of relying on the presence of faecal indicators. Using multitarget gene microarray, Weidhaas et al. [94] showed that many waterborne pathogens were still prevalent in a watershed system although the faecal coliforms numbers had fallen to close to the United States recreational water criterion standards.

8.3.3.7 Molecular beacons

Molecular beacons (MBs) are short singled stranded nucleic acid sequences (30–50 bases) that designed to have a unique sequence flanked by indirect repeats so that a stem-loop structure is formed. Each end of the structure is labelled,

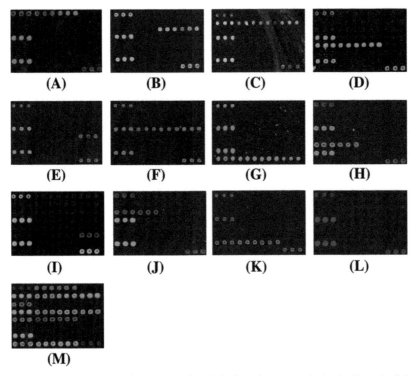

FIG. 8.8 Microarray analysis. Panels A through L show the pattern obtained with each of the nine individual pathogens and panel M shows the pattern obtained when all nine are present together. Panels: A, *Aeromonas hydrophila*; B, *Klebsiella pneumoniae*; C, *Legionella pneumophila*; D, *Pseudomonas aeruginosa*; E, *Salmonella*; F, *Shigella*; G, *Staphylococcus aureus*; H, *Vibrio cholerae*; I, *Vibrio parahaemolyticus*; J, *Yersinia enterocolitica*; K, *Leptospira interrogans*; L, *Legionella bozemanii*; M, *A. hydrophila, P. aeruginosa, Shigella, S. aureus,* and *L. pneumophila.(Reproduced with permission from G. Zhou, et al., Development of a DNA microarray for detection and identification of Legionella pneumophila and ten other pathogens in drinking water, Int. J. Food Microbiol. (2011) https://doi.org/10.1016/j.ijfoodmicro.2011.01.014.)*

one with a fluorophore (5') and the other (3') with a universal quencher. When the stem and loop is formed the quencher masks the fluorescence however if the unique sequence hybridises to a complementary (target) sequence, the stem and loop structure is broken, removing the quencher from the fluorescent dye thereby causing fluorescence to occur. MBs have been found to be extremely specific where even a single base mismatch did not cause fluorescence [95, 96]. MBs have been used successfully to detect *Candida* species in clinical samples [97] *E. coli* from food [98], *E. coli* and *L. monocytogenes* from milk and milk products [99, 100], and more recently *E. coli* from fresh produce and water [101] and *Salmonella* from surface and potable water in North India [102]. Jyoti et al. [102] found that the MB assay could detect the *inv A* gene in *Salmonella* strains at 10 CFU/PCR reaction (10^4 CFU/mL) from spiked water samples which was 100 times more sensitive than conventional PCR using the same oligomers. Sandhya et al. [101] reported that the *uid A* gene in *E. coli* could be detected down to 10 CFU/mL. They claim that MBs were extremely useful since their assay can be used to eliminate false positives results generated from PCR amplification.

For viral detection, MBs hold great promise as a method for viral gene detection in living host cells [96] however the inability to grow many of the enterovirues (*Norwalk, Adenovirus, Astrovirus,* etc.) in culture limits the utility of the method for testing environmental samples.

8.3.3.8   FRET

Fluorescence resonance energy transfer (FRET) is a phenomenon in which energy is transferred from an excited fluorophore, the donor, to a light absorbing molecule, the acceptor. Like MBs it uses fluorophores and quenchers; however, it is designed to use two primers which bind to their target so that the distance between the donor dye on one primer is in close proximity to the acceptor dye on the other primer (within 10 nm) to cause a fluorescent reaction that can be detected (Fig. 8.9). The extreme sensitivity of the efficiency of the energy transfer from donor to acceptor has proven to be a valuable means for studying protein–protein interaction as well as the proteolysis of viral replication in living cells [91] and has the potential to be used to detect pathogenic bacteria [103].

FRET systems have been used to detect pathogens or toxins [104, 105], however, there are no studies that have solely used a FRET system to detect waterborne pathogens from water sources. Several biosensor platforms are based on FRET-like systems and will be discussed in the biosensor section later in the chapter.

8.3.3.9   Microfluidics

Microfluidic technology emerged in the 1980s and allows a reaction to take place on the surface of very small solid material. The best examples are inkjet print heads. The technology deals with the behaviour, precise control, and manipulation of fluids that are geometrically constrained to a small, typically submillimetre scale. Micropumps supply fluids in a continuous manner or are used for dosing. Microvalves determine the flow direction or the mode of movement of pumped liquids. This technology has allowed research in molecular biology to miniaturise some processes and produce Lab-on-chip or DNA-chip technology. This means that molecular biology techniques such as DNA isolation and PCR from samples could be miniaturised and automated to accomplish detection in far less time and labour than macrotechniques [47].

Han et al. [106] showed that *E. coli* could be detected at less than 40 CFU/mL for viable cells in pure culture using latex immunoagglutination coupled with a microfluidic technology. Microfluidics has also been applied to water samples where Chow and Du [107] devised a microfluidic device that

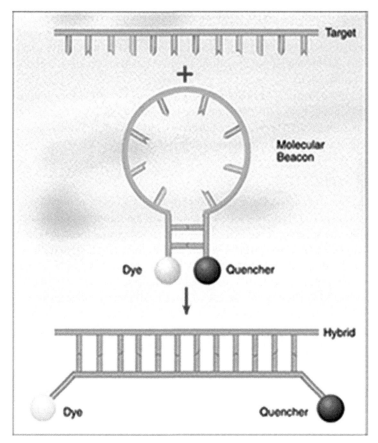

FIG. 8.9 Illustration of the FRET system. Molecular beacons are hairpin shaped molecules with an internally quenched fluorophore. They are designed in such a way that the loop portion of the molecule is a probe sequence complementary to a target nucleic acid molecule. A fluorescent moiety is attached to the end of one arm and a quenching moiety is attached to the end of the other arm. The stem keeps these two moieties in close proximity to each other, causing the fluorescence of the fluorophore to be quenched by energy transfer. When the probe encounters a target molecule, it can form a hybrid that is longer and more stable than the stem. This causes the molecular beacon to undergo a spontaneous conformational change that forces the stem apart, and causes the fluorophore and the quencher to move away from each other leading to fluorescence. *(From http://www.gene-quantitifcation.de/chemistry.html)*

allowed them to trap bacterial cells from a continuous flow stream and differentiate between *E. coli* and *Enterococcus faecalis*. Similarly, Taguchi et al. [108] used a microfluidic device to trap *Cryptosporidium* oocysts which were then subjected to automated FITC labelling and imaging with a detection limit of 36 oocysts/mL. Ramalingam et al. [109] designed a real-time PCR-based microfluidic array chip that was able to simultaneously detect four waterborne pathogens, *Pseudomonas aeruginosa, Aeromonas hydrophila, Klebsiella pneumoniae*, and *Staphylococcus aureus* by using a prototype real-time machine. It is still unknown whether this technology can be reliably applied to raw water samples but it holds much promise as a technology that can reduce the detection time and improve risk assessment. Recently, Chow and Du [107] fabricated a dielectrophoresis-based microfluidic channel to trap bacteria from water samples. They could easily differentiate between polystyrene beads and bacteria using the appropriate operating frequency, however, the ability to differentiate between two types of bacteria was still ambiguous. More recently methods have been based on rapid electrochemical sensors [110] or an immunoagglutination-based protocol coupled with microfluidics [111] to quantify *E. coli* O157:H7 levels directly from water samples. The latter method was fast (within 10 min) and could be adapted to be integrated into a mobile device for rapid and reproducible field detection of waterborne pathogens [111].

Chapter 10 provides a comprehensive overview of the field of microfluidics, as well as a review of how the technology has been applied to various different detection methods for waterborne pathogens.

8.3.3.10 Biosensors

Biosensors are described and reviewed in Chapter 7. That chapter mainly focuses upon detection of whole organisms, while molecular methods are discussed in this chapter. However, many of the biosensor techniques explained in Chapter 7 are applicable to the detection of small molecules. Indeed, the detection of small molecules is often easier and more sensitive than whole-cell detection. Chapter 7 also provides an overview of the types of molecule adopted for surface recognition of the target. Antibodies are widely used for whole-cells but many emerging recognition elements have great promise for highly selective molecular detection on a biosensor.

A biosensor is an analytical device for the detection of an analyte once bound to a biological component (probe molecule) such as an antibody, enzyme, protein, peptide, nucleic acid, etc. For medical applications, analytes can be whole cells, viral capsids, small proteins, nucleic acid, etc. The main advantages of biosensing devices are their specificity, sensitively, simplicity and short detection time [112].

The detector (Fig. 8.10) transforms the signal resulting from the interaction of the analyte with the probe molecule into another signal (i.e. transducers) that can be more easily measured and quantified. Detection methods range from thermal, optical, electrochemical to mass-based. The whole reaction takes place usually on a sensing surface that is able to detectable signals even at low concentrations or frequency of analyte–probe interaction. Bioreceptors are immobilised unto solid phase transducers to form sensing platforms. The immobilisation of the bioreceptor is fundamental for the functionality and integrity of the sensor [112]. Detection of the binding of the analyte to the bioreceptor can be measured many ways including electrochemically, optically, and piezoelectrically.

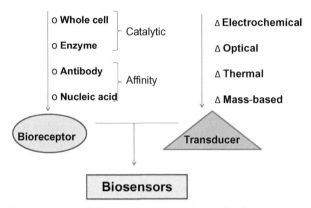

FIG. 8.10 The main types of bioreceptors and transducers that have been used to produce biosensors.

The advent of nanomaterials with advanced chemical and optical properties should increase the ability to detect analyte–bioreceptor interactions.

The use of biosensors for the detection of viruses is relatively new and has been applied mostly to viruses from clinical specimens and not to the detection of viruses from environmental or water samples. The limitation of using biosensors for the detection of viruses from water samples is that many current viral biosensors rely on the culturing of the virus to detect antibody reactions in the host cells. Water samples would also need to be concentrated to overcome low concentrations of pathogen in samples.

The microcystin toxin, produce by species of cyanobacteria such as *Microcystis*, *Oscillatoria*, *Anabaena*, and *Nostoc*, has been found in fresh water bodies worldwide and has been the target of biosensors for cyanobacteria detection. Mycrocystins are monocyclic heptapeptides with many variations in structures. The MC-LR version contains the amino acids leucine and arginine and the most widely associated with toxicity due to water consumption [113]. The target of the microcystin toxin is serine/threonine protein phosphatases type 1 (PP1A) and type 2A (PP2A). This characteristic has been exploited for the development of the biosensors. PP1A bound to optic fibres has been used to determine the amount of MC-LR in a sample by measuring the amount that binds to the PP1A when in direct competition with labelled MC-Cys-FITC (fluorescein isocyanate label). The resultant signals are inversely related to the concentration of the MC-LR in the samples [114]. Electrochemical biosensors which are more cost effective have also used to detect MCs in water samples [115]. The sensitively is limited, however, by the type of electrode used.

ELIZA-based biosensors using monoclonal or polyclonal antibodies have also been tried. These show greater specific binding properties than enzyme-based methods however they can be limited by the need for specific antibodies to be prepared for each compound to be detected, or in this case, each toxin variant [113].

The electrochemical DNA sensors have received much attention in recent years. They are simple, sensitive, specific, and relatively inexpensive. They are based on the detection of DNA hybridisation between the target DNA and the probe DNA. This interaction causes some changes in electrochemical properties that can be measured [110, 116]. Theelectrochemical detection of *Microcystis* spp. was accomplished by immobilising a sequence specific probe unto a transducer surface that is complimentary to a specific gene sequence in *Microcystis* [117]. Methylene blue and ruthenium complex was used as the electrochemical hybridisation indicator. Sensitivity of this method can be increased if samples are first subjected to PCR to increase the species or gene specific sequences that are used as the target sequences.

Aptamers (short single-stranded DNA fragments) that can bind to target molecules such as proteins, amino acids, and other molecules has also been used to develop a surface plasmon resonance (SPR) biosensor. These biosensors can detect binding of MC to the DNA aptamer on the surface optically by measuring the changes in refractive indexes [118].

Currently, no single biosensor has been validated for use in the detection of cyanobacterial presence in water sources. Possibly a combination of immunosensors for the detection of the microcystin toxins released into the environment along with DNA/aptamer biosensors for the analysis of gene sequences from *Microcystis* spp. to detect early algal bloom states would provide a well-coordinated effort to identify cyanobacterial pathogens in water sources.

Biosensor for protozoan pathogens such as *Cryptosporidium* have been developed to detect oocsyts using antibodies to the oocysts and piezoelectric-excited millimetre-sized cantilevers (PEMC) as sensors [119]. They were able to detect between 1 and 1000 oocysts/mL although a decreased in detection was noted in the presence of contaminants in the samples that can interfere with the signal such as 45% milk [120]. Measurement of the impedance instead of the resonant frequency of the cantilever was found to reduce signal to noise ratio and improve the detection limits of these sensors [121]. The use of antibodies for detection however is limited by their specificity and the lack of ability to detect viability of the oocysts.

The use of the expression of the heat shock protein hsp 70 produced by *Cryptosporidium* has examined as a target to detect and determine the viability of oocysts. The mRNA encoding the protein was amplified via NASBA and used as a viability marker [122]. The amplicon was detected with the use of an electrochemical biosensor in sandwich hybridisation assays with capture probe-coated superparamagnetic beads and reporter probe-tagged liposomes. Upon capture the liposomes release the electrochemical markers that are an active redox couple producing a current that can be measured. The mRNA could be detected from as little as 1 oocyst. The difficulty with using mRNA is the extraction of the nucleic acid from the oocysts. Standardisation and validation of the extraction methods used are still pending.

Biosensor development for the detection of bacteria pathogens is more extensive than other groups of waterborne pathogens. Biosensors have been proposed for *E. coli, Salmonella, C. jejuni, Legionella pneumophila*, and *L. monocytogenes* among others.

The most commonly used biosensors are based on peptide-based probes such as antibodies and oligopeptides. Park et al. [123] developed a strip-based biosensor for the detection of *E. coli* O157:H7. It used two antibodies specific to *E. coli* O157:H7 to form a sandwich ELISA. The conjugate carried horseradish peroxidase for colorimetric detection. They could detect 1.8×10^3 CFU/mL within 30 min. A modification of theimmunosensor was developed by Heyduk and Heyduk [103] where they used two specific antibodies against surface proteins on the target pathogen cells. Each antibody is coupled to oligonucleotide that is conjugated to a fluorescent molecule. When both antibodies are bound to the target, the oligos hybridise to each other producing a FRET reaction that can be measured. The reaction could be detected with simple fluorescent microscopy. They were able to detect as little as 300 cells of either *E. coli* or *Salmonella*. The use of antibodies in biosensor applications is limited by their stability in field applications.

Baeumner et al. [124] proposed the use of a capture probe-coated polyether sulfone membrane and reporter probe-tagged liposomes that would capture mRNA of the target organism. This format would determine the gene expression and there by the activity of the cell which can be correlated to viability. They were successful at extracting and amplifying the RNA from *E. coli* by NASBA and detecting the target sequence as low as 5 fmol per sample.

Most biosensors for detection of waterborne pathogens rely on the detection of whole cells. However, with the emergence of nanotechnologies it is becoming increasing possible to detect and quantify protein and DNA from samples rapidly [125]. Quantum dots, gold nanoparticles, magnetic nanoparticles, and nanowires are just some of the materials that are being used for their unique optical and physical properties to develop new diagnostic assays [126]. Combined with microfluidics, nanotechnology-based detection systems have the potential to detect cells down to the single cell level. Important standardisation and validation steps will need to be carried out before implementation of these systems into routine water monitoring.

8.3.3.11 Pyrosequencing

The ability to sequence DNA has provided researchers with one of the most informative resources to study biological systems. The most common DNA sequencing platform utilises dideoxy chain termination technology [127] however, this technology is slow and labour intensive and does not lend itself to rapid and routine use as a pathogen detection method from water samples.

On the other hand, the development of more advanced and high-throughput sequencing methods, collectively known as next generation sequencing (NGS), are suitable for numerous applications. The first of the NGS methods was Pyrosequencing which is based on the synthesis of a new strand, however, the incorporation of the base at each step of synthesis can be detected in real-time via the detection of chemiluminescent signals that allows the determination of the sequence of the template. It eliminates the need for labelled primers, labelled nucleotides and gel-electrophoresis. The technique was developed by PålNyrén and Mostafa Ronaghi at the Royal Institute of Technology in Stockholm in 1996 [128]. The method involves immobilising the template (which can be a PCR product), adding the primer for the target sequence and then synthesising the new strand. Nucleotides are added on at a time, if one is incorporated into the strand by the DNA polymerase, pyrophosphate is released. ATP sulfurylase quantitatively converts PPi to ATP in the presence of adenosine 5′ phosphosulfate. This ATP acts as fuel for the luciferase-mediated conversion of luciferin to oxyluciferin that generates visible light in amounts that are proportional to the amount of ATP produced (Fig. 8.11). The light produced in the luciferase-catalysed reaction is detected and the nucleotide noted. The desired DNA sequence is able to be determined by the fact that only one out of four of the possible A/T/C/G nucleotides are added and available at a time so that only one nucleotide can be incorporated on the single stranded template (Fig. 8.12). The intensity of the light determines if there are more than one of these nucleotides in a row. Unincorporated nucleotides and ATP are degraded by the apyrase before the next nucleotide is added. This process is repeated with each of the four nucleotides until the DNA sequence of the single stranded template is determined [129]. The Roche 454 sequencing platform is able to produce 1 million reads of up to 1000 bases in a 23 h period [16].

So far, pyrosequencing has been limited to the detection of bacterial pathogens in biosolids and sewage treatment plants and human pathogenic viruses in environmental samples based on 16S rRNA sequences [130–132]. This technology is likely to be able to identify novel organisms in samples and correlate illness to the emergence of novel pathogens. The use of unique target sequences for waterborne pathogens [133] could make this method a rapid, accurate technique for detecting currently recognised microbial contamination in water samples in the future.

1. $(NA)_n$ + nucleotide $\xrightarrow{\text{BST polymerase}}$ $(NA)_{n+1}$ + PPi

2. PPi + APS $\xrightarrow{\text{ATP sufurylase}}$ ATP + SO_4^{-2}

3. ATP + luciferin + O_2 $\xrightarrow{\text{Luciferase}}$ AMP + PPi + oxyluciferin + CO_2 + light

FIG. 8.11 Reactions involved in pyrosequencing system. The *Bst* polymerase catalyses the addition of a nucleotide to an existing nucleic acid chain (1). This results in the release of a pyrophosphate molecule (PPi) which can be added to ADP to ATP by ATP sulfurylase (2). Luciferase enzyme can then use the ATP to oxidise luciferin to oxyluciferin which releases light that can be measured (3).

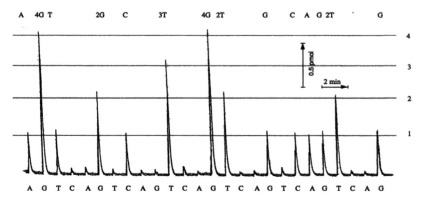

FIG. 8.12   Pyrogram obtained from liquid-phase pyrosequencing. The presence of a signal indicates the presence of the nucleotide and the height of the signals is proportional to the number of bases of the nucleotide that have been incorporated. The sequence of bases added is indicated below the pyrogram and the order of the nucleotides in this sequence in indicated above the pyrogram.*(Reproduced with permission from M. Ronaghi, Pyrosequencing sheds light on DNA sequencing, Genome Res. (2001) https://doi.org/10.1101/gr.11.1.3.)*

8.4.   Current state of pathogen detection in water sources

8.4.1   Sample collection

A standardised collection procedure for water sample analysis is a fundamental requirement for waterborne pathogen detection. Many pathogens are usually present in minute amounts in environmental samples, and always require some form of concentration prior to detection. Many methods for concentration of various target organisms have been published but will need to be validated to be used as routine water detection systems.

Concentration of samples through filter membranes (0.2-0.45 μm) for bacterial detection is usually employed as an initial step in detection. Filter blockage can become a problem if large volumes needed to be concentrated. Centrifugation of the sample has been also used however it can also concentrate contaminates. Hollow-fibre ultrafiltration has been used followed by centrifugation to recover *E. coli* O157 from large volumes of surface water (up to 40 L). Methods that reduce exogenous DNA [63] from the samples are best for reducing false positive reactions.

Enrichment methods for bacterial pathogens have been applied in many cases in order to enhance the concentration of the target organism. However, enrichment of the target organism often allows uncontrollable growth of non-target organisms as well. In such cases, estimation of the precise number of pathogens prior to enrichment is hard to determine. Centrifugation has been employed with large volume samples to increase the probability of detecting rare targets, however it also concentrates other particulate matter in the samples, whereby environmental samples can be affected by the concentration of substances such as humic acids that interfere with downstream detection. More recently, filtration has been studied as possible method for concentration of samples, since it is rapid for processing 1-100 L of water and it may be possible to eliminate exogenous DNA [134]. With filtration, the enrichment step can be eliminated yet still allow the detection and quantification of the pathogen [63].

Overall one of the greatest challenges for the detection of virus in water and environmental samples is the adequate sensitivity due to low concentration, since as few as 1-10 viral particles per gram of soil or litre of water maybe present [135]. Standardisation of units of measurement: PFU, focal forming units, genome equivalents, PCR units, 50% tissue infectivity, also needs to be considered to better compare one method with another. Current research is devoted to improving concentration, purification, and detection; however, there is still a need to have a priori knowledge of the target. Enrichment of viruses from environmental samples can be accomplished by adsorption/elution protocols, entrapment ultrafiltration, ultracentrifugation, or hydroextraction [49]. An overview of the pros and cons of each method are summarised by Bosch et al. [136]. Recovery of norovirus from water samples is especially important since there is no cell culture system for this virus. Recovery of norovirus is dependent on both the filter type (electronegative or electropositive) and sample source (tap, bottles, river water) [137]. Huguet et al. [138] determined that in situ lysis of viral particles and recovery of viral RNA was superior to recovery of intact virus particles. Methods that rely on immunoaffinity with magnetic beads have shown great promise and can simultaneously help purify the sample. However, the cost of this technology is expensive and requires the generation of specific antibodies for each viral variant [136, 139].

Concentration of ooscysts and cysts is also necessary to determine the presence of low numbers of *Cryptosporidium* and *Giardia*, respectively, in environmental samples. Since oocysts and cysts are naturally electronegative, this property can be used to concentrate the organism by both coagulation and filtration [72]. More recently, a microfluidic micromixer device for the effective capture of *Cryptosporidium* oocysts from water is being studied. It has been shown to have a capture efficiency of 96% however it is still in development and not used for routine screening [140].

8.4.2 Molecular extraction

Once concentration has been performed, a standardised robust extraction protocol needs to be adopted to release the target molecules. Many methods have been shown to be successful for bacterial cell lysis but to date none have been accepted as a standard procedure possibly due to the less than optimal reproducibility of the protocols from lab to lab. The challenge of extraction methods is to obtain target molecules free from humic acids, metals and other contaminants that may interfere with subsequent enzymatic reactions and to minimise operator variability. Future automation using microfluidic technology holds the promise of reducing or eliminating the human error factor.

8.4.3 Viruses

Viruses constituent an unique group of infectious agents that are extremely small, have a simple acellular structure, are unable to replicate outside of their host cell, yet can cause a diverse array of illnesses. There are more than 140 enteric viruses known to infect humans, many of which are transmitted via faecal-oral route through infected surface or ground water. The most common enteric viruses associated with human waterborne illnesses are Adenovirus, Astroviruses, Caliciviruses (noroviruses and sapoviruses), Enteroviruses (polioviruses, echoviruses, coxsackie viruses), Hepatitis A and E, and Rotavirus (Table 8.1). They have been detected in sewage, surface water, groundwater, and drinking water sources [2, 170].

Although traditionally cell culture was the most widely used method for the detection of viruses, it is now combined with PCR or immunological testing to improve the sensitivity of detection. The use of molecular techniques can contribute to more accurate assessment of the occurrence and health impact of waterborne viruses by overcoming the need to culture the viruses. However currently none of the molecular tests are practical for routine or large scale monitoring and therefore viral presence is often based on the presence/absence of bacteria indicators such as *E. coli*, total coliforms, *Enterococcus*, *C. perfringens* spores, and bacteriophages [2]. Although enteric viruses cannot multiply outside of their host, they are able to survive longer in water than most intestinal bacteria including indicator organisms therefore the absence of faecal bacteria does not ensure the corresponding absence of the virus.

Detection of viruses in water samples is limited at various stages in the traditional methodology. In samples where viral concentrations are high such as faeces (may excrete up to 10^{11} viral particles per gram of stool), electron microscopy, enzyme immunoassay, latex agglutination or polyacrylamide gel electrophoresis can be used however in water samples where the virus may only reach 1-10 particles per litre, these methods have not been found to be sensitive enough. Therefore, sensitivity requires the concentration of large volumes of the sample to overcome infectious levels of viruses that are often very low. This, in itself, can also cause detection problems since it also concentrates contaminants that may interfere with subsequent testing. Filtration by absorption or size exclusion (ultrafiltration) have both been used successfully [171, 172] although sometime a second concentration step is necessary to improve sensitivity [136]. Once concentrated, the detection efficiency of current methods may vary from only 50% to 95% [48, 170]. Cell cultures of the viruses may still be the most definite test since it can confirm both the presence and infectivity of the viruses however it suffers from the lack of detection of slow-growing viruses, the occasional inability to produce clear cytopathic effects and possible

TABLE 8.1 Major groups of waterborne microbial pathogens with their traditional and proposed detection methods.

Group	Pathogen	Traditional method	Suggested new methods
Viruses	Adenoviridae (Adenovirus)	- Concentration by filtration and detection using mammalian cell cultures [141]	- ICC-PCR [48] - Real-time qPCR detected 1.5 gene copies [142] - RT-qPCR and real-time RT-qPCR with electropositive filtration method [135]
	Astroviridae (Astroviruses)	- Concentration by filtration and detection using mammalian cell cultures [141]	- RT-PCR [143]
	Caliciviruses (Noroviruses, Sapoviruses)	- Concentration by filtration and detection using mammalian cell cultures - No cell culture available for Norovirus so use molecular methods [136]	- Real-time qPCR detected 10 gene copies [142] - RT-qPCR [144] - RT-qPCR and real-time RT-qPCR with electropositive filtration method for Norovirus [135]
	Picornaviridae (Enterovirus–polio, echo, coxsackie)	- Concentration by filtration and detection using mammalian cell cultures [141]	- RT-qPCR using TaqMan could detect 0.8 PFU per reaction [46] - Metagenomic analysis, pyrosequencing [16] - RT-qPCR and real-time RT-qPCR with electropositive filtration method [135]
	Picornaviridiae (Hepatitis A)	- Not available	- ICC-MB assay [96] - Duplex RT-qPCR [145]
	Hepaviridae (Hepatitis E)	- Not available	- Real-time RT-qPCR method [146] - Real-time PCR Taqman assay and real-time PCR ProProET assay [147] - Duplex RT-qPCR [148]

(Continued)

TABLE 8.1 Major groups of waterborne microbial pathogens with their traditional and proposed detection methods—cont'd

Group	Pathogen	Traditional method	Suggested new methods
	Reoviridae (Rotavirus)	- Environmental samples concentrated by filtration and detection using mammalian cell cultures - Human faecal samples analysed by electron microscopy, enzyme immunoassay, latex agglutination and polyacrylamide gel electrophoresis (PAGE) [47]	- Integrated microfluidic-PCR method combined with fluorescence detection [47]
Bacteria	Campylobacter jejuni	- Based on detection of faecal coliform indicator bacteria on microbiological media followed by species specific metabolic tests	- Microaerophilic enrichment at 42°C followed by PCR of 16S rRNA [53] - qPCR of VS1 gene [63]
	Escherichia coli O157:H7	- Based on detection of faecal coliform indicator bacteria on microbiological media, confirmation with MUG	- Microarray [54] - MB-real-time PCR down to 10 CFU/mL after 18 h of enrichment [100, 101] - qPCR using gene detected 10 cells/mL in wastewater samples [63] - FRET [103] - MB microarray using FRET [98] - Magnetic bead-quantum dot assay, could detect 25 CFU/mL [149]
	Legionella pneumophila	- Based on detection of faecal coliform indicator bacteria on microbiological media	- Microarray [93] - qPCR combined with propidium monoazide (PMA) [150]
	Salmonella enterica	- Based on detection of faecal coliform indicator bacteria on microbiological media	- MB based PCR assay for invA gene detected 10 genome equivalents in 3 h [102] - RNA aptamers [151]
	Shigella spp.	- Based on detection of faecal coliform indicator bacteria on microbiological media	- PCR using primers to tuf gene [62]

Vibrio cholera	- The 'gold-standard' is considered to be culturing on selective media, usually after enrichment [152]	- Triplex PCR and *Vibrio*-specific microarray [153] - Multiplex PCR [154] - Real-time PCR and NASBA down to 1 CFU/mL (IMO guidelines) toxigenic *V. cholera* in ballast water [82]
Multiple pathogens		- Microarray [54] - qPCR combined with propidium monoazide (PMA) (review by van Frankenhuyzen et al. [155])
Acanthamoeba castellanii	- Concentration using depth filters followed by microscopic examination of slides	- DNA microarray [71]
Cryptosporidium spp.	- Concentration using depth filters followed by oocyst detection by immunofluorescent antibody (IFA) and conformation by DIC microscopy: USEPA method 1623 [156]	- Nested PCR-RFLP [157, 158] - Multiplex PCR with Luminex assay [159] - Microarray [71, 77] - LAMP [86] - Microfluidics and biosensors (review by Bridle et al. [160])
Cyclospora spp.	- Concentration using depth filters followed by microscopic examination of slides	- Nested-PCR [161] - PCR-OLA [161]
Entamoeba histolytica	- Concentration using depth filters followed by microscopic examination of slides	- Microarray [77]
Microsporidium (*Enterocytozoon* spp., *Encephalitozoon* spp.)	- Concentration using depth filters followed by microscopic examination of slides using Weber's stain [162]	- PCR using SSU-rRNA [72]

Protozoa

(Continued)

TABLE 8.1 Major groups of waterborne microbial pathogens with their traditional and proposed detection methods—cont'd

Group	Pathogen	Traditional method	Suggested new methods
	Giardia lamblia	- Concentration using depth filters followed by microscopic examination of slides and/or immunological tests: USEPA method 1623 [156]	- PCR of 18S rRNA gene, tpi gene, elongation factor 1α gene [163, 164] - DNA microarray [71, 77]
	Naegleria fowleri	- Concentration using depth filters followed by microscopic examination of slides	- qPCR of 18S rRNA gene [165]
Cyanobacteria	*Anabaena*	- Water samples for cyanobacteria analysis were preserved with Lugol's solution and species identified by light microscopy [166] - Mouse bioassays for neurotoxins [166]	- UPLC/MS/MS method determines low concentrations of cylindrospermopsin, anatoxin-a, and several icrocystin congeners [167] - LC–MS/MS [168]
	Microcystis	- Based on cell numbers not on toxin levels	- qPCR [169] - Review of current methods with coordinated approach involving more than one type of biosensor [113]

inability to distinguish between plaques produced by multiple viruses [173, 174]. Although the detection of *Adenoviruses*, *Enteroviruses*, *Astroviruses*, and *Rotaviruses* have been achieved with appropriate cell lines [48, 136], it is not as a realistic approach for Hepatitis A and Norovirus due to assay complexity [136]. Furthermore, cell cultures can give false positives due to contamination that causes cells to change morphologically [139].

For viruses where antibodies are available, immunomagnetic separation (IMS) can be used to concentrate the viruses from samples. In this method, antibody coated paramagnetic beads are used to bind to antigens present on the surface of the target virus. A magnetic is then used to attract the beads to one side of the tube thereby concentrating the virus. This method can be combined with PCR to provide comparable results to plaque assays. Viruses such as Enteroviruses, hepatitis A, Rotavirus, and Norwalk virus have been successfully detected from sewage and artificially contaminated environmental water samples using this technology [175–177]. Drawbacks of the method are the availability of an antibody that can target all strains of a particular virus, the cost of the magnetic beads depending on the reagents coupled to them, and the conditions of the environmental sample since pH and colloidal particles may interfere with the initial binding step between the antibody and antigen [178, 179].

Molecular methods based on PCR have been developed for the key enteric viruses. With the exception of adenoviruses which contain dsDNA as their nucleic acid, the rest are RNA viruses with all of them containing ssRNA with the exception of Rotaviruses which possess a dsRNA genome. RNA viruses represent a major challenge in the development of molecular methods since their genomes are replicated with error prone polymerases that promote high mutation rates and recombination events. Targets for PCR primers or microarray probes need to be carefully selected to ensure they target highly conserved regions.

The advantage of PCR based methods is that amplification of the target sequence can increase the sensitivity of the assay and eliminate the need to reconcentrate samples. Direct PCR, nested PCR and real-time quantitative PCR (qPCR) were able to provide rapid sensitive detection of adenovirus in wastewater, drinking water, recreational waters and rivers [48, 180, 181] with qPCR providing some measure of the concentration of the virus. Jothikumar et al. [182] designed a broadly reactive TaqMan assay for the detection of all adenovirus and a specifically reactive assay, using fluorescence resonance energy transfer (FRET) probes, to successfully detect the adenovirus fibre gene in AdV40 and AdV41. They were able to detect as low as five to eight copies of AdV40/41 with TaqMan and three to five copies with FRET-PCR. Melting curve analysis allowed them to differentiate between AdV40 and AdV41 [182]. Like immunological techniques, PCR methods can also be inhibited by environmental contaminants causing false negatives in this case [139].

Real-time quantitative PCR has also been successfully applied to the detection of hepatitis A [183], hepatitis E [146], *Norwalk*-like virus [184], and norovirus [185]. Recently, Jothikumar et al. [46] developed a RNAX buffer that allows improved isolation of RNA from environmental sample. Using the buffer, they found that contaminants in water not as much of a problem when applied in a RT-qPCR assay.

The major limitation of PCR based method is the lack of correlation between the detected genome copy numbers of the target viral pathogen and the infectivity of the virus detected. Newer PCR methods are being studied that may be able to detect the integrity of the viral genome as a measure of the infectivity. Whereas PCR is usually based on target sequences that are well conserved, unique to the pathogen in question and short, the use of longer sequence targets could give more information of the possible degradation in the viral nucleic acid. It is hypothesised that infectivity will be correlated to degradation of the genome and resultant shorter amplicons then expected will signify degradation of the genomes. Although Simonet and Gantzer [186] saw some correlation between degradation by UV light and fragment size, viruses can lose infectivity without a corresponding deduction in fragment size [186]. This would cause over estimations of risk due to viral contamination and limit the general applicability of this long chain PCR method as an indicator of viral infectivity [50].

The combination of integrated cell culture techniques along with PCR (ICC/PCR) allows both semiquantitative and qualitative information on the measure of infectious virus [47, 48, 55]. This method involves the inoculation of viral samples unto the cell cultures, followed by nucleic acid isolation for the cell cultures and PCR for detection of target sequences. It does not require CPE effects to be visible which decreases the wait time for diagnosis. Reynolds et al. [187, 188] were able to detect enteroviruses as soon as 1 day postinoculation while Jiang et al. [189] was able to detect 1 infective unit of hepatitis A in environmental samples.

The development of microfluidics as applied to nucleic acid analysis is a promising rapid detection system. Miniaturisation of the biological, chemical, or biochemical test allows assay times to be significantly reduced. If it is combined with an integrated product detection system such as capillary electrophoresis separation and fluorescent detection of the amplified products [190], it can be automated with no human intervention which reduces time and contamination. Such systems have been reported for virus detection from clinical samples. Li et al. [47] showed that *Rotavirus* could be detected within 1 h at LOD's of the RNA concentration be 3.6×10^4 copies/μL. Very low levels of virus in water samples, however, still presents a problem of concentration prior to use of the microfluidic device.

The most promising future method is viral metagenomics. The method involves sequence-independent amplification, subcloning and sequencing of purified viral nucleic acids followed by in silico searches for sequence similarities to known viruses [191]. This approach allows novel and emerging viral contaminants to be detected without bias towards any particular viral group. However, little research involving environmental samples has yet been completed using this technology. As of now many environmental samples show no significant nucleotide similarity to any sequences in the database [192].

8.4.4   Bacteria

More than 80% of diarrheal disease each year is attributed to drinking water contaminated with waterborne pathogens. Currently, pathogenic bacteria in water sources are the cause of the majority of waterborne disease outbreaks in the United States [193]. Some of the most prevalent bacterial pathogens belong to pathogenic strains of *E. coli*, *Campylobacter* spp., *Helicobacter pylori*, *Klebsiella* spp., *Salmonella* spp., *Shigella* spp., *Legionella*, *Vibrio* spp., and *Yersinia* spp. Routine monitoring of drinking and recreational water sources for pathogenic bacteria is currently still based on the presence of faecal indicator organisms (FIBs) [194]. Specifically microbiological quality is measured by testing drinking water for *E. coli*, a bacterium that is always present in the intestines of humans and animals and whose presence indicates the contamination of the water source with faecal matter. The current guideline indicates that for the water to be safe for consumption there should be no *E. coli* detected per 100 mL.

The drawback of using FIBs is the lack of correlation with many waterborne pathogens including other bacterial contaminates. Moreover, the lack of identification of the actual pathogen makes it hard to predict pathogens occurrence and public health risk. Therefore, many other methods have been developed and are being tested for their usefulness in detecting and tracking the actual pathogens in water sources. The ideal method would identify the actual pathogen causing the outbreak, be specific enough to resolve the organism down to a taxonomic significant level, be sensitive enough to determine disease relevant levels, and be affordable for standard routine use. Although many techniques have been tested, most of the research to detect bacterial pathogens has concentrated on DNA-based assays. They range from FISH to PCR, qPCR to RT-qPCR, and microarrays to pyrosequencing and biosensor technology.

Regardless of the method employed they all follow a common approach. The sample to be tested usually needs to be processed in a way that ensures the pathogen in question can be detected which often requires a concentration step for pathogens found at low infectious levels. Second, the nucleic acid needs to be extracted. Third, the actual detection of the pathogen needs to be carried out. Ideally the pathogen should also be quantified so that risk from the pathogen can be ascertained. Finally, the method should allow an assessment of the virulence of the pathogen.

Detection of single pathogen(s) by most molecular methods requires the use of primers that have previously been shown to be specific for the pathogens either at the genus or species level. However, the selection of primers for currently recognised waterborne bacterial pathogens is an area where standardisation has not been established. Many of the primers target the 16S rRNA gene which is present in all organisms and contains hypervariable regions that can be targeted down to the genus or species level. PCR has been used to detect *E. coli* [56, 195], *Salmonella* [66, 196, 197], *Campylobacter* [198, 199], *Legionella* [200], and *Bacteroides* [201] from food, water or water-related sources. Chakravorty et al. [51] demonstrated that although no single hypervariable region is able to distinguish among all bacteria, analysis of 9 variable regions determined that using multiple targets, hypervariable regions V2 (nucleotides 137–242), V3 (nucleotides 433–497), and V6 (nucleotides 986–1043), contained enough discriminatory power to distinguish between the 110 species they studied.

Although 16S rRNA is an excellent target due to its high degree of conservation at the genus level, it often fails to discriminate between species or strains level and therefore the 23S rRNA gene [54] and the ITS region [202] between the 16S rRNA and the 23S rRNA regions have been studied for the direct detection of waterborne pathogens. Furthermore, individual genes have been selected for various pathogens that are able to differentiate closely related species such as the gyrB (gyrase subunit B) for detection of *Enterobacter* spp. in infant formula [77]. Many of these targets need to be assessed for application to waterborne pathogens.

Khan et al. [56] designed a qPCR assay using the conserved flanking regions of the 16S rRNA gene, the internal transcribed spacer region (ITS) and the 23S rRNA gene targets for the detection of *E. coli* in the spiked agricultural water samples. It was capable of detecting and quantifying the *E. coli* with a minimum detection limit of 10 cells/mL. This was found to be significantly higher than the culture-based assay on the same samples. Combining RT-PCR assay, in a molecular beacon format, Ram et al. [195] was able to detect 4 CFU/mL of ETEC from water samples spiked by a nonpathogenic *E. coli*. The assay was found to be 500 times more sensitive than conventional PCR using the same oligomers.

More recent studies have involved the use of immunological reactions on biosensor platforms. Guan et al. [203] combined antibody-coated beads for the capture of *E. coli* cells from samples and the detection of adenosine triphosphate (ATP) employing the bioluminescence reaction of firefly luciferin-lucifera-ATP on a microfluidic chip. The method allowed reliable detection of *E. coli* O157:H7 concentrations from 3.2×10^1 to 3.2×10^5 CFU/μL within 20 min. Heyduk and Heyduk [103] developed a FRET assay based on two samples of the antibody, each labelled via nanometre size flexible linkers with short complementary oligonucleotides, when in to close proximity, produce a fluorescent signal that can be detected. They found that *E. coli* O157:H7 cells could be detected at levels of 300 cells in approximately 5 min. Although the method has been demonstrated as a proof of concept it has not been demonstrated using environmental water samples. The limitations of the method are at the antibody recognition step and the false detection of endogenous fluorescence in samples. The recent advances in these biosensor platforms not only makes detection very rapid but it also promises to provide an early warning system for water quality.

Repeated detection of *Salmonella* has occurred from rivers, lakes, coastal waters, ground water, as well as sewage and wastewater effluent [204]. The use of FIBs to determine *Salmonella* contamination in water samples recently is questionable and Wilkes et al. [14] suggested that a level of at least 89 CFU/100 mL of *E. coli* was required for correlation with 89% of *Salmonella* positive samples. The use of *Bacteroides* 16S rRNA markers in river and waste-water samples appeared to correlate positively with *Salmonella* especially when the concentration of human *Bacteroides* genetic markers was greater than 10^6copies/100 mL [205]. Although the infectious dose of *Salmonella* is quite high, the emergence of multiple drug resistant strains (MDR) has generated concern over the inability to track *Salmonella* reliably the presence with the use of traditional FIBs. With the introduction of molecular methods for pathogens detection, the anaerobic *Bacteroidales* order have been suggested as alternative biological indicator of faecal pollution according to their host-specific distributions, their short survival rate once released into the natural environment and their abundance in faeces from warm-blooded animals [206]. The considerable advantage of these alternative indicators is their possible application in determining the source of faecal contamination. Host-specific *Bacteroidales* 16S rRNA genetic markers were recently identified and PCR primers have been designed for detection of human (HF183F–Bac708R), ruminant (CF128F–Bac708R), and swine (PF163F–Bac708R) specific faeces [207]. The markers were found to be useful to discriminate between agricultural and human related sources of contamination, however their levels were unable to be correlated to the presence of pathogens such as *Campylobacter*, *Salmonella*, or *E. coli* O157:H7 [208].

Because of poor correlation of *Salmonella* numbers with bacterial indicator both with culture dependent and culture-independent markers, methods that rely on the direct detection of the pathogens is favoured. Using real-time PCR, *Salmonella* could be detected directly from water samples using the invasion associates gene (*invA*) of *S. typhimurium* [14, 63, 67]. Clark et al. [63] found that it was possible to detect as few as 10 *S. typhimurium* cells in pure culture but sensitively decreased once introduced into an artificially contaminated environmental matrix. As with *E. coli*, *Salmonella* detection has also been tested on biosensor platforms with great success in the laboratory [103]. These platforms are being advertised as simple, fast, inexpensive, sensitive, and quantitative ways to detect cells or molecules from a variety of matrices.

Legionella is an ubiquitous bacteria found in many water environments including surface waters, groundwaters, as well as in human-made systems such as cooling towers, hot water tanks, and spas. Currently no suitable indicators are suitable for predicting *Legionella* spp. in samples. There is some support for the correlation of HPC numbers (> 100 CFU/mL) as an indication of the presence, however, the relationship is not consistent and therefore unreliable [209]. A very recent study demonstrated the direct detection of *Legionella* using an oligonucleotide based microarray method based on the *ITS* region and *gyr* B gene [93]. They were able to detect 10^4 CFU/mL of *Legionella* in pure bacterial cultures. Whether this diagnostic test can be useful for water analysis remains to be determined.

Direct culture independent DNA based assays have also been developed for other waterborne pathogens such as *Campylobacter*, *Shigella*, *Vibrio*, and *Yersinia* spp. Multiplex PCR [210] or qPCR [92, 153] and microarrays for multiple pathogens or a single pathogen are more common every year and many DNA based tests have passed the proof of concept step yet none have been adopted as routine water monitoring tests.

8.4.4.1 Protozoa

Waterborne parasitic protozoan diseases are found worldwide and contribute to the 4 billion cases of diarrhoea and 1.6 million deaths per year. The three main protozoan diseases contracted from water are cryptosporidiosis, giardiasis, and amobiasis caused by *Cryptosporidium*, *Giardia*, and *Entamoeba*, respectively [211]. However, numerous other protozoans can cause human infections such as *T. gondii*, *Acanthamoeba* spp., *Cyclospora cayetanensis*, *Microsporidia*, *Isospora*, *Blastocystis hominis*, *Sarcocystis* spp., *Naegleria* spp., and *Balantidium coli* [212].

Current routine monitoring tests, such as immunofluorescence microscopy for the detection of oocysts [156] are only effective for single protozoan pathogens, are time consuming, and labour intensive which leads to under reporting of outbreaks in many countries. The need for improved surveillance of drinking water for protozoan parasites has necessitated the search for faster, more specific testing for these pathogens. More accessible and reliable tools for protozoan detection have been proposed and include, PCR, DNA microarrays, mass spectroscopy, and biosensors [211].

Rochelle et al. [70] described a PCR based assay for the detection of *C. parvum* and *Giardia lamblia* from water samples that was able to detect from 1 to 10 oocysts and 5 to 50 cysts in pure and seeded samples, respectively. To generate this level of sensitively however they performed two successive rounds of amplification. The convenience of detecting more than one pathogen simultaneously lead to the study of multiplex PCR systems [213] however the need to have the different primer sets function in the same conditions limits the specificity of the primers.

The development of DNA microarray systems holds promise for more rapid, sensitive, and specific detection of protozoans in water samples. They have successfully been applied to phylogenetic studies [73] and are now being adopted for detecting and discriminating between the various protozoan pathogens [77]. Lee et al. [71] have described a DNA microarray system that is able to detect approximately 1×10^3 copies of the SSU rRNA gene of *C. parvum* which corresponds to about 50 oocysts, however, this is still way above the infectious dose of 1 oocyst [209]. Miniaturisation of molecular methods holds the promise of being able to attain the single oocyst detection especially when combined with microfluidic systems [160]. However, the speed of the tests cannot jeopardise the sensitivity and specificity that is needed for these methods to be adopted as standard protocols since the approval of regulatory bodies is required.

8.4.4.2   Cyanobacteria

Cyanobacteria (*Anabaena, Aphanizomenon, Microcystis*) and other blue-green algae can form heavy growths on the surface of pond, lakes and other bodies of water. These water blooms appear most often during hot weather and where the water body is eutrophic. Toxins produced by these organisms are potentially hepatotoxic, nephrotoxoic, and carcinogenic [113]. Guidelines have been proposed for toxin levels for consumption but not for the potential risk for cancer although microcystin toxins are likely to be tumour promoters. In the United States the exposure limits have been set to 1.0 µg/L, in Canada they are 1.5 µg/L, in Australia they are 1.3 µg/L, and the World Health Organization has set theirs at 1.0 µg/L [214]. The route of exposure is likely inhalation and there is no standardised correlation data on *Microcystis* levels in the water and their corresponding levels in air.

Methods currently available for detection of microcystin include whole cell bioassay in mice that detect levels of 25-150 µg/kg, MALDI-TOF which is able to detect microcystin at levels of 1µg/L, and LC–MS and GC–MS which are able to detect levels of 0.02 µg/L and 0.0043 µg/L, respectively. All these methods are expensive and time consuming and not suited for routine water monitoring. Many newer techniques have been based on biosensor technology using a wide range of analytes, including enzyme based biosensors, optical biosensors, electrochemical biosensors, immunosensors, and nucleic acid sensors (see Ref. [108] for review). Recently, Al-Tebrineh et al. [169] described a PCR assay for the detection of hepatotoxigenic *Cyanobacteria* that was quantitative.

Currently, no standardised protocol has been approved for routine detection of cyanobacterial toxins in drinking or recreational waters although cyanotoxins, microcystin, anatoxin-a, and cylindrospermopsin have been put on the Contaminate Candidate List III (CCL III) by the United States Environmental Protection Agency (USEPA) based on occurrence and prevalence research.

8.5.   Faecal source tracking

In order to adequately assess human health risks and for drinking water monitoring it is necessary to know not only the pathogens present but the sources of the pathogen contamination. This information can assist in developing guidelines and management policies to protect watersheds and reduce waterborne transport of enteropathogens worldwide [215]. To date, microbial source tracking (MST) using indicator organisms has been the standard. Besides monitoring for total and faecal coliforms (*E. coli*), *Clostridium* and faecal Enterococci have been used to determine the source of the faecal contamination. The ratios of *E. coli* to *Enterococci* have been used to predict human or animal source with mixed success since some human pathogens such as *Giardia* and *Cryptosporidium* are often associated with animal contamination. Therefore, although the presence of FIBs may be able to shade some light on the source of the faecal contamination, it cannot accurately predict the pathogen or the risk associated with the levels of the pathogen. More information about the route of infection and the risk assessment for human death would benefit from some of the more current protocols for identifying pathogens in the water and water related samples.

All of the methods that have been designed and developed to date need to be assessed before they can be accepted as a routine means of MST. To be recognised as a new method of faecal source tracking, each step of a new protocol needs to be verified and the overall test needs to meet regulatory standards. Field and Samadpour [3] have identified six stages in the testing of faecal source tracking methods: 1. Proof of concept, 2. Feasibility and biological likelihood the new method can detect likely sources, 3. Application of the test to novel samples, 4. Comparison to other methods, 5. Testing with blind samples, and 6. Verifying the test leads for improvement of water quality [3]. The current culture based indicator tests would be comparators in stages 4 and 5.

Culture based methods to determine faecal sources have scored low at identifying novel pathogens or blind samples and are limited by the ability of the pathogens to be isolated on media. Techniques such as antibiotic resistance patterns, carbon utilisation profiling, and phage typing methods are time consuming and often require prior knowledge of the sample source. Other phenotypic methods such as FAME have passed the proof of concept and feasibility stages but rely on libraries of information to predict the contaminate and therefore lack in ability to identify isolates from outside of the library [3].

Culture independent methods are fast growing into the methods of choice for identifying pathogen content in samples and possibly as a new standard for MST. They rely on extraction of a genetic marker from a sample and detection of host specific marker genes. The greatest asset of these methods is their direct detection of a pathogen attribute and their speed of delivery. They are able to detect unculturable isolates as long as some knowledge of the genome of the suspected contaminate is known. Most of these methods, PCR-based, microarray, and biosensors have shown proof of concept and feasibility as MST methods [3] however, they are still lacking from method testing in real life applications. Adoption of one method will require extensive testing from both a comparison point of view to current methods and a guarantee that the method will supply a technique that has ease of use and is financially feasible.

Enteric virus detection is among the promising library independent, culture independent tools that can be used for MST due to their presence in host faeces and their host specificity [19]. Enteric viruses are frequently detected in the environment and are easily differentiated based on sequence differences in genus-common genes. Furthermore, it may be possible to use different enteric virus targets to differentiate between on-going and recent faecal contamination [216]. The persistence of these viruses relative to bacterial indicators could be advantageous when tracking faecal contamination at a distance from the source. Advances in viral detection by next generation sequencing and metagenomics has enhanced the utility of enteric viruses for MST [19]. More recently, detection of adenoviruses by qPCR have been shown to correlate well with somatic coliphages presence and they could possibly by used as an index pathogen for Quantitative Microbial Risk Assessment [181].

8.6. Summary and future outlook

Waterborne pathogens continue to infect and cause illnesses worldwide. Routine analysis of water samples has paved the way to supplying safe water for all to drink. Traditional techniques suffer from limitations, including laborious sample preparation, bulky instrumentation, and slow data readout. In view of the urgency for sensitive, specific, robust, and rapid diagnostics, numerous advancements have been made in the area of diagnostics.

The application of molecular methods to pathogen detection is major breakthrough in our ability to demonstrate the responsibility of specific agents as causes of disease. The use of culture independent testing methods for the detection of pathogens is well established in the clinical setting [217] but for environmental samples, the methodology lags behind due to sample complexity, multiple organism contamination and the inability to culture the pathogens at all. For this reason, detection of faecal indicator bacteria on selective culture media has been the gold standard. However, the inconsistency between waterborne pathogen presence and FIBs has paved the way to alternative testing methods. Theoretically molecular methods have been advertised as being able to provide concise qualitative and quantitative data on the presence of any microbial pathogenic organism in drinking water, wastewater, and sewage samples. This review has summarised some of the methods that have proposed as alternatives to indicator bacteria testing (Table 8.1).

First of all, regardless of method to be utilised, the state of the sample can affect the outcome of the testing. Environmental samples often contain chemical contaminants such as humic acids and metals that need to be diluted or removed prior to extraction of DNA/RNA, proteins or fatty acids since they can interfere with subsequent detection methods. This is complicated by the fact that pathogens can be present at very low levels requiring the samples to be concentrated which increases the likelihood of inhibition due to contaminates. This is particularly true with samples needing to be monitored for virus and protozoa since samples of to 100 L may need to be concentrated to ensure detection since infectivity of the pathogen can be as low as one particle.

For viral sample concentration methods, environmental virology has focused on filtration methods that can maximise viral trapping within the filter during filtration and maximise viral detachment from the filter for recovery [19]. The recent increase in more robust and reliable concentration methods will increase the likelihood of detecting of all types of waterborne pathogens.

Once the sample is prepared, isolation of the macromolecules requires specialised protocols for both the matrix in question and the pathogens to be detected. There has been much success with extraction methods but still one has not been universally accepted as the method to follow as a routine testing protocol. High quality extraction of the macromolecules from samples is highly variable and dependent of the composition (organic matter, humic acids, detergents, pesticides, herbicides, pharmaceuticals, etc.) of target water samples. Validation of the method for reproducibility between operators and laboratories also requires cooperation between research and testing labs.

Low pathogen numbers equate generate low molecule concentrations that can cause random fluctuations in PCR efficiency of detection. Since sample may contain a high variety of pathogens (virus, bacteria, protozoa), competition for nucleic acid templates may interfere with single pathogen detection and confirmatory results may require multiple analysis.

One direction that current research has taken in order to improve concentration, purification and detection of waterborne pathogens is to use immunomagnetic separation. Although it stills requires a priori knowledge of the target, it can bind specifically and concentrate the pathogen while leaving the contaminates behind.

Many methods have been designed and developed for detection of pathogens targets over the last decade yet in the case of DNA based methods no consensus on an universal primer set for the detection of common waterborne pathogens have been established nor have the target genes been agreed on. In the end, the primers sets will need to consistently amplify targets with high reproducibility, sensitivity and specificity. One concern is the ability to establish the physiological condition of the target organism: dead or alive, infectious or noninfectious. Targets that have developed around rRNA genes have had the most success at determining the infectivity of the pathogen since RNA degrades faster than DNA after the death of a cell. The lack of RNA detection however may still occur in viable but not culturable cells and initiate incorrect risk assessment decisions.

On the positive side, most waterborne pathogens can be removed from drinking water with standard drinking water treatment measures. Most if not all outbreaks are tracked back to untreated or inadequately treated waters.

The future of pathogen detection follows the culture-independent, library-independent trend for microbial target detection. An ideal method would allow for a real-time flow thru system to monitor water systems on a continuous basis that could be miniaturised and automated to allow for ease of use and economy of costs. Recently it has been shown that the technical problems with PCR can be overcome by the development and use of isothermal techniques for the nucleic acid amplification such as LAMP (loop-mediated isothermal amplification). This technology when coupled to a photomultiplier tube for fluorescent detection and a microfluidic control module was able to simultaneously detection 4 different pathogens from fish within 65 min with a detection limit of 20 copies [89]. Detection of multiple pathogens can also be accomplished with next-generation sequencing and metagenomic analysis. This approach provides a platform to characterise complex microbial communities although bioinformatic profiles of pathogens in the metagenomic databases is limiting and it is still a challenge to link metagenomic sequences as seen with viral diversity directly to disease [218, 219].

Cell-based sensors are also an emerging frontier in the area of nanodiagnostics [126]. The use of cells as sensors is a very attractive way to devise sensitive biochemical detectors. The main advantages of whole cells as biosensors are that cells have built-in natural selectivity to biologically active chemicals and they can react to analytes in a physiologically relevant way. The transduction of the cell sensor signals maybe achieved by the measurement of transmembrane and cellular potentials, impedance changes, metabolic activity, analyte inducible emission of genetically engineered reporter signals, and optically by means of fluorescence or luminescence [126]. Significant challenges exist for long-term operation since the cells need to be kept alive and healthy under various harsh operating conditions. Much work has been done towards this front, as this technology has been extended to demonstrate automated portable cell based biosensors platform that have been field tested [220]. Furthermore, direct measurement of current through ion channels in the cells has also been used to develop on-chip patch clamp devices, which can potentially be very sensitive to changes in the ambient conditions of the cells. Lastly, many of the new innovative approaches are utilising the unique properties of nanomaterials in order to achieve detection of infectious agents. So far however, most of the research has concentrated on diagnosis of infectious diseases and not on detection of pathogens from water sources [38].

Although humans have been protected from consuming infectious particles from water bodies through water treatment and faecal source tracking using indicator organism, the paradigm has shifted to being able to directly detect the pathogen. With biosensors, microfluidic processing and nanotechnology advancing in leaps and bounds, it is reasonable to assume that this vision is not too far in the future.

References

[1] Health Canada, Canadian Drinking Water Guidelines, Gov Canada, 2012.

[2] Health Canada, Guidelines for Canadian Drinking Water Quality-Summary Table, Gov Canada, 2019. https://www.canada.ca/content/dam/hc-sc/migration/hc-sc/ewh-semt/alt_formats/pdf/pubs/water-eau/sum_guide-res_recom/sum_guide-res_recom-eng.pdf.

[3] K.G. Field, M. Samadpour, Fecal source tracking, the indicator paradigm, and managing water quality, Water Res. (2007), https://doi.org/10.1016/j.watres.2007.06.056.

[4] Health Canada, *Escherichia coli* in Drinking Water, 2019, Available at: https://www.canada.ca/en/health-canada/programs/consultation-e-coli-drinking-water/document.html.

[5] R. Girones, et al., Molecular detection of pathogens in water—the pros and cons of molecular techniques, Water Res. (2010), https://doi.org/10.1016/j.watres.2010.06.030.

[6] U.S. Environmental Protection Agency, Method 1106.1: Enterococci in Water by Membrane Filtration Using Membrane-Enterococcus-Esculin Iron Agar (mE-EIA), in: September, United States Environmental Protection Agency, 16, 2002.

[7] C.P. Gerba, Indicator microorganisms, in: Environmental Microbiology, third ed., 2014, https://doi.org/10.1016/B978-0-12-394626-3.00023-5.

[8] N.H. Tran, K.Y.-H. Gin, H.H. Ngo, Fecal pollution source tracking toolbox for identification, evaluation and characterization of fecal contamination in receiving urban surface waters and groundwater, Sci. Total Environ. 538 (2015) 38–57.

[9] A. Casanovas-Massana, et al., Predicting fecal sources in waters with diverse pollution loads using general and molecular host-specific indicators and applying machine learning methods, J. Environ. Manag. 151 (2015) 317–325.

[10] V.J. Harwood, et al., Performance of viruses and bacteriophages for fecal source determination in a multi-laboratory, comparative study, Water Res. (2013), https://doi.org/10.1016/j.watres.2013.04.064.

[11] V.J. Harwood, C. Staley, B.D. Badgley, K. Borges, A. Korajkic, Microbial source tracking markers for detection of fecal contamination in environmental waters: relationships between pathogens and human health outcomes, FEMS Microbiol. Rev. (2014), https://doi.org/10.1111/1574-6976.12031.

[12] W. Ahmed, S. Sawant, F. Huygens, A. Goonetilleke, T. Gardner, Prevalence and occurrence of zoonotic bacterial pathogens in surface waters determined by quantitative PCR, Water Res. (2009), https://doi.org/10.1016/j.watres.2009.03.041.

[13] W. Ahmed, F. Huygens, A. Goonetilleke, T. Gardner, Real-time PCR detection of pathogenic microorganisms in roof-harvested rainwater in Southeast Queensland, Australia, Appl. Environ. Microbiol. (2008), https://doi.org/10.1128/AEM.00331-08.

[14] G. Wilkes, et al., Seasonal relationships among indicator bacteria, pathogenic bacteria, Cryptosporidium oocysts, Giardia cysts, and hydrological indices for surface waters within an agricultural landscape, Water Res. (2009), https://doi.org/10.1016/j.watres.2009.01.033.

[15] K. Lemarchand, P. Lebaron, Occurrence of Salmonella spp. and Cryptosporidium spp. in a French coastal watershed: relationship with fecal indicators, FEMS Microbiol. Lett. (2003), https://doi.org/10.1016/S0378-1097(02)01135-7.

[16] T.G. Aw, J.B. Rose, Detection of pathogens in water: from phylochips to qPCR to pyrosequencing, Curr. Opin. Biotechnol. (2012), https://doi.org/10.1016/j.copbio.2011.11.016.

[17] T.M. Straub, D.P. Chandler, Towards a unified system for detecting waterborne pathogens, J. Microbiol. Methods (2003), https://doi.org/10.1016/S0167-7012(03)00023-X.

[18] M.J. Allen, J.L. Clancy, E.W. Rice, Plain, hard truth about pathogen monitoring, J. Am. Water Works Assoc. (2000), https://doi.org/10.1002/j.1551-8833.2000.tb09005.x.

[19] K. Wong, T.T. Fong, K. Bibby, M. Molina, Application of enteric viruses for fecal pollution source tracking in environmental waters, Environ. Int. (2012), https://doi.org/10.1016/j.envint.2012.02.009.

[20] J. Jofre, A.R. Blanch, Feasibility of methods based on nucleic acid amplification techniques to fulfil the requirements for microbiological analysis of water quality, J. Appl. Microbiol. (2010), https://doi.org/10.1111/j.1365-2672.2010.04830.x.

[21] B. Slabbinck, B. De Baets, P. Dawyndt, P. De Vos, Towards large-scale FAME-based bacterial species identification using machine learning techniques, Syst. Appl. Microbiol. (2009), https://doi.org/10.1016/j.syapm.2009.01.003.

[22] A. Lipski, E. Spieck, A. Makolla, K. Altendorf, Fatty acid profiles of nitrite-oxidizing bacteria reflect their phylogenetic heterogeneity, Syst. Appl. Microbiol. (2001), https://doi.org/10.1078/0723-2020-00049.

[23] A.M. Glucksman, H.D. Skipper, R.L. Brigmon, J.W.S. Domingo, Use of the MIDI-FAME technique to characterize groundwater communities, J. Appl. Microbiol. 88 (4) (2000) 711–719.

[24] S. Parveen, N.C. Hodge, R.E. Stall, S.R. Farrah, M.L. Tamplin, Phenotypic and genotypic characterization of human and nonhuman *Escherichia coli*, Water Res. (2001), https://doi.org/10.1016/S0043-1354(00)00269-4.

[25] A.G. Werker, J. Becker, C. Huitema, Assessment of activated sludge microbial community analysis in full-scale biological wastewater treatment plants using patterns of fatty acid isopropyl esters (FAPEs), Water Res. (2003), https://doi.org/10.1016/S0043-1354(02)00625-5.

[26] H. Moura, M. Ospina, A.R. Woolfitt, J.R. Barr, G.S. Visvesvara, Analysis of four human microsporidian isolates by MALDI-TOF mass spectrometry, J. Eukaryot. Microbiol. (2003), https://doi.org/10.1111/j.1550-7408.2003.tb00110.x.

[27] M.J. Donohue, A.W. Smallwood, S. Pfaller, M. Rodgers, J.A. Shoemaker, The development of a matrix-assisted laser desorption/ionization mass spectrometry-based method for the protein fingerprinting and identification of Aeromonas species using whole cells, J. Microbiol. Methods (2006), https://doi.org/10.1016/j.mimet.2005.08.005.

[28] K. De Bruyne, et al., Bacterial species identification from MALDI-TOF mass spectra through data analysis and machine learning, Syst. Appl. Microbiol. (2011), https://doi.org/10.1016/j.syapm.2010.11.003.

[29] N. Singhal, M. Kumar, P.K. Kanaujia, J.S. Virdi, MALDI-TOF mass spectrometry: an emerging technology for microbial identification and diagnosis, Front. Microbiol. (2015), https://doi.org/10.3389/fmicb.2015.00791.

[30] J. Gopal, N. Hasan, H.F. Wu, Fabrication of titanium based MALDI bacterial chips for rapid, sensitive and direct analysis of pathogenic bacteria, Biosens. Bioelectron. (2013), https://doi.org/10.1016/j.bios.2012.06.036.

[31] M. Hausner, et al., The use of immunological techniques and scanning confocal laser microscopy for the characterization of *Agrobacterium tumefaciens* ad *Pseudomonas fluorescens*, Atrizine-utilizing biofilms, in: H.-C. Flemming, U. Szewzyk, T. Griebe (Eds.), Biofilms: Investigative Methods and Applications, Technomic Publ., Lancaster, Basel, 2000, pp. 143–154.

[32] P.V. Hornbeck, Enzyme-linked immunosorbent assays, Curr. Protoc. Immunol. (2015), https://doi.org/10.1002/0471142735.im0201s110.

[33] P.Y. Bin, Y.H. Cho, Y.M. Jee, G.P. Ko, Immunomagnetic separation combined with real-time reverse transcriptase PCR assays for detection of norovirus in contaminated food, Appl. Environ. Microbiol. (2008), https://doi.org/10.1128/AEM.00013-08.

[34] S. Park, Y.T. Kim, Y.K. Kim, Optical enzyme-linked immunosorbent assay on a strip for detection of *Salmonella typhimurium*, Biochip J. (2010), https://doi.org/10.1007/s13206-010-4204-y.

[35] D.R. Shelton, et al., Estimation of viable *Escherichia coli* O157 in surface waters using enrichment in conjunction with immunological detection, J. Microbiol. Methods (2004), https://doi.org/10.1016/j.mimet.2004.03.017.

[36] L.M. Fincher, C.D. Parker, C.P. Chauret, Occurrence and antibiotic resistance of *Escherichia coli* O157:H7 in a watershed in north-central Indiana, J. Environ. Qual. (2009), https://doi.org/10.2134/jeq2008.0077.

[37] J. Brunt, M.D. Webb, M.W. Peck, Rapid affinity immunochromatography column-based tests for sensitive detection of *Clostridium botulinum* neurotoxins and *Escherichia coli* O157, Appl. Environ. Microbiol. (2010), https://doi.org/10.1128/AEM.03059-09.

[38] S.B. Shinde, C.B. Fernandes, V.B. Patravale, Recent trends in in-vitro nanodiagnostics for detection of pathogens, J. Control. Release (2012), https://doi.org/10.1016/j.jconrel.2011.11.033.

[39] E.B. Setterington, E.C. Alocilja, Rapid electrochemical detection of polyaniline-labeled *Escherichia coli* O157:H7, Biosens. Bioelectron. (2011), https://doi.org/10.1016/j.bios.2010.09.036.

[40] M. Haffar, K.A. Gilbride, The utility and application of real-time PCR and FISH in the detection of single-copy gene targets in *Escherichia coli* O157:H7 and *Salmonella typhimurium*, Can. J. Microbiol. 56 (3) (2010), https://doi.org/10.1139/W09-126.

[41] C. García-Aljaro, M. Muniesa, J. Jofre, A.R. Blanch, Prevalence of the stx2 gene in coliform populations from aquatic environments, Appl. Environ. Microbiol. (2004), https://doi.org/10.1128/AEM.70.6.3535-3540.2004.

[42] R.S. Quilliam, A.P. Williams, L.M. Avery, S.K. Malham, D.L. Jones, Unearthing human pathogens at the agricultural-environment interface: a review of current methods for the detection of *Escherichia coli* O157 in freshwater ecosystems, Agric. Ecosyst. Environ. (2011), https://doi.org/10.1016/j.agee.2011.01.019.

[43] R. Awais, K. Miyanaga, H. Unno, Y. Tanji, Occurrence of virulence genes associated with enterohemorrhagic *Escherichia coli* in raw municipal sewage, Biochem. Eng. J. 33 (1) (2007) 53–59.

[44] M.M. Moritz, H.C. Flemming, J. Wingender, Integration of *Pseudomonas aeruginosa* and *Legionella pneumophila* in drinking water biofilms grown on domestic plumbing materials, Int. J. Hyg. Environ. Health (2010), https://doi.org/10.1016/j.ijheh.2010.05.003.

[45] R. Saiki, et al., Primer-directed enzymatic amplification of DNA with a thermostable DNA polymerase, Science (80-) (1988), https://doi.org/10.1126/science.239.4839.487.

[46] N. Jothikumar, M.D. Sobsey, T.L. Cromeans, Development of an RNA extraction protocol for detection of waterborne viruses by reverse transcriptase quantitative PCR (RT-qPCR), J. Virol. Methods (2010), https://doi.org/10.1016/j.jviromet.2010.06.005.

[47] Y. Li, C. Zhang, D. Xing, Integrated microfluidic reverse transcription-polymerase chain reaction for rapid detection of food- or waterborne pathogenic rotavirus, Anal. Biochem. (2011), https://doi.org/10.1016/j.ab.2011.04.026.

[48] Y. Dong, J. Kim, G.D. Lewis, Evaluation of methodology for detection of human adenoviruses in wastewater, drinking water, stream water and recreational waters, J. Appl. Microbiol. (2010), https://doi.org/10.1111/j.1365-2672.2009.04477.x.

[49] P. Wyn-Jones, The detection of waterborne viruses, in: Perspectives in Medical Virology, 2007, https://doi.org/10.1016/S0168-7069(07)17009-9 (Chapter 9).

[50] I.A. Hamza, L. Jurzik, K. Überla, M. Wilhelm, Methods to detect infectious human enteric viruses in environmental water samples, Int. J. Hyg. Environ. Health (2011), https://doi.org/10.1016/j.ijheh.2011.07.014.

[51] S. Chakravorty, D. Helb, M. Burday, N. Connell, D. Alland, A detailed analysis of 16S ribosomal RNA gene segments for the diagnosis of pathogenic bacteria, J. Microbiol. Methods (2007), https://doi.org/10.1016/j.mimet.2007.02.005.

[52] I.U.H. Khan, et al., A methods comparison for the isolation and detection of thermophilic Campylobacter in agricultural watersheds, J. Microbiol. Methods (2009), https://doi.org/10.1016/j.mimet.2009.09.024.

[53] I.U.H. Khan, A. Loughborough, T.A. Edge, DNA-based real-time detection and quantification of aeromonads from fresh water beaches on Lake Ontario, J. Water Health (2009), https://doi.org/10.2166/wh.2009.041.

[54] D.Y. Lee, K. Shannon, L.A. Beaudette, Detection of bacterial pathogens in municipal wastewater using an oligonucleotide microarray and real-time quantitative PCR, J. Microbiol. Methods (2006), https://doi.org/10.1016/j.mimet.2005.09.008.

[55] D.Y. Lee, H. Lauder, H. Cruwys, P. Falletta, L.A. Beaudette, Development and application of an oligonucleotide microarray and real-time quantitative PCR for detection of wastewater bacterial pathogens, Sci. Total Environ. (2008), https://doi.org/10.1016/j.scitotenv.2008.03.004.

[56] I.U.H. Khan, et al., Development of a rapid quantitative PCR assay for direct detection and quantification of culturable and non-culturable *Escherichia coli* from agriculture watersheds, J. Microbiol. Methods (2007), https://doi.org/10.1016/j.mimet.2007.02.016.

[57] C.K. Lin, H.Y. Tsen, Comparison of the partial 16S rRNA gene sequences and development of oligonucleotide probes for the detection of *Escherichia coli* cells in water and milk, Food Microbiol. (1999), https://doi.org/10.1006/fmic.1999.0277.

[58] A.K. Bej, R.J. Steffan, J. DiCesare, L. Haff, R.M. Atlas, Detection of coliform bacteria in water by polymerase chain reaction and gene probes, Appl. Environ. Microbiol. 56 (2) (1990) 307–314.

[59] A.K. Bej, J.L. DiCesare, L. Haff, R.M. Atlas, Detection of *Escherichia coli* and Shigella spp. in water by using the polymerase chain reaction and gene probes for uid, Appl. Environ. Microbiol. 57 (4) (1991) 1013–1017.

[60] D. Juck, J. Ingram, M. Prévost, J. Coallier, C. Greer, Nested PCR protocol for the rapid detection of *Escherichia coli* in potable water, Can. J. Microbiol. (1996), https://doi.org/10.1139/m96-110.

[61] L. Heijnen, G. Medema, Quantitative detection of *E. coli*, *E. coli* O157 and other shiga toxin producing *E. coli* in water samples using a culture method combined with real-time PCR, J. Water Health (2006), https://doi.org/10.2166/wh.2006.026.

[62] A.F. Maheux, et al., Analytical comparison of nine PCR primer sets designed to detect the presence of *Escherichia coli*/Shigella in water samples, Water Res. (2009), https://doi.org/10.1016/j.watres.2009.04.017.

[63] S.T.S.T. Clark, et al., Evaluation of low-copy genetic targets for waterborne bacterial pathogen detection via qPCR, Water Res. 45 (11) (2011) 3378–3388.

[64] K. Horáková, H. Mlejnková, P. Mlejnek, Evaluation of methods for isolation of DNA for polymerase chain reaction (PCR)-based identification of pathogenic bacteria from pure cultures and water samples, Water Sci. Technol. (2008), https://doi.org/10.2166/wst.2008.453.

[65] S. Bonetta, et al., Development of a PCR protocol for the detection of *Escherichia coli* O157:H7 and Salmonella spp. in surface water, Environ. Monit. Assess. (2011), https://doi.org/10.1007/s10661-010-1650-x.

[66] P.F.G. Wolffs, K. Glencross, R. Thibaudeau, M.W. Griffiths, Direct quantitation and detection of salmonellae in biological samples without enrichment, using two-step filtration and real-time PCR, Appl. Environ. Microbiol. (2006), https://doi.org/10.1128/AEM.02112-05.

[67] A. Thompson, G. Rowley, M. Alston, V. Danino, J.C.D. Hinton, Salmonella transcriptomics: relating regulons, stimulons and regulatory networks to the process of infection, Curr. Opin. Microbiol. (2006), https://doi.org/10.1016/j.mib.2005.12.010.

[68] J. Theron, D. Morar, M. Du Preez, V.S. Brözel, S.N. Venter, A sensitive seminested PCR method for the detection of Shigella in spiked environmental water samples, Water Res. (2001), https://doi.org/10.1016/S0043-1354(00)00348-1.

[69] H. Fan, Q. Wu, X. Kou, Co-detection of five species of water-borne bacteria by multiplex PCR, Life Sci. J. 54 (4) (2008) 47–54.

[70] P.A. Rochelle, R. De Leon, M.H. Stewart, R.L. Wolfe, Comparison of primers and optimization of PCR conditions for detection of *Cryptosporidium parvum* and *Giardia lamblia* in water, Appl. Environ. Microbiol. (1997).

[71] D.Y. Lee, P. Seto, R. Korczak, DNA microarray-based detection and identification of waterborne protozoan pathogens, J. Microbiol. Methods (2010), https://doi.org/10.1016/j.mimet.2009.11.015.

[72] F. Izquierdo, et al., Detection of microsporidia in drinking water, wastewater and recreational rivers, Water Res. (2011), https://doi.org/10.1016/j.watres.2011.06.033.

[73] T.M. Straub, et al., Genotyping *Cryptosporidium parvum* with an hsp70 single-nucleotide polymorphism microarray, Appl. Environ. Microbiol. (2002), https://doi.org/10.1128/AEM.68.4.1817-1826.2002.

[74] I. Bertrand, C. Gantzer, T. Chesnot, J. Schwartzbrod, Improved specificity for *Giardia lamblia* cyst quantification in wastewater by development of a real-time PCR method, J. Microbiol. Methods (2004), https://doi.org/10.1016/j.mimet.2003.11.016.

[75] A.J. Baeumner, M.C. Humiston, R.A. Montagna, R.A. Durst, Detection of viable oocysts of *Cryptosporidium parvum* following nucleic acid sequence based amplification, Anal. Chem. (2001), https://doi.org/10.1021/ac001293h.

[76] A. de Lucio, et al., Environment Agency: the microbiology of drinking water (2010)—part 14—methods for the isolation, identification and enumeration, Water Res. 9 (1) (2017) 2006–2008.

[77] Z. Wang, G.J. Vora, D.A. Stenger, Detection and genotyping of *Entamoeba histolytica*, *Entamoeba dispar*, *Giardia lamblia*, and *Cryptosporidium parvum* by oligonucleotide microarray, J. Clin. Microbiol. (2004), https://doi.org/10.1128/JCM.42.7.3262-3271.2004.

[78] D. Wiemer, et al., Real-time multiplex PCR for simultaneous detection of *Campylobacter jejuni*, Salmonella, Shigella and Yersinia species in fecal samples, Int. J. Med. Microbiol. (2011), https://doi.org/10.1016/j.ijmm.2011.06.001.

[79] Y. Cao, M.R. Raith, J.F. Griffith, Droplet digital PCR for simultaneous quantification of general and human-associated fecal indicators for water quality assessment, Water Res. 70 (2015) 337–349.

[80] M.B. Cooley, D. Carychao, L. Gorski, Optimized co-extraction and quantification of DNA from enteric pathogens in surface water samples near produce fields in California, Front. Microbiol. (2018), https://doi.org/10.3389/fmicb.2018.00448.

[81] N. Cook, The use of NASBA for the detection of microbial pathogens in food and environmental samples, J. Microbiol. Methods (2003), https://doi.org/10.1016/S0167-7012(03)00022-8.

[82] E.M. Fykse, et al., Real-time PCR and NASBA for rapid and sensitive detection of *Vibrio cholerae* in ballast water, Mar. Pollut. Bull. (2012), https://doi.org/10.1016/j.marpolbul.2011.12.007.

[83] J. Jean, B. Blais, A. Darveau, I. Fliss, Simultaneous detection and identification of hepatitis A virus and rotavirus by multiplex nucleic acid sequence-based amplification (NASBA) and microtiter plate hybridization system, J. Virol. Methods (2002), https://doi.org/10.1016/S0166-0934(02)00096-4.

[84] M.D. Wyer, et al., Relationships between human adenoviruses and faecal indicator organisms in European recreational waters, Water Res. (2012), https://doi.org/10.1016/j.watres.2012.04.008.

[85] P. Karanis, J. Ongerth, LAMP—a powerful and flexible tool for monitoring microbial pathogens, Trends Parasitol. (2009), https://doi.org/10.1016/j.pt.2009.07.010.

[86] M.A. Bakheit, et al., Sensitive and specific detection of Cryptosporidium species in PCR-negative samples by loop-mediated isothermal DNA amplification and confirmation of generated LAMP products by sequencing, Vet. Parasitol. (2008), https://doi.org/10.1016/j.vetpar.2008.09.012.

[87] M.F. Lee, Y.H. Chen, C.F. Peng, Evaluation of reverse transcription loop-mediated isothermal amplification in conjunction with ELISA-hybridization assay for molecular detection of *Mycobacterium tuberculosis*, J. Microbiol. Methods (2009), https://doi.org/10.1016/j.mimet.2008.10.005.

[88] Y. Suzuki, et al., Comparison of real-time reverse-transcription loop-mediated isothermal amplification and real-time reverse-transcription polymerase chain reaction for detection of noroviruses in municipal wastewater, J. Biosci. Bioeng. (2011), https://doi.org/10.1016/j.jbiosc.2011.06.012.

[89] W.H. Chang, et al., Rapid isolation and detection of aquaculture pathogens in an integrated microfluidic system using loop-mediated isothermal amplification, Sensors Actuators B Chem. (2013), https://doi.org/10.1016/j.snb.2011.12.054.

[90] C. Gallas-Lindemann, I. Sotiriadou, M.R. Mahmoodi, P. Karanis, Detection of *Toxoplasma gondii* oocysts in different water resources by Loop Mediated Isothermal Amplification (LAMP), Acta Trop. (2013), https://doi.org/10.1016/j.actatropica.2012.10.007.

[91] C. Maynard, et al., Waterborne pathogen detection by use of oligonucleotide-based microarrays, Appl. Environ. Microbiol. (2005), https://doi.org/10.1128/AEM.71.12.8548-8557.2005.

[92] R.Y.C. Kong, S.K.Y. Lee, T.W.F. Law, S.H.W. Law, R.S.S. Wu, Rapid detection of six types of bacterial pathogens in marine waters by multiplex PCR, Water Res. (2002) 00503–00506, https://doi.org/10.1016/S0043-1354(01.

[93] G. Zhou, et al., Development of a DNA microarray for detection and identification of *Legionella pneumophila* and ten other pathogens in drinking water, Int. J. Food Microbiol. (2011), https://doi.org/10.1016/j.ijfoodmicro.2011.01.014.

[94] J. Weidhaas, A. Anderson, R. Jamal, Elucidating waterborne pathogen presence and aiding source apportionment in an impaired stream, Appl. Environ. Microbiol. 84 (6) (2018), https://doi.org/10.1128/AEM.02510-17.

[95] S. Tyagi, F.R. Kramer, Molecular beacons: probes that fluoresce upon hybridization, Nat. Biotechnol. (1996), https://doi.org/10.1038/nbt0396-303.

[96] H.Y. Yeh, M.V. Yates, W. Chen, A. Mulchandani, Real-time molecular methods to detect infectious viruses, Semin. Cell Dev. Biol. (2009), https://doi.org/10.1016/j.semcdb.2009.01.012.

[97] S. Park, et al., Rapid identification of *Candida dubliniensis* using a species-specific molecular beacon, J. Clin. Microbiol. (2000).

[98] H. Kim, et al., A molecular beacon DNA microarray system for rapid detection of *E. coli* O157:H7 that eliminates the risk of a false negative signal, Biosens. Bioelectron. (2007), https://doi.org/10.1016/j.bios.2006.04.032.

[99] J.L. McKillip, M. Drake, Molecular beacon polymerase chain reaction detection of *Escherichia coli* O157:H7 in milk, J. Food Prot. (2000), https://doi.org/10.4315/0362-028X-63.7.855.

[100] J. Singh, V.K. Batish, S. Grover, Molecular beacon based real-time PCR assay for simultaneous detection of *Listeria monocytogenes* and Salmonella spp. in dairy products, Dairy Sci. Technol. (2011), https://doi.org/10.1007/s13594-011-0007-8.

[101] S. Sandhya, W. Chen, A. Mulchandani, Molecular beacons: a real-time polymerase chain reaction assay for detecting *Escherichia coli* from fresh produce and water, Anal. Chim. Acta (2008), https://doi.org/10.1016/j.aca.2008.03.026.

[102] A. Jyoti, et al., Contamination of surface and potable water in South Asia by Salmonellae: culture-independent quantification with molecular beacon real-time PCR, Sci. Total Environ. (2010), https://doi.org/10.1016/j.scitotenv.2009.11.056.

[103] E. Heyduk, T. Heyduk, Fluorescent homogeneous immunosensors for detecting pathogenic bacteria, Anal. Biochem. (2010), https://doi.org/10.1016/j.ab.2009.09.039.

[104] R.H. Kimura, E.R. Steenblock, J.A. Camarero, Development of a cell-based fluorescence resonance energy transfer reporter for *Bacillus anthracis* lethal factor protease, Anal. Biochem. (2007), https://doi.org/10.1016/j.ab.2007.05.014.

[105] T. Li, J.Y. Byun, B.B. Kim, Y.B. Shin, M.G. Kim, Label-free homogeneous FRET immunoassay for the detection of mycotoxins that utilizes quenching of the intrinsic fluorescence of antibodies, Biosens. Bioelectron. (2013), https://doi.org/10.1016/j.bios.2012.10.085.

[106] J.H. Han, B.C. Heinze, J.Y. Yoon, Single cell level detection of *Escherichia coli* in microfluidic device, Biosens. Bioelectron. (2008), https://doi.org/10.1016/j.bios.2007.11.013.

[107] K.S. Chow, H. Du, Dielectrophoretic characterization and trapping of different water-borne pathogen in continuous flow manner, Sensors Actuators A Phys. (2011), https://doi.org/10.1016/j.sna.2011.03.053.

[108] T. Taguchi, et al., Detection of *Cryptosporidium parvum* oocysts using a microfluidic device equipped with the SUS micromesh and FITC-labeled antibody, Biotechnol. Bioeng. (2007), https://doi.org/10.1002/bit.21104.

[109] N. Ramalingam, et al., Real-time PCR-based microfluidic array chip for simultaneous detection of multiple waterborne pathogens, Sensors Actuators B Chem. (2010), https://doi.org/10.1016/j.snb.2009.11.025.

[110] M.G. Beeman, et al., Electrochemical detection of *E. coli* O157:H7 in water after electrocatalytic and ultraviolet treatments using a polyguanine-labeled secondary bead sensor, Sensors (Switzerland) (2018), https://doi.org/10.3390/s18051497.

[111] T.F. Wu, et al., Rapid waterborne pathogen detection with mobile electronics, Sensors (Switzerland) (2017), https://doi.org/10.3390/s17061348.

[112] R.L. Caygill, G.E. Blair, P.A. Millner, A review on viral biosensors to detect human pathogens, Anal. Chim. Acta (2010), https://doi.org/10.1016/j.aca.2010.09.038.

[113] S. Singh, et al., Recent trends in development of biosensors for detection of microcystin, Toxicon (2012), https://doi.org/10.1016/j.toxicon.2012.06.005.

[114] O.A. Sadik, F. Yan, Novel fluorescent biosensor for pathogenic toxins using cyclic polypeptide conjugates, Chem. Commun. (2004), https://doi.org/10.1002/chin.200432204.

[115] M. Campàs, D. Szydłowska, M. Trojanowicz, J.L. Marty, Enzyme inhibition-based biosensor for the electrochemical detection of microcystins in natural blooms of cyanobacteria, Talanta (2007), https://doi.org/10.1016/j.talanta.2006.10.012.

[116] A. Liu, et al., Development of electrochemical DNA biosensors, TrAC Trends Anal. Chem. (2012), https://doi.org/10.1016/j.trac.2012.03.008.

[117] A. Erdem, et al., DNA biosensor for Microcystis spp. sequence detection by using methylene blue and ruthenium complex as electrochemical hybridization labels, Turk. J. Chem. (2002).

[118] C. Nakamura, T. Kobayashi, M. Miyake, M. Shirai, J. Miyake, Usage of a DNA aptamer as a ligand targeting microcystin, Mol. Cryst. Liq. Cryst. Sci. Technol. Sect. A Mol. Cryst. Liq. Cryst. (2001), https://doi.org/10.1080/10587250108024762.

[119] G.A. Campbell, R. Mutharasan, Near real-time detection of *Cryptosporidium parvum* oocyst by IgM-functionalized piezoelectric-excited millimeter-sized cantilever biosensor, Biosens. Bioelectron. (2008), https://doi.org/10.1016/j.bios.2007.10.017.

[120] S. Xu, R. Mutharasan, Detection of *Cryptosporidium parvum* in buffer and in complex matrix using PEMC sensors at 5 oocysts mL-1, Anal. Chim. Acta (2010), https://doi.org/10.1016/j.aca.2010.04.056.

[121] R.S. Lakshmanan, S. Xu, R. Mutharasan, Impedance change as an alternate measure of resonant frequency shift of piezoelectric-excited millimeter-sized cantilever (PEMC) sensors, Sensors Actuators B Chem. (2010), https://doi.org/10.1016/j.snb.2009.12.048.

[122] S.R. Nugen, P.J. Asiello, J.T. Connelly, A.J. Baeumner, PMMA biosensor for nucleic acids with integrated mixer and electrochemical detection, Biosens. Bioelectron. (2009), https://doi.org/10.1016/j.bios.2008.12.025.

[123] S. Park, H. Kim, S.H. Paek, J.W. Hong, Y.K. Kim, Enzyme-linked immuno-strip biosensor to detect *Escherichia coli* O157:H7, Ultramicroscopy (2008), https://doi.org/10.1016/j.ultramic.2008.04.063.

[124] A.J. Baeumner, R.N. Cohen, V. Miksic, J. Min, RNA biosensor for the rapid detection of viable *Escherichia coli* in drinking water, Biosens. Bioelectron. (2003), https://doi.org/10.1016/S0956-5663(02)00162-8.

[125] N. Sanvicens, C. Pastells, N. Pascual, M.P. Marco, Nanoparticle-based biosensors for detection of pathogenic bacteria, TrAC Trends Anal. Chem. (2009), https://doi.org/10.1016/j.trac.2009.08.002.

[126] J. Theron, T. Eugene Cloete, M. De Kwaadsteniet, Current molecular and emerging nanobiotechnology approaches for the detection of microbial pathogens, Crit. Rev. Microbiol. (2010), https://doi.org/10.3109/1040841X.2010.489892.

[127] F. Sanger, S. Nicklen, A.R. Coulson, DNA sequencing with chain-terminating inhibitors, Proc. Natl. Acad. Sci. U. S. A. (1977), https://doi.org/10.1073/pnas.74.12.5463.

[128] M. Ronaghi, S. Karamohamed, B. Pettersson, M. Uhlén, P. Nyrén, Real-time DNA sequencing using detection of pyrophosphate release, Anal. Biochem. (1996), https://doi.org/10.1006/abio.1996.0432.

[129] M. Ronaghi, Pyrosequencing sheds light on DNA sequencing, Genome Res. (2001), https://doi.org/10.1101/gr.11.1.3.

[130] K. Bibby, E. Viau, J. Peccia, Pyrosequencing of the 16S rRNA gene to reveal bacterial pathogen diversity in biosolids, Water Res. (2010), https://doi.org/10.1016/j.watres.2010.05.039.

[131] K. Bibby, E. Viau, J. Peccia, Viral metagenome analysis to guide human pathogen monitoring in environmental samples, Lett. Appl. Microbiol. (2011), https://doi.org/10.1111/j.1472-765X.2011.03014.x.

[132] L. Ye, T. Zhang, Pathogenic bacteria in sewage treatment plants as revealed by 454 pyrosequencing, Environ. Sci. Technol. (2011), https://doi.org/10.1021/es201045e.

[133] A. Zahedi, et al., Identification of eukaryotic microorganisms with 18S rRNA next-generation sequencing in wastewater treatment plants, with a more targeted NGS approach required for Cryptosporidium detection, Water Res. (2019), https://doi.org/10.1016/j.watres.2019.04.041.

[134] V.R. Hill, et al., Development of a rapid method for simultaneous recovery of diverse microbes in drinking water by ultrafiltration with sodium polyphosphate and surfactants, Appl. Environ. Microbiol. (2005), https://doi.org/10.1128/AEM.71.11.6878-6884.2005.

[135] X.L. Pang, et al., Pre-analytical and analytical procedures for the detection of enteric viruses and enterovirus in water samples, J. Virol. Methods (2012), https://doi.org/10.1016/j.jviromet.2012.05.014.

[136] A. Bosch, S. Guix, D. Sano, R.M. Pintó, New tools for the study and direct surveillance of viral pathogens in water, Curr. Opin. Biotechnol. (2008), https://doi.org/10.1016/j.copbio.2008.04.006.

[137] E. Haramoto, H. Katayama, E. Utagawa, S. Ohgaki, Recovery of human norovirus from water by virus concentration methods, J. Virol. Methods (2009), https://doi.org/10.1016/j.jviromet.2009.05.002.

[138] L. Huguet, C. Carteret, C. Gantzer, A comparison of different concentration methods for the detection of viruses present in bottled waters and those adsorbed to water bottle surfaces, J. Virol. Methods (2012), https://doi.org/10.1016/j.jviromet.2012.01.005.

[139] T.R. Julian, K.J. Schwab, Challenges in environmental detection of human viral pathogens, Curr. Opin. Virol. (2012), https://doi.org/10.1016/j.coviro.2011.10.027.

[140] L. Diéguez, et al., Disposable microfluidic micromixers for effective capture of *Cryptosporidium parvum* oocysts from water samples, J. Biol. Eng. 12 (2018) 4, https://doi.org/10.1186/s13036-018-0095-6.

[141] V.R. Hill, et al., Comparison of hollow-fiber ultrafiltration to the USEPA VIRADEL technique and USEPA method 1623, J. Environ. Qual. (2009), https://doi.org/10.2134/jeq2008.0152.

[142] R.A. Rodríguez, L. Thie, C.D. Gibbons, M.D. Sobsey, Reducing the effects of environmental inhibition in quantitative real-time PCR detection of adenovirus and norovirus in recreational seawaters, J. Virol. Methods (2012), https://doi.org/10.1016/j.jviromet.2012.01.009.

[143] C. Logan, J.J. O'Leary, N. O'Sullivan, Real-time reverse transcription PCR detection of norovirus, sapovirus and astrovirus as causative agents of acute viral gastroenteritis, J. Virol. Methods (2007), https://doi.org/10.1016/j.jviromet.2007.05.031.

[144] J.B. Gregory, L.F. Webster, J.F. Griffith, J.R. Stewart, Improved detection and quantitation of norovirus from water, J. Virol. Methods (2011), https://doi.org/10.1016/j.jviromet.2010.12.011.

[145] S. Martin-Latil, C. Hennechart-Collette, L. Guillier, S. Perelle, Duplex RT-qPCR for the detection of hepatitis E virus in water, using a process control, Int. J. Food Microbiol. (2012), https://doi.org/10.1016/j.ijfoodmicro.2012.05.001.

[146] N. Jothikumar, T.L. Cromeans, B.H. Robertson, X.J. Meng, V.R. Hill, A broadly reactive one-step real-time RT-PCR assay for rapid and sensitive detection of hepatitis E virus, J. Virol. Methods (2006), https://doi.org/10.1016/j.jviromet.2005.07.004.

[147] P. Gyarmati, et al., Universal detection of hepatitis E virus by two real-time PCR assays: TaqMan® and Primer-Probe Energy Transfer, J. Virol. Methods. (2007), https://doi.org/10.1016/j.jviromet.2007.07.014.

[148] S. Martin-Latil, C. Hennechart-Collette, L. Guillier, S. Perelle, Comparison of two extraction methods for the detection of hepatitis A virus in semi-dried tomatoes and murine norovirus as a process control by duplex RT-qPCR, Food Microbiol. (2012), https://doi.org/10.1016/j.fm.2012.03.007.

[149] G.Y. Kim, A. Son, Development and characterization of a magnetic bead-quantum dot nanoparticles based assay capable of *Escherichia coli* O157:H7 quantification, Anal. Chim. Acta (2010), https://doi.org/10.1016/j.aca.2010.07.046.

[150] M.A. Yáñez, et al., Quantification of viable *Legionella pneumophila* cells using propidium monoazide combined with quantitative PCR, J. Microbiol. Methods (2011), https://doi.org/10.1016/j.mimet.2011.02.004.

[151] J.Y. Hyeon, et al., Development of RNA aptamers for detection of *Salmonella enteritidis*, J. Microbiol. Methods (2012), https://doi.org/10.1016/j.mimet.2012.01.014.

[152] A. Huq, et al., Detection of *Vibrio cholerae* O1 in the aquatic environment by fluorescent-monoclonal antibody and culture methods, Appl. Environ. Microbiol. 56 (8) (1990) 2370–2373.

[153] R.Y.C. Kong, M.M.H. Mak, R.S.S. Wu, DNA technologies for monitoring waterborne pathogens: a revolution in water pollution monitoring, Ocean Coast. Manag. (2009), https://doi.org/10.1016/j.ocecoaman.2009.04.011.

[154] J.F. Mehrabadi, P. Morsali, H.R. Nejad, A.A. Imani Fooladi, Detection of toxigenic *Vibrio cholerae* with new multiplex PCR, J. Infect. Public Health (2012), https://doi.org/10.1016/j.jiph.2012.02.004.

[155] J.K. van Frankenhuyzen, J.T. Trevors, H. Lee, C.A. Flemming, M.B. Habash, Molecular pathogen detection in biosolids with a focus on quantitative PCR using propidium monoazide for viable cell enumeration, J. Microbiol. Methods (2011), https://doi.org/10.1016/j.mimet.2011.09.007.

[156] USEPA, Method 1623.1: Cryptosporidium and Giardia in Water by Filtration/IMS/FA, United States Environmental Protection Agency, 2012.

[157] N.J. Ruecker, et al., The detection of Cryptosporidium and the resolution of mixtures of species and genotypes from water, Infect. Genet. Evol. (2013), https://doi.org/10.1016/j.meegid.2012.09.009.

[158] N.J. Ruecker, et al., Molecular and phylogenetic approaches for assessing sources of Cryptosporidium contamination in water, Water Res. (2012), https://doi.org/10.1016/j.watres.2012.06.045.

[159] W. Li, et al., A novel multiplex PCR coupled with Luminex assay for the simultaneous detection of Cryptosporidium spp., *Cryptosporidium parvum* and *Giardia duodenalis*, Vet. Parasitol. (2010), https://doi.org/10.1016/j.vetpar.2010.05.024.

[160] H. Bridle, et al., Detection of Cryptosporidium in miniaturised fluidic devices, Water Res. (2012), https://doi.org/10.1016/j.watres.2012.01.010.

[161] S.m. Xiao, G.q. Li, W.h. Li, R.q. Zhou, J.w. Yang, Development and application of nested PCR assay for detection of dairy cattle-derived Cyclospora sp, Agric. Sci. China (2007), https://doi.org/10.1016/S1671-2927(08)60015-2.

[162] R. Weber, et al., Improved light-microscopical detection of microsporidia spores in stool and duodenal aspirates. The Enteric Opportunistic Infections Working Group, N. Engl. J. Med. (1992), https://doi.org/10.1056/NEJM199201163260304.

[163] J.B. Weiss, H. van Keulen, T.E. Nash, Classification of subgroups of *Giardia lamblia* based upon ribosomal RNA gene sequence using the polymerase chain reaction, Mol. Biochem. Parasitol. (1992), https://doi.org/10.1016/0166-6851(92)90096-3.

[164] P.J. Adams, P.T. Monis, A.D. Elliot, R.C.A. Thompson, Cyst morphology and sequence analysis of the small subunit rDNA and ef1α identifies a novel Giardia genotype in a quenda (*Isoodon obesulus*) from Western Australia, Infect. Genet. Evol. (2004), https://doi.org/10.1016/j.meegid.2004.05.003.

[165] T. Le Calvez, et al., Detection of free-living amoebae by using multiplex quantitative PCR, Mol. Cell. Probes (2012), https://doi.org/10.1016/j.mcp.2012.03.003.

[166] R.J.R. Molica, et al., Occurrence of saxitoxins and an anatoxin-a(s)-like anticholinesterase in a Brazilian drinking water supply, Harmful Algae (2005), https://doi.org/10.1016/j.hal.2004.11.001.

[167] S.A. Oehrle, B. Southwell, J. Westrick, Detection of various freshwater cyanobacterial toxins using ultra-performance liquid chromatography tandem mass spectrometry, Toxicon (2010), https://doi.org/10.1016/j.toxicon.2009.10.001.

[168] F.A. Dörr, et al., Methods for detection of anatoxin-a(s) by liquid chromatography coupled to electrospray ionization-tandem mass spectrometry, Toxicon (2010), https://doi.org/10.1016/j.toxicon.2009.07.017.

[169] J. Al-Tebrineh, M.M. Gehringer, R. Akcaalan, B.A. Neilan, A new quantitative PCR assay for the detection of hepatotoxigenic cyanobacteria, Toxicon (2011), https://doi.org/10.1016/j.toxicon.2010.12.018.

[170] P. Payment, A. Berte, M. Prévost, B. Ménard, B. Barbeau, Occurrence of pathogenic microorganisms in the Saint Lawrence river (Canada) and comparison of health risks for populations using it as their source of drinking water, Can. J. Microbiol. (2000), https://doi.org/10.1139/w00-022.

[171] C. Beuret, A simple method for isolation of enteric viruses (noroviruses and enteroviruses) in water, J. Virol. Methods (2003), https://doi.org/10.1016/S0166-0934(02)00194-5.

[172] L.M. Villar, V.S. de Paula, L. Diniz-Mendes, E. Lampe, A.M.C. Gaspar, Evaluation of methods used to concentrate and detect hepatitis A virus in water samples, J. Virol. Methods (2006), https://doi.org/10.1016/j.jviromet.2006.06.008.

[173] P.F.M. Teunis, W.J. Lodder, S.H. Heisterkamp, A.M. De Roda Husman, Mixed plaques: statistical evidence how plaque assays may underestimate virus concentrations, Water Res. (2005), https://doi.org/10.1016/j.watres.2005.08.012.

[174] T.M. Straub, et al., In vitro cell culture infectivity assay for human noroviruses, Emerg. Infect. Dis. (2007), https://doi.org/10.3201/eid1303.060549.

[175] N. Casas, E. Sunén, Detection of enteroviruses, hepatitis a virus and rotaviruses in sewage by means of an immunomagnetic capture reverse transcription-PCR assay, Microbiol. Res. (2002), https://doi.org/10.1078/0944-5013-00152.

[176] S.G. Gilpatrick, K.J. Schwab, M.K. Estes, R.L. Atmar, Development of an immunomagnetic capture reverse transcription-PCR assay for the detection of Norwalk virus, J. Virol. Methods (2000), https://doi.org/10.1016/S0166-0934(00)00220-2.

[177] M. Myrmel, E. Rimstad, Y. Wasteson, Immunomagnetic separation of a Norwalk-like virus (genogroup I) in artificially contaminated environmental water samples, Int. J. Food Microbiol. (2000), https://doi.org/10.1016/S0168-1605(00)00262-2.

[178] C. Monceyron, B. Grinde, Detection of hepatitis A virus in clinical and environmental samples by immunomagnetic separation and PCR, J. Virol. Methods (1994), https://doi.org/10.1016/0166-0934(94)90100-7.

[179] B. Grinde, T. Jonassen, H. Ushijima, Sensitive detection of group A rotaviruses by immunomagnetic separation and reverse transcription-polymerase chain reaction, J. Virol. Methods (1995), https://doi.org/10.1016/0166-0934(95)00070-X.

[180] S. Choi, S.C. Jiang, Real-time PCR quantification of human adenoviruses in urban rivers indicates genome prevalence but low infectivity, Appl. Environ. Microbiol. (2005), https://doi.org/10.1128/AEM.71.11.7426-7433.2005.

[181] M. Verani, I. Federigi, G. Donzelli, L. Cioni, A. Carducci, Human adenoviruses as waterborne index pathogens and their use for Quantitative Microbial Risk Assessment, Sci. Total Environ. (2019), https://doi.org/10.1016/j.scitotenv.2018.09.295.

[182] N. Jothikumar, et al., Quantitative real-time PCR assays for detection of human adenoviruses and identification of serotypes 40 and 41, Appl. Environ. Microbiol. (2005), https://doi.org/10.1128/AEM.71.6.3131-3136.2005.

[183] H.A. Brooks, R.M. Gersberg, A.K. Dhar, Detection and quantification of hepatitis A virus in seawater via real-time RT-PCR, J. Virol. Methods (2005), https://doi.org/10.1016/j.jviromet.2005.03.017.

[184] H. Katayama, A. Shimasaki, S. Ohgaki, Development of a virus concentration method and its application to detection of enterovirus and Norwalk virus from coastal seawater, Appl. Environ. Microbiol. (2002), https://doi.org/10.1128/AEM.68.3.1033-1039.2002.

[185] L. Baert, et al., Detection of murine norovirus 1 by using plaque assay, transfection assay, and real-time reverse transcription-PCR before and after heat exposure, Appl. Environ. Microbiol. (2008), https://doi.org/10.1128/AEM.01039-07.

[186] J. Simonet, C. Gantzer, Inactivation of poliovirus 1 and F-specific RNA phages and degradation of their genomes by UV irradiation at 254 nanometers, Appl. Environ. Microbiol. (2006), https://doi.org/10.1128/AEM.01106-06.

[187] K.A. Reynolds, C.P. Gerba, I.L. Pepper, Detection of enteroviruses in marine waters by direct RT-PCR and cell culture, Water Sci. Technol. (1995), https://doi.org/10.2166/wst.1995.0635.

[188] K.S. Reynolds, C.P. Gerba, I.L. Pepper, Rapid PCR-based monitoring of infectious enteroviruses in drinking water, Water Sci. Technol. (1997), https://doi.org/10.1016/S0273-1223(97)00297-7.

[189] Y.J. Jiang, et al., Detection of infectious hepatitis A virus by integrated cell culture/strand-specific reverse transcriptase-polymerase chain reaction, J. Appl. Microbiol. (2004), https://doi.org/10.1111/j.1365-2672.2004.02413.x.

[190] X. Huang, J. Ren, Chemiluminescence detection for capillary electrophoresis and microchip capillary electrophoresis, TrAC Trends Anal. Chem. (2006), https://doi.org/10.1016/j.trac.2005.07.001.

[191] E.L. Delwart, Viral metagenomics, Rev. Med. Virol. (2007), https://doi.org/10.1002/rmv.532.

[192] K. Rosario, M. Breitbart, Exploring the viral world through metagenomics, Curr. Opin. Virol. (2011), https://doi.org/10.1016/j.coviro.2011.06.004.

[193] J. Yoder, et al., Surveillance for waterborne disease and outbreaks associated with drinking water and water not intended for drinking—United States, 2005-2006, Morb. Mortal. Wkly. Rep. Surveill. Summ. 57 (9) (2008) 39–62.

[194] APHA, Standard Methods for the Examination of Water and Wastewater, twenty-second ed., American Public Health Association, American Water Works Association, Water Environment Federation, 2012, https://doi.org/10.1520/E0536-16.2.

[195] S. Ram, P. Vajpayee, R. Shanker, Rapid culture-independent quantitative detection of enterotoxigenic *Escherichia coli* in surface waters by real-time PCR with molecular beacon, Environ. Sci. Technol. (2008), https://doi.org/10.1021/es703033u.

[196] R. Ferretti, I. Mannazzu, L. Cocolin, G. Comi, F. Clementi, Twelve-hour PCR-based method for detection of Salmonella spp. in food, Appl. Environ. Microbiol. (2001), https://doi.org/10.1128/AEM.67.2.977-978.2001.

[197] A. Fey, et al., Establishment of a real-time PCR-based approach for accurate quantification of bacterial RNA targets in water, using Salmonella as a model organism, Appl. Environ. Microbiol. (2004), https://doi.org/10.1128/AEM.70.6.3618-3623.2004.

[198] C. Yang, Y. Jiang, K. Huang, C. Zhu, Y. Yin, Application of real-time PCR for quantitative detection of *Campylobacter jejuni* in poultry, milk and environmental water, FEMS Immunol. Med. Microbiol. (2003), https://doi.org/10.1016/S0928-8244(03)00168-8.

[199] C. Yang, et al., A real-time PCR assay for the detection and quantitation of *Campylobacter jejuni* using SYBR green I and the LightCycler, Yale J. Biol. Med. 1–6 (2004) 125–132.

[200] B. Chang, et al., Specific detection of viable legionella cells by combined use of photoactivated ethidium monoazide and PCR/real-time PCRδ, Appl. Environ. Microbiol. (2009), https://doi.org/10.1128/AEM.00604-08.

[201] C.A. Kreader, Persistence of PCR-detectable *Bacteroides distasonis* from human feces in river water, Appl. Environ. Microbiol. (1998).

[202] B. Cao, et al., 16S-23S rDNA internal transcribed spacer regions in four Proteus species, J. Microbiol. Methods (2009), https://doi.org/10.1016/j.mimet.2009.01.024.

[203] X. Guan, H.J. Zhang, Y.N. Bi, L. Zhang, D.L. Hao, Rapid detection of pathogens using antibody-coated microbeads with bioluminescence in microfluidic chips, Biomed. Microdevices (2010), https://doi.org/10.1007/s10544-010-9421-6.

[204] C. Levantesi, et al., Salmonella in surface and drinking water: occurrence and water-mediated transmission, Food Res. Int. (2012), https://doi.org/10.1016/j.foodres.2011.06.037.

[205] O. Savichtcheva, N. Okayama, S. Okabe, Relationships between Bacteroides 16S rRNA genetic markers and presence of bacterial enteric pathogens and conventional fecal indicators, Water Res. (2007), https://doi.org/10.1016/j.watres.2007.03.028.

[206] O. Savichtcheva, S. Okabe, Alternative indicators of fecal pollution: relations with pathogens and conventional indicators, current methodologies for direct pathogen monitoring and future application perspectives, Water Res. (2006), https://doi.org/10.1016/j.watres.2006.04.040.

[207] L.K. Dick, et al., Host distributions of uncultivated fecal Bacteroidales bacteria reveal genetic markers for fecal source identification, Appl. Environ. Microbiol. (2005), https://doi.org/10.1128/AEM.71.6.3184-3191.2005.

[208] B. Fremaux, J. Gritzfeld, T. Boa, C.K. Yost, Evaluation of host-specific Bacteroidales 16S rRNA gene markers as a complementary tool for detecting fecal pollution in a prairie watershed, Water Res. (2009), https://doi.org/10.1016/j.watres.2009.06.045.

[209] Health Canada, Guidelines for Canadian Drinking Water Quality. Guideline Technical Document. Enteric Protozoa: Giardia and Cryptosporidium, 2012.

[210] D.D. Tambalo, B. Fremaux, T. Boa, C.K. Yost, Persistence of host-associated Bacteroidales gene markers and their quantitative detection in an urban and agricultural mixed prairie watershed, Water Res. 46 (9) (2012) 2891–2904.

[211] M. Bouzid, D. Steverding, K.M. Tyler, Detection and surveillance of waterborne protozoan parasites, Curr. Opin. Biotechnol. 19 (3) (2008) 302–306.

[212] S. Baldursson, P. Karanis, Waterborne transmission of protozoan parasites: review of worldwide outbreaks—an update 2004-2010, Water Res. (2011), https://doi.org/10.1016/j.watres.2011.10.013.

[213] W. Quintero-Betancourt, E.R. Peele, J.B. Rose, *Cryptosporidium parvum* and *Cyclospora cayetanensis*: a review of laboratory methods for detection of these waterborne parasites, J. Microbiol. Methods (2002), https://doi.org/10.1016/S0167-7012(02)00007-6.

[214] M.D. Burch, M.D. Burch, A.W.Q. Centre, A.W.Q. Centre, Effective doses, guidelines & regulations, Adv. Exp. Med. Biol. 841 (2007) 819–841 (Chapter 36).

[215] T.K. Graczyk, D.B. Conn, Molecular markers and sentinel organisms for environmental monitoring, Parasite (2008), https://doi.org/10.1051/parasite/2008153458.

[216] E.K. Lipp, J. Carrie Futch, D.W. Griffin, Analysis of multiple enteric viral targets as sewage markers in coral reefs, Mar. Pollut. Bull. (2007), https://doi.org/10.1016/j.marpolbul.2007.08.001.

[217] C.D. Sibley, G. Peirano, D.L. Church, Molecular methods for pathogen and microbial community detection and characterization: current and potential application in diagnostic microbiology, Infect. Genet. Evol. (2012), https://doi.org/10.1016/j.meegid.2012.01.011.

[218] I.A. Hamza, K. Bibby, Critical issues in application of molecular methods to environmental virology, J. Virol. Methods 266 (2019) 11–24.

[219] N.J. Ashbolt, Microbial contamination of drinking water and human health from community water systems, Curr. Environ. Health Rep. (2015), https://doi.org/10.1007/s40572-014-0037-5.

[220] T.H. Rider, et al., A B Cell-Based Sensor for Rapid Identification of Pathogens, Science (80-) (2003), https://doi.org/10.1126/science.1084920.

Chapter 9

Nanotechnology for detection of waterborne pathogens

Helen Bridle

Institute of Biological Chemistry, Biophysics and Bioengineering, Heriot-Watt University, Edinburgh, Scotland

9.1. Introduction

Nanotechnology is the characterisation, fabrication, and/or manipulation of structures, devices, or materials between 1 and 100 nm. The properties observed in materials at this size range are different from those found in the bulk material, mainly due to a large surface-to-volume ratio. This results in enhanced surface reactivity, quantum confinement effects, enhanced electrical conductivity, and enhanced magnetic properties, among others. For these reasons, nanotechnology has recently been exploited in a wide range of different applications including biotechnology, drug delivery, and other biomedical applications, catalysis, electronics, food, cosmetics, and design of new fabrics and materials, for example construction or sports articles.

In addition to these other applications, nanotechnology presents a great opportunity to develop fast, accurate, and cost-effective diagnostics for the detection of pathogenic infectious agents [1]. For a review of nanotechnology in the identification of microbial pathogens, we recommend this 2010 article [1] and a 2017 one focussing on foodborne pathogens, of which many are the same as waterborne ones [2]. Nanomaterials have also been widely used in treatment processes as recently reviewed (2020 book chapter by Ankita Ohja [3]) and are commercially available as point-of-use devices like LifeStraw and WaterStick as well as incorporated into various membrane products like NanoCeram, ArsenXnp, Karofi, TATA Swach, though further work is necessary to ascertain safe levels of nanomaterial incorporation into such products, considering leaching and following proper toxicity studies [4].

This chapter aims to provide an introduction to nanotechnology for monitoring applications. Particular attention is devoted to examples concerning waterborne pathogens, given in each of the categories of waterborne pathogens. Firstly, an overview of the types of nanomaterials that have been employed in monitoring is given. Secondly, this chapter will focus on how nanotechnology can enhance the sample processing and detection methods presented in earlier chapters.

Waterborne Pathogens: Detection Methods and Applications. https://doi.org/10.1016/B978-0-444-64319-3.00009-5
293

9.2. Background

This section introduces the nanoscale and the types of nanomaterials that exist. Fig. 9.1 indicates the typical size of some of the nanomaterials commonly used in monitoring applications. The figure compares this to the size of pathogens, showing how nanomaterials can be of similar size or even smaller than the detection target to which they are applied. Fig. 9.2 shows schematics representing the structure of different types of nanomaterials.

9.2.1 Gold nanoparticles

Gold nanoparticles (AuNPs) are a diverse group of nanomaterials ranging in size from 5 to 110 nm which can take a variety of different forms including spheres, cubes, nanorods, and nanoribbons. The advantages of AuNPs are their ability to resonantly scatter light, their high chemical stability, their excellent conductivity, and their colour change upon aggregation.

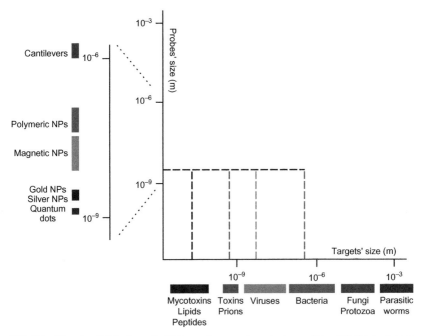

FIG. 9.1 The size distribution of widely used nanosystems in comparison with the most common types of infectious disease agents. A particular nanoparticle (represented by a black horizontal dashed line) could interact differently with targets of various sizes, such as peptides, toxins, viruses, and bacteria (blue, grey, green, and orange dashed lines, respectively). In this particular case, the nanoparticle has a size of 100 nm, whereas virulence factors and disease markers are smaller (e.g. lipids, peptides, DNA, toxins), and pathogens can be of roughly equal (e.g. most viruses) or bigger size than the nanoparticle (e.g. bacteria, fungi, etc.). *Scheme 1 from [1].*

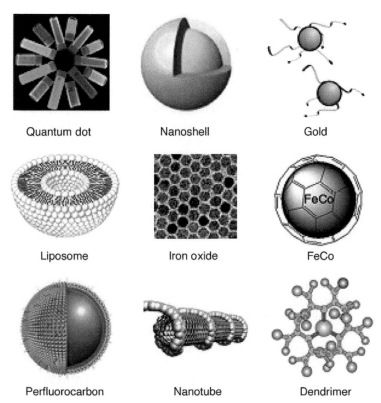

Quantum dot	Nanoshell	Gold
Liposome	Iron oxide	FeCo
Perfluorocarbon	Nanotube	Dendrimer

FIG. 9.2 Nanoparticles currently synthesised for biomedical applications. These include metallic, semiconductor, and organic molecule nanomaterials of a variety of shapes, sizes, and structures. *(Fig. 5 from [5]. Reprinted with permission. Originally reprinted with permission of the Wiley Publishing Company, Cai et al. Small 3 (2007) 1840–1854).*

When dispersed in solution AuNPs are red due to strong localised surface plasmon resonance. However, the colour is highly dependent upon the interparticle distance. A distinct colour change from red to blue is observed when particles aggregate such that the NPs are separated by less than 2.5 times their diameter due to interparticle plasmon coupling. This property has been widely exploited for colorimetric biodetection [6].

The colour change of AuNPs is not the only property making them useful for biological detection. In addition to their unique optical properties, these NPs offer biocompatibility, low toxicity, and easy modification of gold structures to incorporate biofunctionality.

The synthesis of AuNPs was pioneered by Turkevich et al. in 1951 and later refined by Frens et al. in 1973 [7]. The typical method involves the chemical reduction of gold salts, for example reduction of hydrogen tetrachloroaurate ($HAuCl_4$) using citrate, or sodium borohydrate [8], as the reducing agent.

9.2.2 Silver nanoparticles

Like AuNPs, silver nanoparticles (AgNPs) have a typical size of 10–100 nm and can exhibit a wide variety of shapes. The advantages of AgNPs include chemical stability, good conductivity, and excellent optical properties. AgNPs have been employed for signal enhancement of AuNP detection schemes (see Tallury [9]) though they have attracted most interest due to their optical properties since they are extraordinarily efficient at absorbing and scattering light. Example applications include metal enhanced fluorescence and surface-enhanced Raman scattering. In addition, like AuNPs, AgNPs display surface plasmon resonance which can be used for colorimetric detection and quantification of targets [1].

AgNPs are also well known for their antimicrobial properties. Silver has been used for antimicrobial applications for thousands of years and recently AgNPs have been incorporated into consumer products, primarily due to these antibacterial and antifungal properties, such as clothes made of AgNP-containing fabric [10] and personal hygiene products [11]. Others are incorporated into wound dressings [12], nano-silver toothpaste, and colloidal silver suspensions, designed as nutritional supplements [13]. There has also been considerable interest in the use of nanotechnology in water purification. In particular, the use of nanomaterials in small-scale, point-of-use, or emergency response treatment systems has been proposed and investigated. For example, silver-embedded ceramic filters have been trialled in several developing countries and bactericidal silver nanoparticle paper was recently reported [14].

The synthesis of AgNPs is commonly performed via the reduction of silver salts with reducing agents such as sodium borohydrate in the presence of a colloidal stabiliser, including polyvinyl alcohol, bovine serum albumin, or citrate. Green methods of synthesis have also been recently developed [15].

9.2.3 Quantum dots

Fluorescent quantum dots (QDs) are colloidal semiconductors with unique optical properties when compared to conventional semiconductors, organic fluorescent dyes, and proteins. Specifically, these materials offer excellent photostability as well as tuneable emission spectra. For a review of, and introduction to QDs see the work by Arya [16].

A QD often consists of a semiconductor core such as cadmium, mixed with selenium or tellurium, surrounded by a semiconductor outer shell, often zinc sulphide, and a layer of polymer to stabilise the QD and allow (bio)chemical functionalisation. The typical size of the core is 2–10 nm, with a total diameter of 10–20 nm greater than that accounting for the outer shell and polymer. QDs are commonly prepared using colloidal synthesis although other options are molecular beam epitaxy or electron beam lithography.

The emission spectra of QDs are tunable from the ultraviolet (UV) to the near-infrared (NIR) region simply by modulating their size. Simultaneous excitation at a single wavelength of different sized QDs is possible due to the broad Stokes shift of the QDs thus allowing for simple multiplex detection. Besides, QDs are highly bright and photostable, meaning they avoid the problems of self-absorption, self-fluorescence, and photobleaching from which traditional fluorophores suffer. Therefore, it is unsurprising that they have found applications within many different areas, including waterborne pathogen detection.

However, one of the major future challenges is the development and large-scale production of stable and reproducible QDs with biocompatible coatings.

9.2.4 Fluorescent polymer nanoparticles

Polymeric nanoparticles encapsulate a fluorophore within a polymeric coating. Examples include fluorescent silica nanoparticles and these types of material have been applied for detection using fluorescence spectrometers or fluorescent reading microtitre plate readers as well as in flow cytometry. These nanoparticles are synthesised using a linear or branched polymer that promotes the encapsulation of a fluorophore within the nanoparticle's cavity or hydrophobic microdomains. The protection of the fluorophore by the polymer provides high stability under a range of conditions proving superior to organic dyes in terms of photostability and versatility [1].

9.2.5 Magnetic nanoparticles

Magnetic NPs are made of compounds of magnetic elements such as iron, nickel, and cobalt and range in size from 1 to 100 nm. They display excellent conductivity and are also often used to concentrate samples. One example is of iron oxides which have been utilised as contrast agents in magnetic resonance imaging (MRI) as well as the immunomagnetic separation of, and detection of, numerous pathogens. Iron oxides are often synthesised via alkaline precipitation of iron salts and control of nanoparticle shape and magnetic properties are possible via different synthetic strategies, for example temperature, choice of reagents, and reaction times [1].

The heterogeneity of magnetic NPs is one challenge in the further application of these materials. Another is the difficulty in precisely controlling the number of functional molecules added at an NP. Therefore, research is required to develop improved ways to produce magnetic NPs focussing upon ensuring precise composition, uniform surface modification, and reproducible functionalisation [6].

9.2.6 Carbon nanotubes

Vertically aligned carbon nanofibres (VACNFs) have received a lot of research attention as one-dimensional nanoscale electrodes for biosensing since they offer high electrical and thermal conductivities, superior mechanical strength, flexible surface chemistry, biocompatibility, and a wide electrochemical potential window. Carbon nanotubes are an allotrope of carbon in which planar sheets of carbon atoms (graphene) are rolled up into cylindrical tubes (a single-walled carbon nanotube). Multi-walled carbon nanotubes can also be formed. Typical sizes are 0.4–3 nm in diameter and 2–100 nm in length.

Carbon nanotubes (CNTs) are semiconductors with excellent electrical properties and this property is often exploited for signal amplification or transducers for the detection of pathogens. Another possible use for CNTs is the protection of molecular probes, which could be particularly useful in the molecular detection of pathogens in environmental matrices, where DNA probes can easily be degraded [17].

Arumugam reported in 2009 on an improved fabrication method for VACNFs [18], which is one of the main challenges for this type of nanotechnology. In addition to the problems of the fabrication of CNTs, questions have also been raised about their potential toxicity [17].

9.2.7   Polysaccharide nanoparticles

Recently, there has been a greater interest in the use of naturally-derived materials like chitosan. Chitosan is the most important derivative of chitin, which is obtained from crustacean shells or the cell wall of fungi. Nanoparticles can be produced by a variety of means, the most of common of which are ionotropic gelation, microemulsion, emulsification solvent diffusion, and emulsion-based solvent evaporation, and nanoparticles sizes between 10 and 900 nm have been reported; an overview of the fabrication methods and other applications of chitosan nanoparticles was reviewed in 2017 [19]. There are many other natural polysaccharide nanomaterials, the applications of which are being explored, and we recommend a recent review for more details [20].

Chitosan nanoparticles have been produced and shown to be effective against a variety of waterborne pathogens, with promising results and the low-cost and generally nontoxic properties being attractive for use in water treatment technologies. Chitosan has been used with *Cryptosporidium* [21], *Giardia* [22], and *Toxoplasma* [23] as well as been shown to have antibacterial effects, in combination with other compounds [24]. Curcumin nanoparticles have also been used, against *Giardia* [22].

9.3.   Nanotechnology in sample processing

Nanotechnology has not been extensively applied to the challenge of waterborne pathogen sample processing. This section presents the examples found in the literature, which are mainly concerned with viruses and bacteria. Examples solely focussed on water treatment using either filter modified with nanomaterials [25] or those formed from carbon nanotubes [26] are excluded.

In 2008, Wegmann and colleagues modified the surface of microporous ceramic filters by dip-coating with zirconia nanopowder, changing the surface charge, and increasing the surface area at least six-fold. The outcome was a large increase in the retention of the virus-like MS bacteriophage from 75% to over 99.99% (7 log removal) [27].

In 2010, Li and co-workers reported the use of nanoalumina fibres in a disc filter format for the sample processing of viruses [28]. The nanofibres were spatially constrained onto a microglass surface to create the filter. As the nanoalumina fibres are electropositive, negatively charged viruses are attracted to them and captured. A higher recovery rate was reported compared to the Millipore HAWP filter for adenovirus, assessed by flow cytometry. Additional advantages were that no prefiltration or prior adjustment of pH or addition of cations was required and that the filters proved good at rapidly processing large volumes of different environmental samples. These disc filters are also cheaper than the NanoCeram cartridge filters, made of the same material (see Chapter 4 for more details). However, when assessing recovery by q-PCR, the nanoalumina fibres performed less well than the HAWP filter, suggesting that inhibitors of the PCR process are also concentrated by this filter, which is a major disadvantage if the intended detection approach is a molecular method.

In 2014, Jiang et al. developed an impedance-based detection system for waterborne bacteria and utilised a nanoporous paper as part of the concentration step, where a 60 mL sample could be passed through the device using a syringe and bacteria captured on the nanoporous paper in a volume of just 30 μL for direct impedance detection [29]. In 2016, a porous silcon nanowire forest was created inside a microfluidic device for label-free virus capture and release [30].

Depletion flocculation of bacteria using rod-shaped nanoparticles, with a length of 90 nm and a width of 8 nm, was described in 2012 by Sun et al. [31]. The depletion of nonadsorbing polymers for the aggregation of bacterial suspensions has not been well exploited although bacteria can be considered to be a dispersion of negatively charged particles surrounded by nonadsorbing polyelectrolytes, which arise from the bacterially secreted extracellular polymeric substances (EPS). The theory is that colloidal particles have a surrounding depletion zone into which nonadsorbing polymers cannot penetrate. When two particles approach each other, the depletion layers overlap and an imbalance in osmotic pressure, as well as polymer exclusion from between particles maximising entropy, drive coagulation of particles. For more details see [31] and references within.

Aggregation of *Escherichia coli* bacteria through depletion interactions in the presence of nonadsorbing polymer, sodium polystyrene sulfonate has only been reported recently [32]. Sun et al. found that using their cellulose nanocrystals (CNC) flocculation was very effective even at concentrations of 0.1%. They explain the mechanism as follows 'Both bacteria and CNC were negatively charged but the CNC had a higher negative zeta potential than the bacteria. As bacteria get close, CNC rods can no longer enter the gap between the big bacteria particles. Then, CNC rods push bacteria together. Hence, CNC particles would be repelled from bacterial cells and could be expected to flocculate the

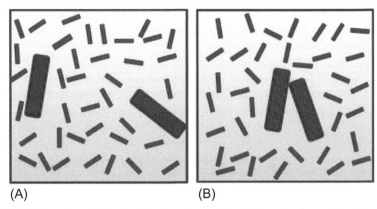

(A)   (B)

FIG. 9.3   Illustration of depletion effect in a mixture of two big bacteria plus small rodlike cellulose nanocrystals. As the bacteria approach each other (*a* and *b*), rod-like nanoparticles cannot enter the gap between the bacteria and bacteria flocculates. *(Fig. 5 from [31]. Reprinted with permission.*

bacteria by depletion' (Fig. 9.3). In 2016, a porous silicon nanowire forest was created inside a microfluidic device for label-free virus capture and release [30].

   There are many reports of the use of antibody-coated magnetic nanoparticles (MNP) for separations linked to different detection formats for foodborne pathogens [33]. However, this strategy is less common for waterborne pathogens. MNPs have a large surface to volume ratio so allow for high capture efficiencies. For *E. coli* immunocapture with these beads was reported to reach 80% compared with 60% for ordinary paramagnetic beads. Varshney et al. reported capture efficiencies of 94% in ground beef samples [34]. The strategy has also been used for seafood samples, though it was combined with a sample preenrichment for the best results [35] as well as for *Salmonella* in a microfluidic device integrated with antibody functionalised magnetic nanoclusters [36]. For foodborne pathogens, simultaneous capture of pathogens as well as the use of apatmer functionalised MNPs have been reported, and have been linked to various methods of detection including IR spectroscopy [37] and bioluminescence [38] as well as smartphone imaging of fluorescent signals [39]. MNPs have also been applied as the separation and concentration element in a detection method called 'biobarcodes' described in Section 9.3.1. One of the major challenges with the MNP approach is the development of robust surface chemistries for the linkage of the MNPs to recognition elements [17].

   MNPs have also been applied in conjunction with microfluidics, particularly with a circular microchannel set-up [40]. The immunosensor chip reported by Agrawal et al. in 2012 was fabricated with microdimensional copper wire and a permanent magnet. The device had two stages; the first being sample isolation and concentration using antibodies conjugated to the MNPs, and the second being detection with fluorescent QDs. The authors stated that the use of the MNPs facilitated the capture of the antigen in a confined space thus enhancing the subsequent fluorescence signal. Cd-Te QDs with different emission wavelengths were conjugated to capture *E. coli* and *S. typhimurium*, respectively. Detection was possible in the range of 10^3–10^7 cfu/mL for a 20 μL sample. The recovery rate of the concentration stage was not reported.

9.4. Nanotechnology in pathogen detection

Most reviews of nanotechnology for pathogen detection consider a wide range of applications without a specific focus upon waterborne pathogens. Biomedical applications seem to have received the most attention [1, 5]. Some 2010 reviews by Hauck [5], and also by Tallury [41], are particularly interesting since they focus on the application of nanotechnology as a means of improving detection for developing country settings [5, 41]. A table adapted from the second review gives an overview of the uses of nanotechnology in the detection of a range of pathogens, many of them waterborne (Table 9.1). Herein, we review how nanotechnology has been applied to enhance the detection of waterborne pathogens, or to enable new detection schemes.

9.4.1 Optical detection

Improved optical properties are a well known feature for a variety of different nanomaterials, in particular gold and silver NPs and QDs. Section 9.3.1 describes the application of these nanomaterials to enhance fluorescent detection schemes, including in flow cytometry, the use of AuNPs in the colorimetric assays enabled by the unique colour change properties of AuNPs and the use of NPs to enhance Raman detection through surface-enhanced Raman spectroscopy (SERS). The topic of optical detection/biosensors for waterborne pathogen applications was reviewed in 2019 with the authors concluding that good progress has been achieved with the best-reported technology able to detect at a level of 2 cfu/mL in 30 min though it is critical to focus on solving issues of potential toxicity, stability in aqueous conditions, and integration with sample processing along with multiplexing [56]. The integration of nanotechnology with aptamers is also a recently growing trend [47, 49], though little demonstrated as yet with protozoa [57].

One strategy employing AuNPs for detection is through fluorescence quenching. For example, Philipps and colleagues achieved a LOD of 10^2 bacteria/mL in the solution for *E. coli* using enzyme-linked AuNPs [48]. The cationic AuNPs were functionalised with quaternary amine groups and electrostatically bound to the enzyme, β-galactosidase. Upon binding of the bacteria, the enzyme is released, restoring its activity and amplifying the signal (Fig. 9.4). The LOD increased to 10^4 bacteria/mL when in a test strip format. A similar approach was taken by Jung et al., who utilised a graphene oxide (GO) surface, and a secondary AuNP label, which binds to captured rotavirus and quenches the GO surface fluorescence through fluorescence resonance energy transfer (FRET) [58]. AuNPs have also been used for the detection of protozoa such as *Cryptosporidium*, offering 10 oocyst/mL visual determination of protozoan

TABLE 9.1 Overview of how nanotechnology has been applied to the detection of waterborne pathogens.

Pathogen	Nanomaterial	Recognition	Detection method	Efficiency/detection limit	Reference
E. coli O157:H7	Qdots	Biotinylated antibody	Fluorescence microscopy	Two orders more sensitive than conventional dyes	[14]
	Qdots and magnetic NPs	Antibody	Fluorometry	100 times more sensitive than FITC	[12]
	Qdots	Fim-H mannose-specific lectin	Fluorometry	10^4 bacteria/mL	[18]
	Dye-doped silica NPs	Antibody	Plate counting/flow cytometry	1–400 E. coli within 20 min	[26]
			Surface plating	1.6×10^1 to 7.2×10^7 cfu/mL	[42]
	Magnetic NPs	Antibody	IR spectroscopy	10^4–10^5 cfu/mL	[43]
			ATP bioluminescence	20 cfu/mL	[44]
	Au NPs	Antibody	Microscope and visual	10 ng	[45]
S. saprophyticus	Magnetic NPs	vancomycin	MALDI-MS	7×10^4 cfu/mL in urine	[46]
S. aureus	Magnetic NPs	Vancomycin	MALDI-MS	7×10^4 cfu/mL in urine	[46]
	CCMV	Antibody	Test strip	Visual detection	[36]
	Au NPs	Antibody	I.C. Assay Test device	100% Sensitivity	[47]
Cholera toxin	Liposomes	Gangliosides	Fluoroimmunoassay	1 nM	[30]
	Au NPs	Thiolated lactose	Test strip	10 fg/mL in 20 min	[31]
			Visual and UV–Vis	10 min	[48]
Salmonella	Au NPs	Antibody	Test strip	Red dots appearance in 2 h	[49]

Salmonella enteritidis	Au and magnetic NPs	DNA assay	Fluorescence	1 ng/mL	[50]
H. pylori	Au NPs	Antibody	SEM	10 ng	[45]
Gram-negative bacteria	Magnetic NPs	Vancomycin	Fluorescence microscopy	4 cfu/mL	[51]
E. faecalis, S. epidermidis	Magnetic NPs	Vancomycin	Plate counting	–	[52]
P fimbriated E. coli	Magnetic NPs	Pigeon ovalbumin	MALDI-MS	~9.6×10^4 cfu/0.5 mL	[53]
Multiplexed					
E. coli, Salmonella	Qdots	Antibody	Fluorescence spectroscopy	10^4 cfu/mL	[13]
Cholera toxin, ricin, Shinga-like toxin1, staphylococcal enterotoxin B	Qdots	Antibody	Fluroimmunoassay	In ng/mL quantity	[15]
E. coli, Salmonella, S. aureus	Dye-doped silica NPs	Antibody	Luminophore immunoassay	within 20 min	[24]
Virus					
Rotavirus	Ag nanorod array	–	SERS	–	[54]
Influenza virus	Liposomes	Sialic acid on glycoproteins	Colorimetry	1 HAUs	[32]
Adenovirus	Eu doped polystyrene NPs	Antibody	Fluoroimmunoassay	800-fold more sensitive than conventional immunoassays	[34]
Protozoan parasite					
Giardia lamblia	Au NPs/Ag staining	Antibody	UV-Vis spectroscopy	1.088×10^3 cells/mL	[55]

I.C., immunochromatographic; HAUS, haemagglutinating units.
Table 1 from Ref. [7]. Reproduced with permission.

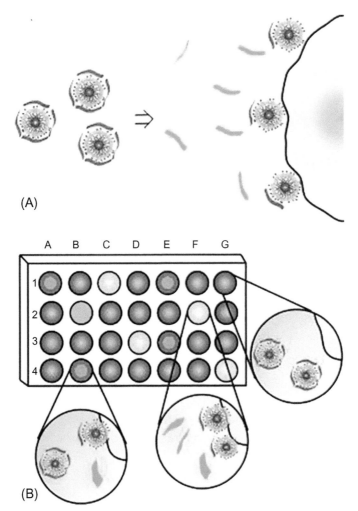

FIG. 9.4 Design of the nanoparticle–conjugated polymer sensor array. (A) Schematic representation of the displacement of anionic conjugated polymers from cationic nanoparticles by negatively charged bacterial surfaces. (B) Schematic illustration of fluorescence pattern generation on a microplate. In the case of release from the nanoparticle, the initially quenched PPEs regain their fluorescence. The fluorescence response is dependent upon the level of displacement determined by the relative nanoparticle—PPE binding strength and bacteria—nanoparticle interactions. By modulating such interactions, the sensor array may generate distinct response patterns against different bacteria. In the diagram, A–G on the microplate represents bacteria of different types, and codes 1–4 represent the PPE—nanoparticle constructs. *(Reproduced with permission from [48].)*

presence [59]. Graphene nanomaterials have also been utilised in fluorescent-based detection via noncovalent self-assembly of a DNAzyme onto the graphene surface; exposure to *E. coli* samples releases the DNAzyme and leads to the cleavage-mediated production of a fluorescent signal [60]. The approach could easily be adapted for other waterborne pathogens.

Both quantum dots and fluorescent nanoparticles are less prone to the problems of thermal fluctuations, self-absorption, self-fluorescence, and photobleaching that traditional fluorescent dyes suffer from. QDs have the additional advantage of a broad excitation spectrum with narrow emission bands. Therefore, multiplexed detection of a range of pathogens, using one excitation source is enabled. QDs have been functionalised with a range of recognition elements and applied to the detection of viruses, bacteria, and protozoa [1, 17, 41, 55]. However, we found no reports of QDs specifically applied to the detection of waterborne viruses.

QDs have been used in immunoassays for a range of foodborne bacteria, many of which are the same as waterborne pathogens, for example *E. coli*, *Salmonella*, and *Shigella* [17]. In one case, QDs facilitated a 16-fold decrease in sensitivity compared to FITC for the detection of *E. coli* [61]. The major advantage that QDs offer of easy multiplexed detection has also been exploited for foodborne pathogens, with LODs on the order of 10^2–10^3 cfu/mL comparable to those recorded in single pathogen assays.

Single-cell *E. coli* detection with QDs was reported by Hahn et al. in 2005, claiming 2 orders of magnitude greater sensitivity than with traditional organic dyes [62]. In 2006, Edgar et al. used QDs to specifically detect *E. coli* in mixed bacterial samples to a LOD of 10 cells/mL [45]. In the same year, Yang and Li reported simultaneous detection of *E coli* and *Salmonella* with a LOD of 10^4 cfu/mL in under 2 h [61]. In 2007, the Chan group demonstrated a multiplexed high throughput analysis microfluidic system capable of rapidly detecting both *E. coli* and hepatitis, as a model virus, in less than 1 h and with 50 times greater sensitivity than existing methods [63]. The detection system employed QD biobarcodes.

Su and Li adopted magnetic particles for sample processing in their 2004 study detecting *E. coli* with QDs [64]. The LOD was at least 100 times lower than with fluorescein isothiocyanate (FITC) with a detection time of under 2 h. Zhao et al. who developed the multiplex foodborne pathogen QD detection described earlier used silica-coated MNPs for the enrichment of bacteria before detection [54], as did Agrawal in their circular microfluidic QD detection set-up [40].

Dwarakanath and colleagues utilised a different approach involving QDs for the detection of pathogens, quantifying the blue shift exhibited upon bacteria binding [65]. The authors explained this shift by physical deformation of the QD conjugates, as well as changes in the chemical environment, upon binding to bacteria. The intensity of the shifted peak grows in correlation with the number of bacteria while the original QD emission peak reduces (Fig. 9.5.). Another alternative strategy was developed by Mukhopadhyay et al. who used QDs to induce *E. coli* aggregation with a LOD of 10^4 bacteria/mL [66]. This method will work well at high concentrations though not be so applicable to approach single-cell LODs. Another approach combined two differently coloured QDs with CNTs in a dual-FRET detection scheme for foodborne pathogens [67]. Foodborne studies are more prevalent than water examples, and a couple of different 2015 articles detected *Salmonella* in microfluidic systems with

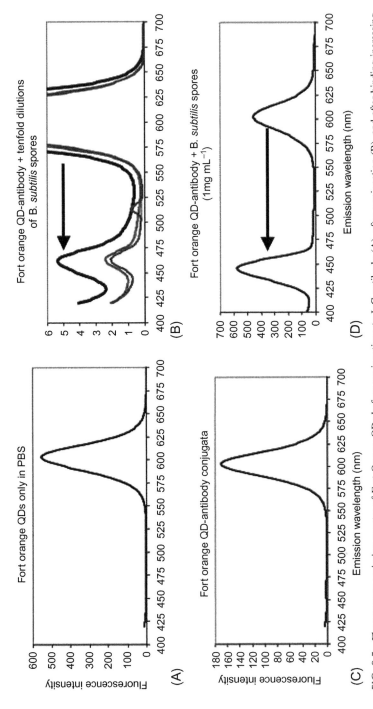

FIG. 9.5 Fluorescence emission spectra of Fort Orange QDs before conjugations to IgG antibody (A), after conjugation (B), and after binding increasing amounts of bacterial spores (C, D). *(Adapted from Ref. [38] and reproduced in Ref. [71].)*

QDs [68, 69] for food applications with one study from the same year using carbon dots and aptamers for tap water studies [70]. Besides, in 2015, CdSe/ZnS QDs were used for *E. coli* O157:H7 detection with an LoD of 10 cfu/mL [71]. In 2016, chitosan-coated CdSe QDs were used for with a 30 cfu/mL LOD [42]. In 2017, a QD approach was used to discriminate between different species as well as identify antimicrobial resistance genes, though the study was not focussed on waterborne pathogens [43].

In 2004, Zhu et al. reported the use of QDs to detect *Cryptosporidium* and *Giardia* in water samples [44]. The advantages of QDs over traditional fluorophores were a higher signal to noise ratios, greater photostability, and dual-colour detection of these pathogens.

QDs have also been used in conjunction with flow cytometry. In 2008, Hahn and co-workers reported the use of QDs for the flow cytometric detection of *E. coli* O157:H7, claiming brighter fluorescent intensities, lower detection limits, and greater specificity than traditional probes [72]. The LOD was 10^6 cells/mL, in a background of 100 times as many cells of another *E. coli* strain. In contrast, in 2007, Ferrari et al. found that flow cytometric detection of *Cryptosporidium* with QDs was worse than with organic fluorophores, reporting both a lower fluorescence intensity from QDs and greater nonspecific binding [73]. Given their finding a year later that the minimum fluorophore concentration required for detection was 100-fold less than FITC [50], this suggests that optimisation of the QD functionalisation procedure to attach antibodies against *Cryptosporidium* is needed.

QDs offer great photostability, enhanced sensitivity of detection, and easy simultaneous detection of multiple pathogens. They have been widely applied to a range of pathogens, detecting mainly nuclei acids but also whole pathogens and even toxins, though they appear to have been less applied in the study of waterborne pathogens. Disadvantages that have been noted are interactions with other compounds decreasing the sensitivity or quenching the fluorescence [17], which might explain why QDs have not been applied in complex environmental water samples. However, the main reason might be to do with their cost, which is higher than traditional dyes. Concerns have also been raised about their size, which is an order of magnitude greater than most dyes, and might pose problems for biorecognition in multiplex assays [17].

Dye-doped silica NPs were employed by Zhao et al. to achieve single-cell *E. coli* O157: H7 detection in processed beef samples. The samples were incubated with the NPs and unbound NPs were removed via centrifugation. Total detection time was less than 20 min and the result could be read out using a spectrofluorometer or a simple flow cytometer. Each antibody against the bacteria was bound to an NP containing 1000 dye molecules inside a silica matrix thus resulting in a 1000 times amplification of fluorescent signal compared to traditional antibody-fluorophore complexes. Therefore, this method offered single-cell resolution without the need for time-consuming culture steps.

Adenovirus has been detected though the use of europium(III)-chelate doped NPs to a limit of 5000 virions per millilitre via a sandwich-based assay [52].

This study, which used patient samples rather than water samples, reported an 800-fold improvement in detection compared to existing immunofluorometric assays. We found no reports of fluorescent NPs applied to the detection of other waterborne viruses or waterborne protozoa. Fluorescein labelled trimethyl chitosan NPs have been used for bacterial detection though, using an aptamer for specific *E. coli* recognition [20]. Another chitosan-based approach was colorimetric allowing for the detection of bacteria in 10 min though the LOD was 10^4 cfu/mL by the naked eye and 10^2 cfu/mL with a spectrophotometer [74]. Antibody-linked fluorescent graphene QDs have been used for the specific detection of *E. coli* O157: H7 at a LOD of 100 cfu/mL [46].

AuNPs have been used in a resonance light scattering (RLS) technique for the detection of pathogen antigens. The technique was developed by Ray and co-workers who exploited the nonlinear optics of AuNPs and applied hyper-Rayleigh scattering to detect binding of oligonucleotide probes [75]. Since double-stranded oligonucleotides have different electrostatic properties the binding of the target to the AuNP conjugated probe leads to dissociation of the probe from the NP, removing the fluorescence quenching of the probe, and also initiating aggregation and colorimetric change of the AuNPs. This was applied by Lin et al. for the detection of *E. coli* antigen to a LOD of 10 ng [51]. Other examples based solely on the colorimetric aspects of AuNPs are the detection of *Salmonella* [53, 76], Shiga toxin producing *E. coli* [77] and the cholera toxin [78]. The latter provided quantitative detection within 10 min. Magnetic iron oxide NPs were used with a fluorescent polymer for multiplexed bacterial detection for the low-cost screening of environmental samples [79].

AuNP aggregation approach has not been attempted often for the detection of waterborne protozoa. One potential reason for this is that on the micrometre scale of for example a bacteria, AuNPs tend to bind to the surface rather than aggregating [80]. For viruses, the first paper to demonstrate virus detection using this approach was published by Lee et al. in early 2013, though the technique has been previously applied to the detection of viral nucleic acids, e.g. with the ability to perform single mismatch detection or direct detection without amplification [81]. Lee et al. stressed the need for an excess of AuNP to allow for sufficient binding to occur to achieve AuNPs in close enough proximity to trigger the colorimetric change. The one example found from 2018 used colistin and AuNPs to allow detection down to 10 bacterial cells per millilitre in 5 min; the approach detects the presence of bacteria as colistin binds to the bacteria and, therefore, is not available in solution to form a complex with the AuNPs (the aggregated complex would exhibit a blue colour), which are thus free in solution and red [80]. Another competitive binding assay used polyethyleneimine-coated AuNPs and exploited the fact that in the presence of bacteria these AuNPs will preferentially bind to bacteria rather than an enzyme and thus in the presence of bacteria enzyme activity is increased and can be colorimerically determined. The LoD was 10 cfu/mL and results could be obtained with an optical reader within 10 min and 2–3 h with the naked eye [82].

With the colorimetric detection approach, silver can be applied for signal enhancement. The AuNPs act as nucleation sites for the silver. Metallic silver has a highly intense black colour, easily visible by eye. This has been exploited for *Salmonella* detection to reach a LOD of 5 cfu/mL [83] and has also been applied for the amplification of *Giardia* detection [84].

So far the colorimetric assay approach has been little applied for the detection of waterborne pathogens but offers a promising and relatively simple detection approach. This is, especially advantageous for the low income of field testing settings as no microscope is required for the visible by eye colorimetric change. As the authors of the paper reporting the first detection of a whole virus via this technique point out an excess of AuNPs is essential for the method to work. This is, therefore, most likely best-suited for the detection of viruses at lower numbers. Work over the next few years will be needed to extend the technique to a wider range of pathogens, and appropriate surface functionalisation will be key to achieving this goal. Integration with appropriate detection systems will also be vital, and one example from 2015 shows how integrating nanoparticles (in this example graphene oxide was used) with paper microfluidics allowed for a 10 cfu/mL detection of *E. coli* in bottled water [79].

As was discussed in Chapter 5, surface-enhanced Raman scattering uses the surface plasmons on roughened metallic surfaces to enhance the Raman effect [85]. SERS has been applied to the detection of waterborne viruses [86], bacteria [87, 88], and protozoa [89]. However, one of the main challenges of SERS detection is obtained multiplex detection with high specificity, due to a high degree of similarity between SERS spectra, and the variability of SERS spectra with species, viability and even age of the sample. The technique is highly sensitive with the potential to offer a wealth of information if accurate discrimination between all these possible states can be realised.

In some cases, nanostructures are fabricated on the Raman substrate and in other nanoparticles are employed. For example, AgNPs have been used to distinguish between *Listeria* species [90], detect *Salmonella* [91], and AuNPs integrated with a Raman reporter have been employed as pathogen-specific SERS probes in a study that detected multiple foodborne pathogens [92]. In this latter work, silica-coated magnetic immunoprobes were utilised for initial pathogen capture. A filter like surface made of AuNPs and mesoporous silica demonstrated a 900 fold enhancement in signal for bacteria [93]. Magnetic AuNPs have also been utilised [93]. When using nanoparticles, one further challenge is the batch-to-batch variability of nanoparticle size, which is crucial to achieve a reproducible and quantitative assay [17].

9.4.2 Electrical

One example of the use of electrical methods in the detection of pathogens with nanotechnology is a method developed by Dobozi-King et al., which exploits the changes in electrical conductivity of the sample solution upon bacteriophage lysis of bacteria [94]. Another is the work by Maalouf and co-workers using MNPs to capture and transport bacterial pathogens to a gold surface at which impedimetric measurements were undertaken [95]. Impedance was also used in combination with an AuNP covered graphene paper sensor system, with immobilised antibodies for *E. coli* O157:H7 [96]. The system was trialled in food samples with a LOD of 1.5×10^2 cfu/mL.

There is a growing interest in the use of nanomaterials to enhance impedimetric methods of detection for pathogens, particularly with examples of bacteria in foodborne applications [97]. One such example is the incorporation of AuNPs to enhance the performance via the AuNP-bacteria interaction providing an electron transfer pathway over the insulating self-assembled monolayer on the electrode surface, which allowed for detection of *E. coli* O157: H7 at a LOD of 100 cfu/mL [98]. Another is the use of MNPs in an assay that combined impedance with magnetohydrodynamic analysis, distinguishing types of bacteria based on differences in surface charge and mass [99]. Cyclic voltammetry and impedance were used for *Salmonella* detection down to 25 cfu/mL in a set-up that combined a nanocomposite of reduced graphene oxide and multiwalled CNTs with aptamer recognition on a glassy carbon electrode [100]. The most successful report achieved rapid (25 min), single colony-forming units per millilitre detection of *E. coli* in potable water, using screen-printed electrodes and magnetite NPs and electrical impedance spectroscopy (see Fig. 9.6) [101]. Otherwise, the area of electrical methods combined with nanotechnology has not been much applied for waterborne pathogens, even though there are several examples of virus detection schemes based on the integration of NPs with electrical approaches, while the use of electrochemical biosensors in a popular topic (covered in Section 9.4.3) [102].

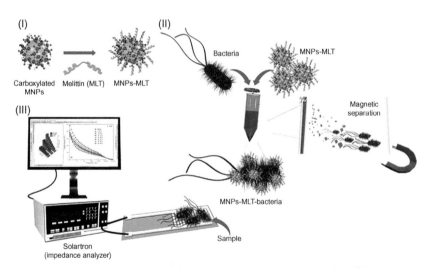

FIG. 9.6 Schematic of the detection scheme. *(Reproduced with permission from [101].)*

9.4.3   Biosensors

There are three main ways of biosensor signal transduction, optical, electrical, and mass-sensitive as described in Chapter 7. The magnetic approaches described in Section 9.4.5 represent an alternative transduction approach, with the examples described there enabled by nanomaterials. Nanotechnology can act to improve the performance in different ways, which is illustrated for AuNPs in Fig. 9.7., and recent developments in the use of nanomaterials and nanotechnology were reviewed in 2018, though few examples concentrated on waterborne pathogens [103, 104] while foodborne pathogens are regularly targeted [105].

Another 2017 review highlighted examples of research success in utilising nanomaterials in pathogen biosensors but stressed the need for method validation and demonstration in complex matrices in real-world samples [106]. The previous work concentrated mostly on bacteria and nanomaterial biosensors for viruses, none of which were waterborne in the described studies, were specifically reviewed in 2017, though the conclusions were similar to that of the previous bacteria review article [107].

Optical detection methods, including some approaches described by the authors as biosensors, were covered in detail in Section 9.4.1. Photonic crystals are nanoarrays of dielectric scatters and were covered in Chapter 7.

In their 2012 review, Shinde et al. claimed that nanotechnology offers a solution to the classical problems of electrochemical biosensors, for example the limitations of poor sensitivity and false-negative results [102]. There are certainly a large number of electrochemical studies reported, indicating the popularity of this approach. Sensitivity can be improved by the deposition of metal nanoparticles thus increasing the surface area, and enhancing the interaction of the recognition element (Fig. 9.7). CNTs have also been utilised in electrochemical detection protocols.

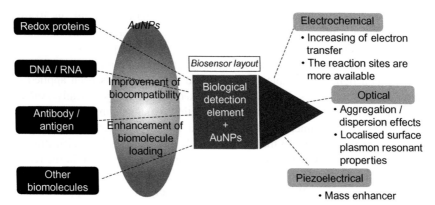

FIG. 9.7   Schematic of how AuNPs can enhance biosensor technology. *(Reproduced from Ref. [53].)*

Early studies of amperometric biosensors for *E. coli* were reported by Hasebe et al. and Brewster and Mazenko. The former achieved a LOD of 10^3–10^4 cells/mL when coupled with a 3 h preenrichment period. The latter authors improved on this by combining the electrochemical detection with bacteria captured on a filter, thus enabling-g detection to a limit of 5000 cells/mL in just 25 min. Later work has improved on these LODs.

In 2006, Viswanathan used liposome-coated CNTs for electrochemical detection of cholera toxin with a LOD of 10–16 g, and in 2012 utilised CNTs with screen printed electrodes for multiple bacteria detection [108]. In 2009, Cheng and colleagues employed CNTs, coated in Fe_3O_4 NPs in an amperometric biosensor for the rapid detection of coliforms. The detection was based upon the determination of the phenol produced by an enzymatic reaction in the bacteria solution. Another enzymatic approach used the activity of β-D-glucuronidase to detect *E. coli*, with a LOD of 104 cfu/mL in 3 h [109].

Another electrochemical method used in conjunction with CNTs is that of field-effect transistors (FETs). FETs fabricated using semiconducting single-walled CNTs can monitor a direct charge transfer between the target and the CNT with single-molecule sensitivity. *E. coli* has been detected using this approach and an aptamer-functionalised CNT: this was achieved by So et al. in 2006. The use of CNTs and electrochemical biosensors was reviewed in 2019 for foodborne pathogens, concluding that critical outstanding issues in the development of this technology are in the stability of recognition elements on CNT surfaces and a deeper understanding of the detection mechanisms [110].

Potentiometric biosensors have achieved good results for bacterial detection with work from 2010 and 2011 reporting LODs of 6 cfu/mL (with CNTs) and 10 cfu/mL (with graphene).

In 2011, Setterington and Alocilja used immunomagnetic separation to isolate *E. coli* for labelling with electroactive polyaniline where magnetic forces were also used to deliver the labelled pathogens to the surface of screen-printed carbon electrodes. This strategy amplified the electrochemical signal generated by the polyaniline enabling detection to a LOD of 7 cfu/mL in 70 min. Polyaniline was more recently used with AuNPs and DropSens carbon screen-printed electrodes for DNA based detection of *E. coli* [111].

Another application of CNTs in potentiometric biosensors is in a technique known as a light addressable potentiometric sensor that has been applied to several microorganisms, including multiplexed detection. Specifically, AuNPs have also been used in electrochemical detection for signal amplification in anodic stripping voltammetry and to achieve a 13-fold increase over traditional screen printed electrodes in the detection of *E. coli* O157: H7. Modification of the electrode with AuNPs in this latter example enabled detection in the range 10^2–10^7 cfu/mL. A LOD of 30 and 10 cfu/mL was achieved for the same bacteria by Guner et al. [112] and Yang et al., respectively, when they applied either hybrid nanocomposites based on AuNPs, CNTs, and chitosan or platinum-coated gold nanoporous electrodes. In 2015, single bacteria level detection was reported using differential pulse voltammetry with an aptamer-based sandwich assay using AgNPs [113]. Cyclic voltammetry was used with silica NPs to enhance *E. coli* detection, with the advantage being simple functionalisation via spin-coating, and the authors hope to build on this proof-of-principle to decrease the LOD from the present 10^3 cfu/mL [114].

As the examples mentioned previously illustrate there are many examples of work, both more recently and over the last couple of decades, applying electrochemical biosensing to bacteria, particularly *E. coli*. This was recently reviewed in 2019, with the main conclusions being that this approach is highly promising due to the high sensitivity [115]. The literature summary indicated that there were reports of single figure colony-forming units per millilitre LODs now being achieved, particularly with Ag and Au NPs. However, this was less often applied to waterborne pathogens, and integration with sample processing is necessary.

In terms of viruses, no examples of waterborne virus detection using nanomaterial enabled electrochemical biosensors were located though many studies are reporting general virus detection, typically utilising affinity interactions between antibody and antigen together with either amperometric or voltammetry [116]. For protozoa, one example from 2015 focused on *C. parvum* detection using an aptamer to capture the oocysts onto an AuNP modified screen-printed electrode and a LOD of 100 was achieved [57]. Concerning antimicrobial resistance, a nanocomposite electrochemical microbial biosensor was used to determine the effectiveness of different antibiotics against different bacteria [117].

Chapter 7 reported very few biosensors for virus detection, with optical methods dominating. One of those, developed by Jung et al., employed AuNPs in a fluorescence quenching set-up for rotavirus [58]. For protozoa, mass-sensitive approaches were found to be the most common type of biosensor in Chapter 7. While nanotechnology can enhance the performance of mass-sensitive biosensors as described earlier this has not yet been applied to the detection of waterborne protozoa. Chapter 7 described many bacterial biosensors, particularly for *E. coli*. Nanoparticles have been utilised to enhance the detection of bacteria using SPR, QCM, and cantilevers.

The sensitivity of SPR is sometimes limited by the inability to measure small changes in refractive index and NPs offer a solution to this problem. Noble metal NPs on the surface of an SPR sensor leads to a local surface plasmon resonance, which exhibits intense absorption and scattering peaks. This signal amplification enhances the sensitivity of SPR and has been applied to detect pathogen RNA and toxins [118]. The former used AuNPs whereas the latter employed hybrid Au-AgNPs. Dendrimers and liposomes (chemical structures on the nanoscale, not described in Section 9.1) have also been employed to improve the detection of *E. coli* and RNA sequences from *E. coli* and *Cryptosporidium* [118]. More recently, the technique has been extended to detect more whole bacteria: combining aptamers with nanomaterials enabled multiplexed detection of three different bacteria [91], and antibody-conjugated AuNPs were also used to detect foodborne pathogens via SPR enhancement [119]. SPR has also been utilised with molecularly imprinted polymer-based nanoparticles for highly sensitive virus detection [120]. Another interesting approach, for bacteria, which combined sample processing and detection; magnetic NPs and AuNPs with a Raman reporter were utilised to capture bacteria in a centrifugal filtration process after which silver intensification was used to enhance the SERS signals [121].

Cantilevers can be considered a form of nanotechnology and were included in the 2010 review by Kaittanis [1]. Using silicon nitride cantilevers, Weeks et al. reported in 2003, the detection of as few as 25 *Salmonella enterica* bacteria, by monitoring the cantilever's surface bending, which was directly associated with the number of bacteria associating on the cantilever [122]. Miniaturisation of cantilevers to nano dimensions is claimed to result in ultrasensitivity, faster detection, and better mass resolution [17]. Static-mode microfluidic cantilevers were also employed for the detection of *Cryptosporidium*, though the LoD was high [123]. Piezoelectric millimetre-sized cantilevers have been used for the detection of molecular products, e.g. DNA and RNA [124].

MNPs have been used to amplify the signal from microcantilevers from the sensitive detection of proteins [125], though not yet for waterborne pathogens. Other NPs have also been shown to offer a means of signal amplification for mass-sensitive biosensors. Wang and colleagues reported an AuNP amplified QCM DNA sensor for *E. coli* O157: H7 detection, with two amplification steps [126]. The first step involved the improved surface area offered by the AuNPs resulting in a higher binding of the target. The second step employed more NPs to further amplify the signal, achieving a LOD of 2×10^3 cfu/mL, an improvement on traditional QCM results. Mao et al. have also used NPs as mass enhancers to amplify the frequency shift for this bacteria reporting a limit of 2.67×10^2 cfu/mL [127]. In terms of detection of bacteria cells, Olsen et al. employed nanosized phages as the recognition element for the rapid detection of *S. typhimurium*, with a detection time of less than 180 s reaching a LOD of 10^2 cells and giving a linear response in the range $10-10^7$ cells/mL [128]. QCM has been used for viruses but a recent review did not report any examples of waterborne viruses or nanomaterial enhancement [129].

9.4.4 Molecular methods

Many of the ways to utilise nanotechnology to enhance the detection of pathogens described in other sections of this chapter can also be applied to the detection of molecular targets and indeed some examples have been given under the other sections, for example optical methods [6, 63], SPR [118], electrochemical detection [111], QCM detection of bacteria [126–128], and magnetic approaches [130]. AuNP interactions with DNA for surface plasmon peak shifts were described in the 1990s as a method of molecular detection [1] and carbon nanotubes have been shown to protect oligonucleotide sequences from degradation. The use of nanomaterials in molecular methods has been harnessed in many examples [1], though as yet the applications to waterborne pathogen detection remains a relatively under-exploited target market.

Biobarcodes are emerging nanobiotechnology for the molecular detection of pathogens [17]. In these assays, signal amplification is achieved using a DNA reporter molecule. Firstly, the target is concentrated with antibody or aptamer coated NPs and secondly, an NP coated in short DNA 'barcode' stretches is applied to facilitate sensitive detection. This is achieved due to the multiple copies of the barcode per nanoparticle and in one report the LOD was lowered by 10 million times compared to a conventional ELISA method [17]. This approach has been applied for the detection of norovirus in food samples [131] as well as *E. coli* O157:H7 in water [132]. In the latter example, a LOD of 25 cfu/mL, or 87 gene copies for double-stranded DNA, was reported. QDs have also been utilised to monitor bacteria without the need for PCR amplification [133], as have metal ions bound to metallothionein in an electrochemical assay [134].

Nanoparticles have also been used for cell lysis and DNA capture. Chitosan covered magnetic nanoparticles were used in combination with mechanical vibrations to cause cell lysis of bacteria followed by the capture of the DNA on the chitosan, enabling a one-step lysis and extraction system. The device was tested with six gram-negative waterborne bacteria and at a vibrational frequency of 180 Hz, the extraction efficiency of 97% was determined [135]. Magnetic nanoparticles have been used in a similar one-step RNA extraction method for SARS-CoV2 [136], demonstrating applicability to viruses. Hydrogel-based nanotraps were used to achieve a 100-fold enrichment of the dengue virus before RT-PCR detection [137]. We (Bridle, Johnston, and Vaidya) have also recently obtained results using nanoparticles for protozoan lysis (unpublished work; Ameya Vaidya thesis), though more work is required to combine this with extraction and DNA purification.

9.4.5 Magnetic detection

Nanotechnology has also enabled novel detection strategies, such as the magnetic relaxation switch approach developed by Perez et al. [138]. In this set-up, antibody carrying dextran-coated iron-oxide NPs alter the spin–spin relaxation of adjacent water molecules upon target recognition. The results can be read out using a benchtop magnetic relaxometry. This has facilitated the detection of as few as 5 viral particles for adenovirus in 10 μL of 25% serum, with the LOD rising to 10 in 100% serum [139]. Interestingly, for targets larger than the NPs, such as bacteria, a nonlinear relationship between the output signal and bacterial concentration is observed with maximal sensitivity for low bacteria concentrations (see Fig. 9.8).

An alternative magnetic approach was developed by Koets and co-workers who used superparamagnetic particles as detection labels for DNA from *E. coli* and *Salmonella* [130]. The magnetic NPs were detected by a Giant Magneto Resistance (GMR) sensor. Using this system 4–250 pM amplicon concentrations

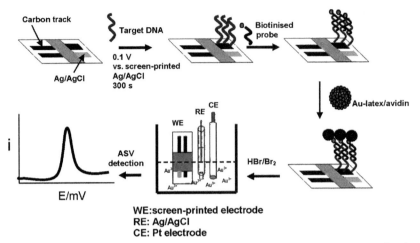

FIG. 9.8 Schematic representation of the procedure for detection of DNA hybridisation using the Au nanoparticle-coated latex labels. *(From Ref. [75].)*

were detected in less than 3 min. Mukika et al. used a magnetostrictive immunosensor to measure *E. coli* O157: H7 in food and clinical samples, where the output signal was the quantification of small magnetic field variations induced by the binding of the MNPs [140]. Maalouf et al. took a different approach using MNPs for *E. coli* detection using the MNPs to attract the pathogens to a gold surface where impedimetric measurements were performed [95]. This yielded detection in the range of $10–10^3$ cfu/mL.

Superconducting quantum interference devices (SQUID) have also been developed using MNPs to detect pathogens. The principle behind this mechanism of detection is the differential oscillations of MNPs in the absence or presence of the target pathogen during exposure to ac magnetic fields. Free, unbound NPs quickly relax by Brownian motion whereas those bound to the target undergo Neel relaxation leading to a gradually dissipating flux. Grossman et al. achieved a LOD of 10^6 cells in a sample volume of $20\,\mu L$ for *L. monocytogenes*. A similar concept was used for the detection of avian flu virus with a LOD of 5 pg/mL. The advantage of this approach is that no washing step to remove unbound MNPs is required. This strategy has been applied to viruses and bacteria though not yet to protozoa or waterborne pathogens in particular (Fig. 9.9).

9.4.6 Mass spectrometry detection

Tap and drinking water samples were analysed using MALDI-MS for bacterial analysis in combination with the use of a ZnO NP and polymethyl methacrylate (PMMA) concentration system. Using the ZnO-PMMA with a dispersive liquid–liquid extraction, the higher sensitivity of MALDI data

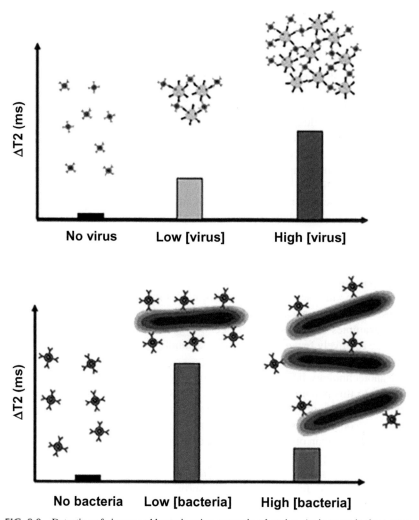

FIG. 9.9 Detection of viruses and bacteria using magnetic relaxation. An increase in the concentration of the virus facilitates the clustering of the nanoparticles and high changes in the spin–spin relaxation times ($\Delta T2$). On the other hand, at low concentrations of bacteria, high $\Delta T2$ are obtained due to nanoparticle assembly on the bacteria surface. $\Delta T2$ decreases as the bacterial concentration increases because the nanosensors switch to a dispersed-like state, reminiscent of the one observed in the sterile medium. *(Scheme 3 from [1]. Reproduced with permission.)*

was obtained although the LoDs were still relatively high and more work is required to increase sensitivity if this kind of approach was to be adapted to a wider range of waterborne pathogens [141]. A MALDI-MS approach was recently combined with a graphene magnetic chitosan nanomaterial to enhance detection [142].

9.5. Summary

In 2009, writing in the Clinical Microbiology Newsletter, Driskell and Tripp predicted nanotechnology for pathogen detection will have a major impact on healthcare, medicine, food and agriculture, biodefence, and the environment. Other authors have added their voice to the view that nanotechnology will revolutionise pathogen diagnostics over this century [17, 118]. Research into nanotechnology has yielded many developments over the past decades; searching the literature has revealed numerous recent articles applying nanomaterials and nanodevices to the detection of waterborne pathogens.

As seen in this chapter, there is a wide variety of different nanomaterials that can be applied in various ways, that is for signal enhancement of most other types of detection protocol or to create novel methods of detection. We have reviewed in this chapter the applications of nanomaterials to the optical, electrical, biosensor, and molecular methods of detection, focusing on examples for waterborne pathogens. While there have been several recent reviews on the topic, these have mainly focussed on medical, bioterrorism, or foodborne pathogens, and especially food, rather than the particular challenges of detecting pathogens in water. However, often the pathogens are similar. Our search of the literature has revealed that *E. coli* has been a popular area of study with many examples of the use of nanotechnology for bacteria detection, whereas there are relatively fewer reports of nanotechnology applied to viruses and protozoan detection in water.

Nanotechnology has many stated advantages although it has not yet fully lived up to this potential, however, there are now studies reporting single bacteria level detection. While many studies report signal enhancement, these are sometimes only one or two orders of magnitude improvements, and in some cases not better than other nonnanotechnology reports of a certain technique. In one case, notably for QDs for *Cryptosporidium* detection with flow cytometry, the use of nanomaterials diminished detection. However, this finding strongly supports the need for research into improved functionalisation chemistries which were thought to be the cause of this failure. In some cases, though single-cell detection is reported, and given that this is a rapidly growing field, further developments will undoubtedly improve nanotechnology performance for waterborne pathogen detection.

Further research must address several challenges facing nanotechnology in this application. For example, as mentioned previously, robust functionalisation chemistries are required, particularly to apply the methods to a wider range of waterborne than thus far studied. Compared to other target analytes, waterborne pathogen detection is less well-studied and methods combining sample processing with nanotechnology-based detection should be a focus of future work. Since the first edition of this book, there has been some progress in this direction with several systems now integrating a preconcentration stage with detection, though typically these are not both nanomaterial-based.

Considering sample processing it is sometimes mentioned as an advantage that nanotechnology enables a reduction in the sample volume. This is not necessarily advantageous for waterborne pathogen detection and so this may place a greater challenge on sample processing. We have seen examples of nanotechnology enhancing filtration or magnetic separations and performances, though both these strategies have drawbacks in terms of the need for subsequent elution or the need for recognition elements conjugated to the magnetic beads. A novel approach to concentration was taken using depletion flocculation and the development of similar schemes could be an interesting route forward. However, a future possibility raised in the first edition of this book for which there has been some development over the last few years is in the integration with microfluidic devices; the use of microfluidics will be considered fully in the next chapter but examples included with nanotechnology included the formation of nano-forests inside channels, cell lysis approaches, and links with on-chip detection protocols.

There are many exciting opportunities for nanomaterials in waterborne pathogen monitoring. In this chapter we've seen how QDs have been widely utilised in optical methods of detection, how metallic nanoparticles have improved the sensitivity of techniques such as SPR and Raman, and the wide exploitation of CNTs in electrochemical biosensors. AuNPs are also employed in colorimetric detection schemes which offer particular advantages for developing country applications, as the result is easily read without expensive equipment. Nanotechnology has also opened up new detection schemes, and as research and development continue more will surely be invented. Going forward translation of technologies developed for food applications into the water sector should be one area of focus along with the design of new approaches for nanotechnology-enabled sample processing steps (from concentration to enrichment and perhaps most likely the lysis, extraction, and purification steps to enable molecular methods) and integration of detection methods with sample processing. Improved understanding of processes and comparisons of different NPs and detection approaches would be useful to facilitate successful comparison and identification of the most promising approaches. While it might take many years, nanotechnology will hopefully deliver on its promise of revolutionising waterborne pathogen detection.

References

[1] C. Kaittanis, S. Santra, J.M. Perez, Emerging nanotechnology-based strategies for the identification of microbial pathogenesis, Adv. Drug Deliv. Rev. 62 (4–5) (2010) 408–423.

[2] N.C. Cady, et al., 12—Micro- and nanotechnology-based approaches to detect pathogenic agents in food, in: A.M. Grumezescu (Ed.), Nanobiosensors, Academic Press, 2017, pp. 475–510.

[3] A. Ojha, Chapter 19—Nanomaterials for removal of waterborne pathogens: opportunities and challenges, in: M.N. Vara Prasad, A. Grobelak (Eds.), Waterborne Pathogens, Butterworth-Heinemann, 2020, pp. 385–432.

[4] S.T. Khan, A. Malik, Engineered nanomaterials for water decontamination and purification: from lab to products, J. Hazard. Mater. 363 (2019) 295–308.

[5] T.S. Hauck, et al., Nanotechnology diagnostics for infectious diseases prevalent in developing countries, Adv. Drug Deliv. Rev. 62 (2010) 438–448.

[6] P.C. Ray, et al., Nanomaterials for targeted detection and photothermal killing of bacteria, Chem. Soc. Rev. 41 (2012) 3193–3209.

[7] V.V. Mody, et al., Introduction to metallic nanoparticles, J. Pharm. Bioallied Sci. 2 (4) (2010) 282–289.

[8] A. Low, V. Bansal, A visual tutorial on the synthesis of gold nanoparticles, Biomed Imaging Interv J. 6 (1) (2010), e9.

[9] P. Tallury, et al., Nanobioimaging and sensing of infectious diseases, Adv. Drug Deliv. Rev. 62 (4–5) (2010) 424–437.

[10] K. Kulthong, et al., Determination of silver nanoparticle release from antibacterial fabrics into artificial sweat, Part. Fibre Toxicol. 7 (8) (2010) 1743–8977.

[11] J.B. Chao, et al., Speciation analysis of silver nanoparticles and silver ions in antibacterial products and environmental waters via cloud point extraction-based separation, Anal. Chem. 83 (2011) 6875–6882.

[12] S. Silver, T. Phung le, G. Silver, Silver as biocides in burn and wound dressings and bacterial resistance to silver compounds, J. Ind. Microbiol. Biotechnol. 33 (2006) 627–634.

[13] B. Nowack, H. Krug, M. Height, 120 years of nanosilver history: implications for policy makers, Environ. Sci. Technol. 45 (2011) 1177.

[14] T.A. Dankovich, D.G. Gray, Bactericidal paper impregnated with silver nanoparticles for point-of-use water treatment, Environ. Sci. Technol. 45 (2011) 1992–1998.

[15] V.K. Sharma, R.A. Yngard, Y. Lin, Silver nanoparticles: green synthesis and their antimicrobial activities, Adv. Colloid Interface Sci. 145 (1–2) (2009) 83–96.

[16] H. Arya, Z. Kaul, R. Wadhwa, K. Taira, T. Hirano, S.C. Kaul, Quantum dots in bio-imaging: revolution by the small, Biochem. Biophys. Res. Commun. 329 (2005) 1173–1777.

[17] N. Gilmartin, R. O'Kennedy, Nanobiotechnologies for the detection and reduction of pathogens, Enzyme Microb. Technol. 50 (2) (2012) 87–95.

[18] P.U. Arumugam, et al., Wafer-scale fabrication of patterned carbon nanofiber nanoelectrode arrays: a route for development of multiplexed, ultrasensitive disposable biosensors, Biosens. Bioelectron. 24 (9) (2009) 2818–2824.

[19] M.A. Mohammed, et al., An overview of chitosan nanoparticles and its application in non-parenteral drug delivery, Pharmaceutics 9 (4) (2017) 53.

[20] F.G. Torres, et al., Natural polysaccharide nanomaterials: an overview of their immunological properties, Int. J. Mol. Sci. 20 (20) (2019) 5092.

[21] S.A. Ahmed, H.S. El-Mahallawy, P. Karanis, Inhibitory activity of chitosan nanoparticles against *Cryptosporidium parvum* oocysts, Parasitol. Res. 118 (7) (2019) 2053–2063.

[22] D.E. Said, L.M. ElSamad, Y.M. Gohar, Validity of silver, chitosan, and curcumin nanoparticles as anti-giardia agents, Parasitol. Res. 111 (2) (2012) 545–554.

[23] T. Aref, et al., Anti-toxoplasma activity of various molecular weights and concentrations of chitosan nanoparticles on tachyzoites of RH strain, Int. J. Nanomedicine 13 (2018) 1341–1351.

[24] A. Shetta, J. Kegere, W. Mamdouh, Comparative study of encapsulated peppermint and green tea essential oils in chitosan nanoparticles: encapsulation, thermal stability, in-vitro release, antioxidant and antibacterial activities, Int. J. Biol. Macromol. 126 (2019) 731–742.

[25] H. Zhang, et al., Direct growth of hierarchically structured titanate nanotube filtration membrane for removal of waterborne pathogens, J. Membr. Sci. 343 (1–2) (2009) 212–218.

[26]   A.S. Brady-Estevez, et al., Impact of solution chemistry on viral removal by a single-walled carbon nanotube filter, Water Res. 44 (13) (2010) 3773–3780.

[27]   M. Wegmann, et al., Modification of ceramic microfilters with colloidal zirconia to promote the adsorption of viruses from water, Water Res. 42 (6–7) (2008) 1726–1734.

[28]   D. Li, H. Shi, S.C. Jiang, Concentration of viruses from environmental waters using nanoalumina fiber filters, J. Microbiol. Methods 81 (2012) 33–38.

[29]   J. Jiang, et al., Smartphone based portable bacteria pre-concentrating microfluidic sensor and impedance sensing system, Sens. Actuators B 193 (2014) 653–659.

[30]   Y. Xia, et al., Label-free virus capture and release by a microfluidic device integrated with porous silicon nanowire forest, Small 13 (6) (2017), 1603135.

[31]   X. Sun, et al., Flocculation of bacteria by depletion interactions due to rod-shaped cellulose nanocrystals, Chem. Eng. J. 198–199 (2012) 476–481.

[32]   K.E. Eboigbodin, et al., Role of nonadsorbing polymers in bacterial aggregation, Langmuir 21 (2005) 12315–12319.

[33]   J. Chen, et al., Integrating recognition elements with nanomaterials for bacteria sensing, Chem. Soc. Rev. 46 (5) (2017) 1272–1283.

[34]   M. Varshney, et al., Magnetic nanoparticle-antibody conjugates for the separation of *Escherichia coli* O157:H7 in ground beef, J. Food Prot. 68 (2005) 1804–1811.

[35]   Y. Liu, et al., A highly sensitive and flexible magnetic nanoprobe labeled immunochromatographic assay platform for pathogen *Vibrio parahaemolyticus*, Int. J. Food Microbiol. 211 (2015) 109–116.

[36]   K. Kant, et al., Microfluidic devices for sample preparation and rapid detection of foodborne pathogens, Biotechnol. Adv. 36 (4) (2018) 1003–1024.

[37]   S.P. Ravindranath, et al., Biofunctionalised magnetic nanoparticle integrated mid-infrared pathogen sensor for food matrixes, Anal. Chem. 81 (2009) 2840–2846.

[38]   Y. Cheng, et al., Combining biofunctional magnetic nanoparticles and ATP bioluminescence for rapid detection of *Escherichia coli*, Talanta 77 (2009) 1332–1336.

[39]   S. Shrivastava, W.-I. Lee, N.-E. Lee, Culture-free, highly sensitive, quantitative detection of bacteria from minimally processed samples using fluorescence imaging by smartphone, Biosens. Bioelectron. 109 (2018) 90–97.

[40]   S. Agrawal, et al., Multiplexed detection of waterborne pathogens in circular microfluidics, Appl. Biochem. Biotechnol. 167 (2012) 1668–1677.

[41]   P. Tallury, et al., Nanobioimaging and sensing of infectious diseases, Adv. Drug Deliv. Rev. 62 (2010) 424–437.

[42]   Ü. Dogan, et al., Rapid detection of bacteria based on homogenous immunoassay using chitosan modified quantum dots, Sens. Actuators B 233 (2016) 369–378.

[43]   K. Cihalova, et al., Antibody-free detection of infectious bacteria using quantum dots-based barcode assay, J. Pharm. Biomed. Anal. 134 (2017) 325–332.

[44]   L. Zhu, S. Ang, W.T. Liu, Quantum dots as a novel immunofluorescent detection system for *Cryptosporidium parvum* and *Giardia lamblia*, Appl. Environ. Microbiol. 70 (1) (2004) 597–598.

[45]   R. Edgar, et al., High-sensitivity bacterial detection using biotin-tagged phage and quantum-dot nanocomplexes, Proc. Natl. Acad. Sci. U. S. A. 103 (13) (2006) 4841–4845.

[46]   X. Yang, L. Feng, X. Qin, Preparation of the Cf-GQDs-*Escherichia coli* O157: H7 bioprobe and its application in optical imaging and sensing of *Escherichia coli* O157: H7, Food Anal. Methods 11 (8) (2018) 2280–2286.

[47]   G. Singh, et al., Novel aptamer-linked nanoconjugate approach for detection of waterborne bacterial pathogens: an update, J. Nanopart. Res. 19 (1) (2016) 4.

[48] R.L. Philipps, et al., Rapid and efficient identification of bacteria using gold-nanoparticle-poly (para-phenylethynylene) constrcuts, Angew. Chem. Int. Ed. 47 (2008) 2590–2594.

[49] J.G. Bruno, Application of DNA aptamers and quantum dots to lateral flow test strips for detection of foodborne pathogens with improved sensitivity versus colloidal gold, Pathogens 3 (2) (2014) 341–355.

[50] R. Ibanez-Peral, et al., Potential use of quantum dots in flow cytometry, Int. J. Mol. Sci. 9 (2008) 2622–2638.

[51] F.Y.H. Lin, et al., Development of a nanoparticle-labeled microfluidic immunoassay for detection of pathogenic microorganisms, Clin. Diagn. Lab. Immunol. 12 (2005) 418–425.

[52] A. Valanne, et al., A sensitive adenovirus immunoassay as a model for using nanoparticle label technology in virus diagnostics, J. Clin. Virol. 33 (2005) 217–223.

[53] S. Wang, et al., Rapid colorimetric identification and targeted photothermal lysis of *Salmonella* bacteria by using bioconjugated oval-shaped Au NPs, Chem. A Eur. J. 19 (2010) 5600–5606.

[54] Y. Zhao, et al., Simultaneous detection of multifoodborne pathogenic bacteria based on functionalised quantum dots coupled with immunomagnetic separation in food samples, J. Agric. Food Chem. 57 (2009) 517–524.

[55] J.D. Driskell, R.A. Tripp, Emerging technologies in nanotechnology-based pathogen detection, Clin. Microbiol. Newsl. 31 (18) (2009) 137–144.

[56] N. Bhardwaj, et al., Optical detection of waterborne pathogens using nanomaterials, TrAC Trends Anal. Chem. 113 (2019) 280–300.

[57] A. Iqbal, et al., Detection of *Cryptosporidium parvum* oocysts on fresh produce using DNA aptamers, PLoS ONE 10 (9) (2015), e0137455.

[58] J.H. Jung, et al., A graphene oxide based immuno-biosensor for pathogen detection, Angew. Chem. Int. Ed. Engl. 49 (2010) 5708–5711.

[59] C. Thiruppathiraja, et al., An advanced dual labeled gold nanoparticles probe to detect *Cryptosporidium parvum* using rapid immuno-dot blot assay, Biosens. Bioelectron. 26 (11) (2011) 4624–4627.

[60] M. Liu, et al., Graphene-DNAzyme-based fluorescent biosensor for *Escherichia coli* detection, MRS Commun. 8 (3) (2018) 687–694.

[61] L. Yang, Y. Li, Simultaneous detection of *E. coli* O157:H7 and *S. typhimurium* using quantum dots as fluorescence labels, Analyst 131 (2006) 394–401.

[62] M.A. Hahn, J.S. Tabb, T.D. Krauss, Detection of single bacterial pathogens with semiconductor quantum dots, Anal. Chem. 77 (15) (2005) 4861–4869.

[63] J.M. Klostranec, et al., Convergence of quantum dot barcodes with microfluidics and signal processing for multiplexed high-throughput infectious disease diagnostics, Nano Lett. 7 (9) (2007) 2812–2818.

[64] X.I. Su, Y. Li, Quantum dot biolabelling coupled with immunomagnetic separation for detection of *Escherichia coli* O157:H7, Anal. Chem. 76 (2004) 4806–4810.

[65] S. Dwarakanath, et al., Quantum dot-antibody conjugates shift fluorescence upon binding bacteria, Biochem. Biophys. Res. Commun. 325 (2004) 739–743.

[66] B. Mukhopadhyay, et al., Bacterial detection using carbohydrate-functionalised CdS quantum dots: a model study exploiting *E. coli* recognition of mannosides, Tetrahedron Lett. 50 (2009) 886–889.

[67] N. Duan, et al., Simultaneous detection of pathogenic bacteria using an aptamer based biosensor and dual fluorescence resonance energy transfer from quantum dots to carbon nanoparticles, Microchim. Acta 182 (5) (2015) 917–923.

[68] G. Kim, et al., A microfluidic nano-biosensor for the detection of pathogenic *Salmonella*, Biosens. Bioelectron. 67 (2015) 243–247.

[69] R. Wang, et al., Immuno-capture and in situ detection of *Salmonella typhimurium* on a novel microfluidic chip, Anal. Chim. Acta 853 (2015) 710–717.

[70] R. Wang, et al., Rapid and sensitive detection of *Salmonella typhimurium* using aptamer-conjugated carbon dots as fluorescence probe, Anal. Methods 7 (5) (2015) 1701–1706.

[71] D.O. Demirkol, S. Timur, A sandwich-type assay based on quantum dot/aptamer bioconjugates for analysis of *E. coli* O157:H7 in microtiter plate format, Int. J. Polym. Mater. Polym. Biomater. 65 (2) (2016) 85–90.

[72] M.A. Hahn, P.C. Keng, T.D. Krauss, ArticleFlow cytometric analysis to detect pathogens in bacterial cell mixtures using semiconductor quantum dots, Anal. Chem. 80 (3) (2008) 864–872.

[73] B.C. Ferrari, P.L. Bergquist, Quantum dots as alternatives to organic fluorophores for *Cryptosporidium* detection using conventional flow cytometry and specific monoclonal antibodies: lessons learned, Cytometry A 71A (4) (2007) 265–271.

[74] T.N. Le, T.D. Tran, M.I. Kim, A convenient colorimetric bacteria detection method utilizing chitosan-coated magnetic nanoparticles, Nanomaterials (Basel). 10 (1) (2020) 92.

[75] J. Griffin, et al., Sequence-specific HCV RNA quantification using the size-dependent nonlinear optical properties of gold nanoparticles, Small 5 (2009) 839–845.

[76] S.B. Fang, et al., Identification of *Salmonella* using colony print and detection with antibody-coated gold nanoparticles, J. Microbiol. Methods 77 (2009) 225–228.

[77] A. Jyoti, P. Pandey, S.P. Singh, S.K. Jain, R. Shanker, Colorimetric detection of nucleic acid signature of Shiga toxin producing *E. coli* using Au NPs, J. Nanosci. Nanotechnol. 10 (7) (2010) 4154–4158.

[78] C.L. Schofield, R.A. Field, D.A. Russell, Glyconanoparticles for the colorimetric detection of cholera toxin, Anal. Chem. 79 (2007) 1356–1361.

[79] Y. Wan, et al., Quaternized magnetic nanoparticles–fluorescent polymer system for detection and identification of bacteria, Biosens. Bioelectron. 55 (2014) 289–293.

[80] P. Singh, et al., Drug and nanoparticle mediated rapid naked eye water test for pathogens detection, Sens. Actuators B 262 (2018) 603–610.

[81] C. Lee, et al., Colorimetric viral detection based on sialic acid stabilized gold nanoparticles, Biosens. Bioelectron. 42 (2013) 236–241.

[82] R. Thiramanas, R. Laocharoensuk, Competitive binding of polyethyleneimine-coated gold nanoparticles to enzymes and bacteria: a key mechanism for low-level colorimetric detection of gram-positive and gram-negative bacteria, Microchim. Acta 183 (1) (2016) 389–396.

[83] Z. Wang, et al., Ultrasensitive chemiluminescent immunoassay of *Salmonella* with silver enhancement of nanogold labels, Luminescence 2 (2011) 136–141.

[84] X.X. Li, et al., Detection of pathogen based on the catalytic growth of gold nanocrystals, Water Res. 43 (2009) 1425–1431.

[85] V.K.K. Upadhyayula, Functionalized gold nanoparticle supported sensory mechanisms applied in detection of chemical and biological threat agents: a review, Anal. Chim. Acta 715 (2012) 1–18.

[86] C. Fan, et al., Detecting food- and waterborne viruses by surface-enhanced Raman spectroscopy, J. Food Sci. 75 (5) (2010) M302–M307.

[87] M.F. Escoriza, et al., Raman spectroscopy and chemical imaging for quantification of filtered waterborne bacteria, J. Microbiol. Methods 66 (1) (2006) 63–72.

[88] S.P. Ravindranath, Y. Wang, J. Irudayaraj, SERS driven cross-platform based multiplex pathogen detection, Sens. Actuators B 152 (2) (2011) 183–190.

[89] A.E. Grow, et al., New biochip technology for label-free detection of pathogens and their toxins, J. Microbiol. Methods 53 (2) (2003) 221–233.

[90] G.C. Green, et al., Identification of *Listeria* species using a low-cost surface-enhanced Raman scattering system with wavelet-based signal processing, IEEE Trans. Instrum. Meas. 58 (10) (2009) 3713–3722.

[91] S.M. Yoo, D.-K. Kim, S.Y. Lee, Aptamer-functionalized localized surface plasmon resonance sensor for the multiplexed detection of different bacterial species, Talanta 132 (2015) 112–117.

[92] Y. Wang, S. Ravindranath, J. Irudayaraj, Separation and detection of multiple pathogens in a food matrix by magnetic SERS nanoprobes, Anal. Bioanal. Chem. 399 (3) (2011) 1271–1278.

[93] C.-C. Lin, et al., A filter-like AuNPs@MS SERS substrate for *Staphylococcus aureus* detection, Biosens. Bioelectron. 53 (2014) 519–527.

[94] M. Dobozi-King, et al., Rapid detection and identification of bacteria: sensing of phage triggered ion cascade (SEPTIC), J. Biol. Phys. Chem. 5 (2005) 3–7.

[95] R. Maalouf, et al., Comparison of two innovative approaches for bacteria detection: paramagnetic nanoparticle and self assembles multilayer process, Microchim. Acta 163 (2008) 157–161.

[96] Y. Wang, et al., Impedimetric immunosensor based on gold nanoparticles modified graphene paper for label-free detection of *Escherichia coli* O157:H7, Biosens. Bioelectron. 49 (2013) 492–498.

[97] G. Kim, A.S. Om, J.H. Mun, Nano-particle enhanced impedimetric biosensor for detedtion of foodborne pathogens, in: International Conference on Nanoscience and Technology, IOP-Science, Basel, Switzerland, 2007, p. 112.

[98] J. Wan, et al., Signal-off impedimetric immunosensor for the detection of *Escherichia coli* O157:H7, Sci. Rep. 6 (2016) 19806.

[99] Zeeshan, et al., Impedance and magnetohydrodynamic measurements for label free detection and differentiation of *E. coli* and *S. aureus* using magnetic nanoparticles, IEEE Trans. Nanobioscience 17 (4) (2018) 443–448.

[100] F. Jia, et al., Impedimetric *Salmonella* aptasensor using a glassy carbon electrode modified with an electrodeposited composite consisting of reduced graphene oxide and carbon nanotubes, Microchim. Acta 183 (1) (2016) 337–344.

[101] D. Wilson, et al., Electrical detection of pathogenic bacteria in food samples using information visualization methods with a sensor based on magnetic nanoparticles functionalized with antimicrobial peptides, Talanta 194 (2019) 611–618.

[102] S.B. Shinde, C.B. Fernandes, V.B. Patravale, Recent trends in in-vitro nanodiagnostics for detection of pathogens, J. Control. Release 159 (2) (2012) 164–180.

[103] A. Sposito, et al., Application of nanotechnology in biosensors for enhancing pathogen detection, Wiley Interdiscip. Rev. Nanomed. Nanobiotechnol. 10 (2018), e1512.

[104] M. Giovanni, et al., Electrochemical quantification of *Escherichia coli* with DNA nanostructure, Adv. Funct. Mater. 25 (25) (2015) 3840–3846.

[105] F. Dridi, et al., 5—Nanomaterial-based electrochemical biosensors for food safety and quality assessment, in: A.M. Grumezescu (Ed.), Nanobiosensors, Academic Press, 2017, pp. 167–204.

[106] F. Mustafa, Y.A.R. Hassan, S. Andreescu, Multifunctional nanotechnology-enabled sensors for rapid capture and detection of pathogens, Sensors 17 (9) (2017).

[107] A. Mokhtarzadeh, et al., Nanomaterial-based biosensors for detection of pathogenic virus, TrAC Trends Anal. Chem. 97 (2017) 445–457.

[108] S. Viswanathan, C. Rani, J.-a.A. Ho, Electrochemical immunosensor for multiplexed detection of food-borne pathogens using nanocrystal bioconjugates and MWCNT screen-printed electrode, Talanta 94 (2012) 315–319.

[109] M. Rochelet, et al., Rapid amperometric detection of *Escherichia coli* in wastewater by measuring β-D glucuronidase activity with disposable carbon sensors, Anal. Chim. Acta 892 (2015) 160–166.

[110] S. Muniandy, et al., Carbon nanomaterial-based electrochemical biosensors for foodborne bacterial detection, Crit. Rev. Anal. Chem. 49 (6) (2019) 510–533.

[111] N. Shoaie, M. Forouzandeh, K. Omidfar, Voltammetric determination of the *Escherichia coli* DNA using a screen-printed carbon electrode modified with polyaniline and gold nanoparticles, Microchim. Acta 185 (4) (2018) 217.

[112] A. Güner, et al., An electrochemical immunosensor for sensitive detection of *Escherichia coli* O157:H7 by using chitosan, MWCNT, polypyrrole with gold nanoparticles hybrid sensing platform, Food Chem. 229 (2017) 358–365.

[113] A. Abbaspour, et al., Aptamer-conjugated silver nanoparticles for electrochemical dual-aptamer-based sandwich detection of staphylococcus aureus, Biosens. Bioelectron. 68 (2015) 149–155.

[114] M. Mathelié-Guinlet, et al., Silica nanoparticles-assisted electrochemical biosensor for the rapid, sensitive and specific detection of *Escherichia coli*, Sens. Actuators B 292 (2019) 314–320.

[115] R. Pourakbari, et al., Recent progress in nanomaterial-based electrochemical biosensors for pathogenic bacteria, Microchim. Acta 186 (12) (2019) 820.

[116] S.I. Kaya, et al., Chapter 18—Electrochemical virus detections with nanobiosensors, in: B. Han, et al. (Eds.), Nanosensors for Smart Cities, Elsevier, 2020, pp. 303–326.

[117] M. Sedki, et al., Sensing of bacterial cell viability using nanostructured bioelectrochemical system: rGO-hyperbranched chitosan nanocomposite as a novel microbial sensor platform, Sens. Actuators B 252 (2017) 191–200.

[118] N. Sanvicens, et al., Nanoparticle-based biosensors for the detection of pathogenic bacteria, Trends Anal. Chem. 28 (11) (2009) 1243–1250.

[119] N. Verdoodt, et al., Development of a rapid and sensitive immunosensor for the detection of bacteria, Food Chem. 221 (2017) 1792–1796.

[120] Z. Altintas, et al., Detection of waterborne viruses using high affinity molecularly imprinted polymers, Anal. Chem. 87 (13) (2015) 6801–6807.

[121] I.-H. Cho, et al., Membrane filter-assisted surface enhanced Raman spectroscopy for the rapid detection of *E. coli* O157:H7 in ground beef, Biosens. Bioelectron. 64 (2015) 171–176.

[122] B.L. Weeks, et al., A microcantilever-based pathogen detector, Scanning 25 (6) (2003) 297–299.

[123] H. Bridle, et al., Static mode microfluidic cantilevers for detection of waterborne pathogens, Sensors Actuators A Phys. 247 (2016) 144–149.

[124] A.P. Haring, E. Cesewski, B.N. Johnson, Piezoelectric cantilever biosensors for label-free, real-time detection of DNA and RNA, in: B. Prickril, A. Rasooly (Eds.), Biosensors and Biodetection: Methods and Protocols, Volume 2: Electrochemical, Bioelectronic, Piezoelectric, Cellular and Molecular Biosensors, Springer New York, New York, NY, 2017, pp. 247–262.

[125] L. Ma, C. Wang, M. Zhang, Detecting protein adsorption and binding using magnetic nanoparticle probes, Sens. Actuators B 160 (1) (2011) 650–655.

[126] L.J. Wang, et al., The *Escherichia coli* O157:H7 DNA detection on a gold nanoparticle-enhanced piezoelectric biosensor, Chin. Sci. Bull. 53 (8) (2008) 1175–1184.

[127] X. Mao, et al., A nanoparticle amplification based quartz crystal microbalance DNA sensor for detection of *Escherichia coli* O157:H7, Biosens. Bioelectron. 21 (2006) 1178–1185.

[128] E.V. Olsen, et al., Affinity selected filamentous bacteriophage as a probe for acoustic wave biodetectors of *Salmonella typhimurium*, Biosens. Bioelectron. 21 (8) (2006) 1434–1442.

[129] A. Afzal, et al., Gravimetric viral diagnostics: QCM based biosensors for early detection of viruses, Chem. Aust. 5 (1) (2017) 7.

[130] M. Koets, et al., Rapid DNA multi-analyte immunoassay on a magneto-resistance biosensor, Biosens. Bioelectron. 24 (2009) 1893–1898.

[131] M. Adler, R. Wacker, C.M. Niemeyer, Sensitivity by combination: immuno-PCR and related technologies, Analyst (6) (2008) 702–718.

[132] D. Zhang, M.C. Huarng, E.C. Alocilja, A multiplex nanoparticle based biobarcoded DNA sensor for the simultaneous detection of multiple pathogens, Biosens. Bioelectron. (4) (2010) 1736–1742.

[133] T.-Y. Wu, et al., A novel sensitive pathogen detection system based on microbead quantum dot system, Biosens. Bioelectron. 78 (2016) 37–44.

[134] L. Zheng, et al., Simultaneous detection of multiple DNA targets based on encoding metal ions, Biosens. Bioelectron. 52 (2014) 354–359.

[135] V. Kamat, et al., A facile one-step method for cell lysis and DNA extraction of waterborne pathogens using a microchip, Biosens. Bioelectron. 99 (2018) 62–69.

[136] Z. Zhao, et al., A simple magnetic nanoparticles-based viral RNA extraction method for efficient detection of SARS-CoV-2, bioRxiv (2020), 2020.02.22.961268.

[137] J.-H. Lee, Application of Hydrogel Nanoparticles for Detection of Dengue Virus, George Mason University, 2018.

[138] J.M. Perez, et al., Magnetic relaxation switches capable of sensing molecular interactions, Nat. Biotechnol. 20 (2002) 816–820.

[139] J.M. Perez, et al., Viral-induced selfassembly of magnetic nanoparticles allows the detection of viral particles in biological media, J. Am. Chem. Soc. 125 (2003) 10192–10193.

[140] M. Mujika, et al., Magnetoresistive immunosensor for the detection of *Escherichia coli* O157:H7 including a microfluidic network, Biosens. Bioelectron. 24 (2009) 1253–1258.

[141] G. Gedda, et al., ZnO nanoparticle-modified polymethyl methacrylate-assisted dispersive liquid–liquid microextraction coupled with MALDI-MS for rapid pathogenic bacteria analysis, RSC Adv. 4 (86) (2014) 45973–45983.

[142] H.N. Abdelhamid, H.-F. Wu, Multifunctional graphene magnetic nanosheet decorated with chitosan for highly sensitive detection of pathogenic bacteria, J. Mater. Chem. B 1 (32) (2013) 3950–3961.

Chapter 10

Miniaturised detection systems

Helen Bridle

Institute of Biological Chemistry, Biophysics and Bioengineering, Heriot-Watt University, Edinburgh, Scotland

Previous chapters have covered a range of detection techniques, such as optical, electrical and molecular methods as well biosensors. Chapter 9 looked at how nanotechnology could improve these methods. This chapter will focus on the role that miniaturisation, in particular using microfluidic systems, can play in the delivery of 'lab-on-a-chip' devices to perform the detection procedures previously described.

The chapter starts with an introduction to the field of microfluidics, covering microfabrication, the underlying fluid mechanics and the types of components which can be incorporated to deliver more complex fluid handling capability. Next, the chapter presents how microfluidic devices can be used to enhance the performance of the detection methods proposed in Chapters 5–8. Particular attention will be devoted to examples concerning waterborne pathogens.

10.1. Microfluidics

Microfluidic systems, that is, fluid handling systems with dimensions on the micrometre scale, have developed rapidly during the past decade and have found many applications, especially within the chemical analysis and biological assays. This is unsurprising considering their numerous advantages which include reduced sample consumption, increased speed of analysis, improved efficiency and process parallelisation as well as access to phenomena and mechanisms that are not accessible on the macroscopic scale [1].

Microfluidics is an enormous field and thus it is impossible to review fully here, and the reader is referred to recent reviews [2–4]; this chapter describes briefly the microfabrication of microfluidic devices, specifically focussing upon devices manufactured in polydimethylsiloxane (PDMS) as this offers a way of rapid, and relatively cheap prototyping, the fluid mechanics governing device function and the incorporation of components, such as valves, mixers and methods of temperature control (Fig. 10.1).

Waterborne Pathogens: Detection Methods and Applications. https://doi.org/10.1016/B978-0-444-64319-3.00010-1

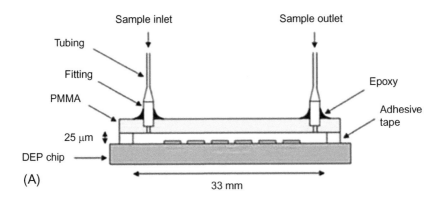

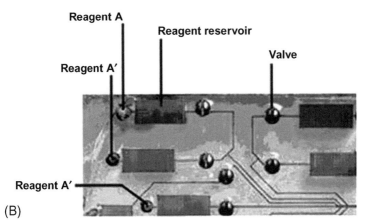

FIG. 10.1 Illustration of microfluidic systems. (A) Cross-sectional schematic of a microchannel. (b) Top view of a microfluidic network of reservoirs, channels and valves. *((A) Fig. 1 from K.S. Chow, H. Du, Dielectrophoretic characterization and trapping of different waterborne pathogen in continuous flow manner, Sensors Actuators A 170 (2011) 24–31. (B) Fig. 1 from J.Y. Yoon, B. Kim, Lab-on-a-chip pathogen sensors for food safety. Sensors 12 (2012) 10713–10741.)*

10.1.1 Microfabrication

Microfabrication techniques, developed for the microelectronics industry, enable the creation of complex systems with dimensions down to tens of nanometres in materials, such as silicon, glass or metals. These techniques include the patterning of substrates by lithography, the deposition of thin films using evaporation, sputtering or oxidation and the removal of material by etching or lift-off processes. Microfabrication is often expensive, time-consuming and requires cleanroom facilities. Since microfluidic systems generally have a larger footprint than microelectronic devices, this means that the cost per device is high. Additionally, for some microfluidic applications, the traditional microfabrication materials are not suitable [1]. Therefore, new alternative methods of microfabrication have been developed that reduce cost by allowing for replication of numerous structures from one master and offer the ability to use polymeric materials. These methods are known as soft lithography. The master is produced using conventional microfabrication and is subsequently negatively replicated by either replica moulding, hot embossing or injection moulding [1] (Fig. 10.2).

Polymers used for microfluidics include, among others, polycarbonate, polymethylacrylate (PMMA), polyethylene and PDMS [9]. Rapid prototyping using replica moulding in PDMS was introduced by Whitesides et al. at the end of the 1990s [10] and since then PDMS has become a key material within microfluidics. PDMS is extensively used since it is inexpensive, optically transparent, flexible, biocompatible, impermeable to water but permeable to gases and has low electrical conductivity as well as high oxidative and thermal stability [11].

Irreversible sealing of devices is easily achieved in PDMS without the need for high temperatures, pressures or voltages. Exposure to oxygen plasma or coronas [12] creates silanol groups at the PDMS surface and −OH-containing groups on another surface, for example, PDMS, glass or silicon, which form

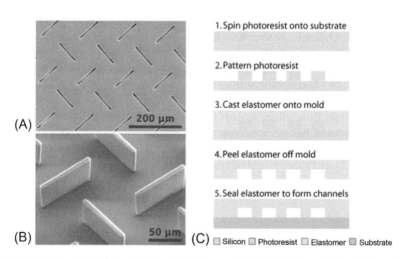

FIG. 10.2 Microfluidic Manufacture. SEM micrograph of (A) PDMS mould for plastic casting and (B) the epoxy chip fabricated by casting; (c) Schematic of the casting method showing an elastomer material poured over a moulding template, peeled off and sealed with an appropriate substrate, such as glass or silicon, to form microfluidic channels. *(Reproduced from Fig. 2 in C. Lui, N.C. Cady, C.A. Batt, Nucleic acid-based detection of bacterial pathogens using integrated microfluidic platform systems, Sensors, 9 (2009) 3713–3744. (A) and (B) Reprinted with permission C. Folk, X. Chen, F. Wudl, C.M. Ho, Hydrogel Microvalves With Short Response Time. ho.seas.ucla. edu/publications/conference/2003/232.pdf, 2003 © 2007 Springer.)*

covalent −O−Si−O-bonds when brought into contact, creating a tight, irreversible seal that can withstand pressures up to 30–50 psi [11]. The ease of bonding also allows for the creation of 3D devices [13]. Additionally, PDMS makes reversible van der Waals contacts to smooth surfaces which can also be used to seal devices and has been used in patterning [9].

The elasticity of PDMS offers several advantages. Firstly, it is easy to remove the PDMS mould from the master which prolongs the lifetime of the master. Secondly, interfacing with external components for, for example, sample introduction, is simply performed using press-fit connections [14]. Thirdly, it allows for the on-chip incorporation of components, such as pumps or valves [15]. However, the elasticity limits the achievable aspect ratio since shrinking or sagging of features can occur [9]. Furthermore, bulging of PDMS microchannels under pressure-driven flow has been reported [16]. These problems may be overcome by using PDMS with a high density of crosslink.

The chemical structure of PDMS has repeating $-O-Si(CH_3)$ units which render the surface hydrophobic. However, as mentioned, exposure to oxygen plasma, or a corona, generates silanol groups and thus creates a hydrophilic surface. If the surface is kept in contact with water or polar solvents the hydrophilicity is maintained. Otherwise, over time uncrosslinked PDMS chains migrate to the surface [9]. The extraction of uncrosslinked PDMS by organic solvents can delay the return to hydrophobicity [17]. The hydrophobic nature of the surface can cause several different problems, for example, nonspecific adsorption [11, 18] and absorption [19], swelling by nonpolar solvents [17] and difficulty in filling the channels. Numerous methods of surface modification have been proposed [20], including self-assembly of charged surfactants or polyelectrolyte layers, chemical vapour deposition or formation of a phospholipid bilayer, to either maintain a hydrophilic surface or to pattern the surface, for example, immunoassays [21]. However, the hydrophobic properties can also be exploited to control the fluid flow within devices [22, 23].

Recently there has been interest in developing lower-cost and more environmentally friendly approaches to manufacture [24], inspiring a new choice of materials [25] and paper-based microfluidics is well-developed [3, 26], including many water testing applications [27].

10.1.2 Fluid mechanics

Fluid flow behaves very differently in microfluidic systems than one would expect from everyday macroscopic observations. Characteristics of microfluidic systems include viscous dominated, turbulence-free laminar flow and high surface-to-volume ratios. To be able to design and understand microfluidic systems, it is crucial to understand the fluid mechanics in small spaces [1]. Fluid mechanics deals with the behaviour of fluids, for example, liquids or gases and how they interact with their surroundings. The equation describing the motion of incompressible Newtonian fluids is the Navier–Stokes equation

$$\rho\left(\frac{\partial v}{\partial t}+v\cdot\nabla v\right)=-\nabla p+\eta\nabla^2 v \qquad (10.1)$$

which is, in principle, a representation of Newton's 2nd law of mechanics for a fluid [28]. The left hand side comprises the density, ρ, multiplied by the acceleration. The two terms in the acceleration are the local acceleration and the convective acceleration, respectively. The convective acceleration arises from spatial variations in the velocity field and is nonlinear; this explains why analytical solutions to general flow using the Navier–Stokes equation are difficult to obtain. The right-hand side of the equation represents the forces acting upon the flow where ∇p is the pressure and $\eta\nabla^2 v$ is the viscous force.

From the Navier–Stokes equation it is possible to derive a dimensionless parameter known as the Reynolds number, defined as

$$\text{Re}=\frac{\rho v l}{\eta} \qquad (10.2)$$

where ρ is the fluid density, v the velocity, l the typical length scale of the system and η the viscosity. The Reynolds number is an estimate of the moment of inertia versus the viscous forces in a fluid system [28]. As the length scales and transport velocities of a system decrease, so does the Reynolds number. When the Reynolds number is very small the nonlinear terms in the Navier–Stokes equation disappear, resulting in linear and predictable Stokes flow. This is the case in typical microfluidic systems; for example, for water, velocities of $1\,\mu\text{m/s}$ to $1\,\text{cm/s}$ and channel radii of 1–$100\,\mu\text{m}$, the Reynolds numbers are in the range 10^{-6}–10. Such low Reynolds numbers mean that the viscous forces dominate the system and that, as mentioned above, flows are laminar and turbulence-free, with diffusion as the primary method of mixing [29]. This enables the design of systems with functionalities that are virtually impossible to obtain at the macroscopic scale.

Diffusion is the transport of matter as a result of random molecular motion which ultimately acts to neutralise concentration gradients. The proportionality of the flux of molecules to the concentration gradient, can be described using Fick's first law, which in one dimension is given by

$$J=-D\frac{dc}{dx} \qquad (10.3)$$

where J is the flux density, that is, the rate of transfer per unit area normal to the x-axis, c is the concentration and D is the diffusion coefficient. This equation is for isotropic diffusion in ideal solutions. Combination of Fick's first law with the continuity equation results in Fick's second law of diffusion,

$$\frac{\partial c}{\partial t}=D\frac{\partial^2 c}{\partial x^2} \qquad (10.4)$$

which describes concentration changes in both time and space [30].

The variance of the solution to Fick's second law in one dimension is given by the Einstein–Smoluchowski relation

$$\left(x(t)-x_0\right)^2 = 2Dt \qquad (10.5)$$

where $<>$ denotes the ensemble average and $(x(t)-x_0)^2$ represents the mean square distance travelled by a particle. This relation illustrates why diffusion is an efficient transport process on small length scales but inefficient over large distances. A typical value of D for a small molecule is $10^{-10}\,\text{m}^2/\text{s}$ and it would take this molecule only 0.5 s to diffuse 10 µm but several days to diffuse 1 cm.

The diffusion coefficient can be estimated for spherical particles using the Stokes–Einstein relation

$$D = \frac{kT}{6\pi\eta a} \qquad (10.6)$$

where k is the Boltzmann constant, T is temperature, η is viscosity and a is the effective hydrodynamic radius. The diffusion coefficient is highly sensitive to changes in temperature since in addition to the explicit inclusion of temperature in the numerator, the viscosity of water is also very dependent upon temperature. Both of these factors contribute to an increase in the diffusion coefficient as temperature increases which should be considered in the design of temperature-controlled microfluidic systems.

10.1.3 Components and operations with microfluidics

10.1.3.1 Transport of fluids

Application of pressure from external components interfaced to the microfluidic system is often used to drive flow through the device. For most channel geometries this results in a parabolic flow profile [29] and the flow rate Q is given by

$$Q = \frac{\Delta P}{R} \qquad (10.7)$$

where ΔP is the pressure difference and R is the channel resistance [11]. The pressure drop can be provided by opening the outlet to atmospheric pressure and applying positive pressure at the inlet, via, for example, a diaphragm pump or gravity, by the suspension of solution reservoirs above the microfluidic system. If the gravitational flow is used the pressure drop is given by

$$\Delta P = \rho g \Delta h \qquad (10.8)$$

where ρ is the density, g is the gravity acceleration and Δh is the height difference. For a circular channel, the resistance is given by

$$R = \frac{8\eta l}{\pi r^4} \qquad (10.9)$$

whereas for a rectangular channel with an aspect ratio close to 1 the resistance
is given by

$$R = \frac{12\eta L}{wh^3}\left[1 - \frac{h}{w}\left(\frac{192}{\pi^5}\sum_{n=1,3,5}^{\infty}\frac{1}{n^5}\tanh\left(\frac{n\pi w}{2h}\right)\right)\right]^{-1} \tag{10.10}$$

(for high or low aspect ratio channels the term within the square brackets vanishes) where η is the fluid viscosity, l is the length, r is the radius, h is the height
and w is the width of the channel [1]. Other pressure-driven means of pumping include different types of peristaltic pumps, fabricated on-chip [31, 32].
Micropumping has even been performed using cardiomyocytes [33].

Electroosmotic pumping is an alternative method of fluid transport [28]. The
main advantages are that fluid flow can be controlled by switching voltages on
and off, thus negating the need for valves, and that the flow profile is plug-like.
However, this method is only compatible with certain solutions, that is, with a
conductive solvent, and requires hydrophilic surfaces. Additionally, demixing
of solutions due to different electrophoretic mobilities can occur.

10.1.3.2 Valves

Valves are essential to be able to control flows within microfluidic systems [1].
The two main strategies for creating valves in elastomeric microfluidic devices
with pressure-driven flows are the use of responsive materials and channel deformation [28]. Responsive materials include conductive polymers [34], microspheres and hydrogels [35]. Samel et al. have shown a single-use valve using
thermally responsive microspheres which increase their volume upon heating
[36]. Hydrogels are polymer gels that undergo reversible volume changes in
response to varying environmental conditions, for example, changes in pH, temperature, light or electric fields [28, 37]. Microfluidic valves have been fabricated from thermally responsive [8] as well as pH sensitive [37, 38] hydrogels
and self-regulating fluidic systems have been created using pH-sensitive hydrogels [39]. The disadvantage of using hydrogels is the long response times, on
the order of tens of seconds.

There are many examples of mechanical valves incorporated into microfluidic systems, most of which involve the use of a deflectable membrane that
closes off a fluidic channel when pressure is applied [28]. Unger et al. showed
pneumatically acutated valves using a cross-channel architecture in which pressure applied to the upper channel deflects a PDMS membrane downward closing the lower channel [31]. Such valves offer a low dead volume, rapid response
times and their small size allows for a high density of valves to be incorporated
on one chip. However, to achieve complete closure of the lower channel bell-shaped channels are required and the manufacturing procedure limits the aspect
ratios possible [40]. In addition, low aspect ratios are necessary since otherwise the actuation pressures required would be impractically high [41]. Li et al.
have developed a normally closed microvalve allowing for vertical walls and

a variety of aspect ratios for portable microfluidic point-of-care devices [40] and Studer et al. have reported on a microfluidic valve requiring lower actuation pressures facilitating the use of different aspect ratios [41]. Ismagilov et al. developed an elastomeric switch involving two crossing channels in different layers [42]. Application of pressure to the crossing affected the aspect ratio and thus the direction of fluid flow could be controlled. As device complexity increases it is essential to be able to independently actuate valves which has led to the development of a multiplexed latching valves system [43].

In all the above systems the valves are incorporated into the microfluidic system which ensures a low dead volume. However, the incorporation of valves inside devices obviously increases the complexity of the system and may complicate fabrication. An alternative strategy is to interface to external valves. External valves integrated with microfluidic flow systems have been used to control and pump fluid flows [32], to generate sequential exposures [44–46] and to mix fluids [32, 44]. Disadvantages with external valves include large dead volumes and thus slow solution exchange times. In this thesis, a system using two external valves was developed to speed up exchange of the dead volume. This system allows for easy interfacing to microfluidic systems without the need for complex microfabrication and, when combined with the particular microfluidic system used in this thesis, for switching between numerous laminar flow patterns where each solution segment can be individually updated over time [47].

10.1.3.3 On-chip mixing

While on-chip mixing is desirable for many applications, it is inherently problematic within microfluidic systems due to low Reynolds and high Peclet numbers, which result in long mixing times [48]. The Peclet number, Pe, indicates the relative importance of convection to diffusion; it is given by

$$Pe = \frac{vl}{D} \tag{10.11}$$

where v is the flow velocity, l is the characteristic length of the system (e.g. channel width) and D is the diffusion coefficient. In microfluidic systems, typical values of the Peclet number are between 10 and 10^5 and thus convection is much faster than molecular diffusion [48]. The time, t, required for diffusional mixing of two adjacent fluid streams in one microchannel is given by

$$t = \frac{l^2}{D} \tag{10.12}$$

Assuming that l is 100 µm, D is 1×10^{-11} m^2/s (typical of a large molecule) and the flow rate is 500 µm/s, t will be 1000 s and the length of microchannel required is 500 cm. Thus, mixing in microfluidic systems can be very slow and require long channels. This can be advantageous and has been exploited in some applications, for example, separations and laminar flow patterns [28]. However, it is problematic if rapid mixing is desired.

Numerous strategies to enhance mixing have been presented [1, 28, 48] and the systems presented below do not represent an exhaustive list. The key to effective mixing is to increase the interfacial area between the fluids [48]. Active mixers exploit external fields to improve mixing. Examples include systems using magnetic stirrers [49], electrokinetic instabilities [50] and disturbances in the method for driving the flow [32, 51]. The incorporation of valves into microfluidic systems has also been exploited to control and improve mixing [40, 52]. Active mixers require interfacing to external components, use of external fields may interfere with detection methods and controlling the flow rate results in an uneven outlet flow. An alternative to active mixers are passive mixing systems, which generally involve complex microchannel geometries that may not be so straightforward to microfabricate. Gradient generating devices have been produced which use repeated lamination and splitting of flows to create gradients spanning several orders of magnitude [53, 54]. However, these gradients are temporally static. Serpentine and twisted channel structures [55], as well as channels containing chevron [56] or grooved structures [57, 58], have also been applied to create chaotic advection and enhance mixing.

10.2. Applications

Microfluidics has found many applications within biology in general, some of which could be extended to waterborne pathogens. For example, there have been developments in using microfluidics to obtain better environmental control over cells during culture, including bacteria [59], and this could be a potential development for traditional culture-based approaches. Microfluidics might also allow the design of environments to promote the culture of microorganisms that have not yet been cultured in the lab. Yoon and Kim, however, are not positive about this approach for foodborne pathogens with the justification that lab-on-a-chip has focussed on rapid methods of detection while culturing is time-consuming [6]. Their review from 2012 is an excellent overview of microfluidic detection methods applied to foodborne pathogens [6], many of which are identical to waterborne pathogens, and Kant et al. add to this covering the last decade [60]. There are other reviews of microfluidics specifically for water monitoring [61–63] and more recently, and broadly, environmental monitoring [64, 65]. For consideration of how microfluidics might apply in developing world settings, the Nature review by Yager is recommended [66].

This section will describe microfluidic applications of the detection methods described in Chapters 5–8. However, one of the main challenges for microfluidic devices in environmental applications such waterborne pathogen monitoring is the large sample volumes and complex matrices. For medical applications, the ability of microfluidics to deal with small sample volumes is an advantage, though here lab-on-a-chip systems are proven to deal with complex samples, like blood. There are, however, many techniques used for cell sorting within microfluidics and this section will start with an introduction to these and their applications to waterborne pathogens, along with other waterborne pathogen sample processing on-chip.

10.2.1 In sample processing

A large number of papers have been published discussing cell sorting in microfluidics, and there are many excellent reviews of the topic including [67, 68]. Methods can be sorted into those which exploit intrinsic properties of the cell, or resort to extrinsic 'labels' for cellular differentiation. Additionally, separation can be performed by a variety of means including optical, electric and magnetic forces, or hydrodynamic forces, and examples of such techniques are shown in Fig. 10.3.

Despite the proliferation of papers dealing with cell sorting, much less work has focussed on sample processing of waterborne pathogens, though this has increased in recent years. Strategies include some form of specific recognition, for example, with antibodies, though this can be expensive and highly sensitive to environmental factors; electrical concentration though this is also susceptible to environmental variations and so far limited to low volumes; physical methods such as acoustic or hydrodynamic sorting which are mainly based on size and thus cannot distinguish to the same degree as antibody or electrical systems; and physical methods using filters or sieves, which allow for high-resolution separation by size though bring manufacturing challenges at smaller sizes and potential for clogging issues. However, sample volumes typically remain small and many devices have been shown only with purified water. Centrifugation on-chip

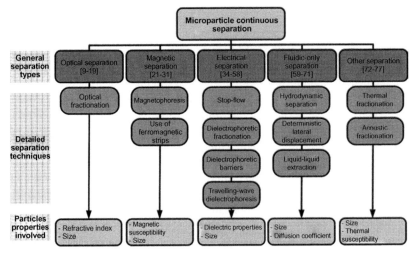

FIG. 10.3 Illustrating different methods of cellular separation in microfluidic systems. *(Reproduced from M. Kersaudy-Kerhoas, R. Dhariwal, M.P.Y. Desmulliez, Recent advances in microparticle continuous separation, IET Nanobiotechnol. 2 (1) (2008) 1–13.)*

has also been shown [69], and was recently applied to waterborne pathogens [70]. Several approaches combine some form of microfluidic enrichment with detection. Some examples of flow cytometry are covered later under optical detection.

Dharmasiri et al. described a cell purification platform for the isolation and washing of *Escherichia coli* in antibody-coated high aspect ratio curved microchannels [71]. Prefiltered lake water was used in the study with the microfluidic device processing 1 mL at a flow rate of around 5 μL/min. A 10^2 concentration factor was achieved with a 71% recovery rate. Microfluidic chips with antibody-coated microbeads inside the microchannels overcome the potential limitations of the low surface area of planar microchannels, and has been applied for the immunoseparation of *E. coli* [72]. A flow-through IMS for *Cryptosporidium* and *Giardia*, developed by Ramadan et al. [73], was described in Chapter 4Agrawal et al. described a PDMS 3D circular microfluidic system with imbedded permanent magnets, designed for multiplex pathogen detection, where the first stage is capture of the bacteria with immunomagnetic nanoparticles [74]. A micromixer set-up was reported in 2018, aiming to enhance the capture efficiency of antibody-functionalised devices, achieving a 96% capture efficiency of *Cryptosporidium* with the ability to subsequently detect on-chip [75].

Early work using DEP to concentrate pathogen samples on a microfluidic chip was performed by Goater et al., as part of an electrororotation detection strategy for *Cryptosporidium* oocysts [76]. Electrokinetic separation of bacteria was reported by Cabrera and Yager [77]. Gomez–Sjoberg reported DEP-based concentration of *E. coli* in 400 pL chamber, achieving concentration factors of 10^4–10^5 [78]. Chow et al. reported a DEP system for trapping and concentrating *E. coli* [5]. The system operated at a flow rate of 1 μL/min and claimed a 100% trapping efficiency if an appropriate DEP force was chosen. Additionally, different voltages could be applied to distinguish between different species and viable and nonviable microorganisms.

Insulator-based DEP (iDEP) was first described in by Lapizco-Encinas [79]. Later iDEP was shown in a plastic chip for the capture of *E. coli* by Cho et al. [80]. At a flow rate of 100 L/min their system achieved a maximal capture efficiency of 66%. This seems relatively low, especially in comparison to IMS though it is comparable to recovery rates following membrane filtration and elution. It is unlikely to be sufficient for reliable detection of single organisms. A 2011 thesis reported the design of 3D iDEP systems for trapping of bacteria, reporting stronger trapping forces at lower temperatures, reducing the risk of thermal damage to the trapped cells, thus potentially enabling further downstream analysis [81]. This work represents the first observation of intraspecies differences in membrane surface properties using iDEP. One potential problem with DEP or iDEP as a concentration and isolation technique is that it operates in batch mode which could complicate integration into continuous flow systems.

Hydrodynamic approaches to the concentration of waterborne pathogens include inertial focussing, a technique in which relatively high flowrates (on the order of mL/min) are used along with geometries that enable exploitation of inertial forces within microfluidics, resulting in size, shape and deformability based separations [82]. Jimenez et al. applied this to waterborne pathogens using a spiral channel inertial focussing approach (Fig. 10.4), reporting a 100% efficiency of *Cryptosporidium* collection from drinking water [83], and subsequently developing the system to process 50 mL of sample with 96% recovery of *Cryptosporidium* and 86% recovery *Giardia* [84]. *Giardia* has also been concentrated from food samples, and purified with removal of small particles to facilitate analysis [85]. Miller et al. described how a parallelised cascaded approach could be adopted to enable the microfluidic processing of 1000 L, thus enabling the application of the inertial focussing technology to the existing requirements for protozoan monitoring [86]; a spin-out company uFraction8 was created though the focus at present is on the algae dewatering market rather than waterborne pathogen monitoring.

Taguchi et al. have investigated the microfluidic trapping of protozoa, for subsequent fluorescence analysis [87]. They developed a micro-well array strategy, consisting of 32×32 microfabricated wells with a 10 or 30 µm diameter and a 10 µm depth, for oocysts capture. After microfabrication, the micro-wells were selectively coated with streptavidin and anti-*C. parvum* antibodies. For capture experiments, 10 mL of a sample mixture of *C. parvum* oocysts (10^7 oocysts/mL) suspended in PBS was simply deposited onto the array and the whole chip rotated horizontally for 1 h.

Refinement of the method to increase the capture efficiency was carried out with the use of a laser-machined stainless steel micromesh consisting of a 10×10 array of 2.7 mm diameter cavities to capture single oocysts incorporated into a PDMS microfluidic device [88]. The maximum flow rate tested was 350 µL/min, so 5 mL could be processed in under 15 min. When loading a 0.5 mL test sample (spiked oocysts in PBS) at a concentration of 36 oocysts/mL a recovery rate of 93% from the mesh was reported, comparable to that achieved by off-chip IMS. Batch processing of the sample occurs in the current design; thus while integration into automated systems would be possible, real-time continuous monitoring would not be.

Work carried out by Liu et al. illustrates an alternative strategy of trapping *Cryptosporidium* oocysts, using sieves or filters [89, 90]. In one example, a weir was created by interfacing a deep channel (50 µm) with a very shallow channel (1 or 2 µm). Using positive pressure, a mixture of protozoa in water was injected into the channel, trapping the cells against the wall of the deep channel. The common disadvantage of sieves or filters systems is their rapid clogging, perhaps due to the weir system. However, by developing a so-called rain drop bypass filter, Liu et al. significantly reduced this issue. The design consists of three prefilters and a wide composite filter structure, which allows alternative fluidic paths and therefore significantly reduces the pressure and the clogging

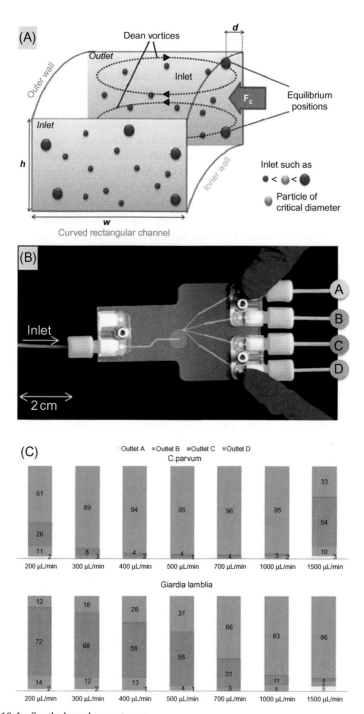

FIG. 10.4 See the legend on next page.

on the filter. The filters are made of fine arrays of raindrop-like-shaped pillars, arranged in gaps ranging from 0.2 to 1 mm, in the trapping zones and coarse arrays in the bypass zones. The device was used for bacterial capture and detection with an limit of detection (LOD) of 10^5 cfu/mL. However, further details regarding the performance (LOD, volumes, etc.) of this device with protozoa were not available to fully analyse its potential.

More recently, Liu et al. have incorporated a membrane into a PDMS microfluidic device for capture of (oo)cysts, reporting 90% detection of targets from 30 mL of drinking water combined with a fluorescent-based direct detection on the membrane [91]. One advantage of the approach is the ability to then release the (oo)cysts for further investigation although this was not trialled and the effectiveness of the release is not known. One of the most successful filter/sieve-based approaches is that of Pires and Dong, capable of operating with 100 L reducing volumes to hundreds of mL with 81% recovery of *Cryptosporidium* and 86% of *Giardia*, even at very low spiking levels and in a range of water turbidities and sample volumes [92].

10.2.2 In optical methods

Several examples of lab-on-a-chip systems exist for optical detection of waterborne pathogens. There are also other potential systems, which have not yet been applied to waterborne pathogens, for example, SERS on-chip, which was reviewed by Chen and Choo [93].

In some of the above examples of (oo)cyst and bacteria capture, the sample processing element of the device was also used for detection, through the addition of fluorescent stains. Two of the above systems represent a microfluidic alternative to the existing IMS and fluorescence detection protocol for (oo) cysts. Taguchi et al. note the advantages of their method include the predefined location of the binding of the oocysts and their good adhesion to the substrate during the washing and staining steps [87]. Their second design also allowed the detection to be done in 60 min compared to 2–3 h claimed for the IMS method (including staining) with automated FITC labelling and imaging was used for

---

FIG. 10.4 Illustration of the use of inertial focussing technology for protozoan concentration. (A) The schematic illustrates the principle of inertial focussing whereby the interplay of different forces within the flow channel acts to locate particles at a particular channel location perpendicular to the flow direction (this is known as the focussing position and is typically closer to the inner wall of a spiral channel). The effect is dependent upon flow rate and particle size and deformability and Fc denotes the centrifugal action acting on the liquid towards the outer wall. (B) The image illustrates the set-up and the size of the microfluidic device. This system can focus on particles ≤2 μm. (C) Recovery rates into the different outlets achieved for *Cryptosporidium parvum* and *Giardia lamblia* at different flow rates. The colour coding in the images corresponds to the different outlet colour-coding as shown in B. The percentage of pathogens found in each outlet at the different flow rates is shown, indicating the flow rate at which maximal recovery rates into one outlet can be achieved, that is, into outlet D 0.7 mL/min for *Cryptosporidium* at 96% and 1.5 mL/min for *Giardia* at 86%.

detection. 36 oocysts/mL was reported as the LOD, above the desired ability to detect single oocysts [88].

In 2001 Stokes et al. developed a sandwich immunoassay in a 0.4 mL reaction chamber on-chip, showing *E. coli* detection. In the same year McClain et al. used a microfluidic flow cytometry set-up for *E. coli* detection. Another microfluidic flow cytometer was developed in 2004 [94]. This device, made from PDMS, hydrodynamically focuses the sample where it is excited and detected using optical fibres. The chip also uses cheap, compact and low-power PIN photodiodes with lock-in amplification to show the first single-cell fluorescence detection using this type of photodiode. Device performance was characterised with beads and yeast cells, with a sample volume of 5 mL/h, but has not yet been applied to pathogen detection. Another sandwich assay presented by Li and Su detected *E. coli* with an LOD of 10–100 cfu/mL in a 1-mL sample in 2 h, claiming microfluidics allowed for an improvement of detection sensitivity due to reduced reagent consumption and increased immunoassay kinetics [95].

Sakamoto et al. also described a microfluidic flow cytometry set-up for *E. coli* detection capable of analysing six 10 µL samples in just 30 min [96]. The system was reported to deliver good agreement with traditional counts in shorter times, and was tested with river water. Flow cytometry on-chip for *E. coli* was also reported by Yamaguchi achieving a good comparison to traditional counting methods, even in pond water [97]. The device integrated on-chip mixing of sample and fluorescent stain to avoid pretreatment steps off-chip and used automated image analysis, delivering results in 1 h. The authors note that using LED light sources could allow for development of a portable system and that integration with sample concentration microfluidics would be an option to reduce the LOD from the reported 10^4 cells/mL. Li and Su produced a microfluidic chip to capture *E. coli* with antibodies for detection by UV/vis spectroscopy.

A microfluidic approach using IMS and chemiluminescence detected *E. coli* O157:H7 down to 34 cells in 90 min, with a sample of 100 µL [98]. *E. coli O157* activity has also been monitored in real time, following immunoseparation by microbeads, with bioluminescence technology [72]. Karsunke et al. reported the detection of *Salmonella*, *E. coli* and *Legionella* using chemiluminescence on-chip, with detection limits of 3×10^6, 1×10^5 and 3×10^3 cells/mL, respectively, for an assay which took just 13 min. A year later the same group slightly redesigned the system, though performance was not improved, with the assay time increasing to 18 min and no large changes in LOD [99].

Single-cell level detection was reported for using a latex immunoagglutination for real-time detection of *E. coli* on a microfluidic chip, reading out the light scattering results using a proximity (i.e. one not integrated on-ship) optical fibre [100]. Single-cell detection per microdevice was reported which increased to 4 cfu upon the addition of an extra washing step to determine only viable bacteria. While the volume processed by each device was 100 µL, later work by the authors connected the microfluidic system to a larger piped network, sampling a fraction of the water supply and achieving a 10 cfu/mL LOD in less than 5 min.

In 2011 *E. coli* from lettuce wash was detected using a particle immunoagglutination assay to an LOD of 10 cfu/mL in just 6 min [101] (Fig. 10.5). The drawback of this system was that it only tested 60 μL of sample. However, the microfluidic chip was integrated into a handheld portable unit. The same year another approach for *E. coli* detection involved incorporation of single microorganisms into droplets, offering the advantages of fast fluorescent tracer dye accumulation within droplets as well as digital counting [102]. Detection was based on metabolism, showing good agreement with traditional colony counts, achievable within 2 h for some bacteria, with full quantification possible in less than 10 h.

Connelly et al. developed a microfluidic system for virus detection, combining preconcentration and signal amplification via liposomes with different detection methods [103]. Fig. 10.6 shows the optical and electrical detection approaches. Liposomes and the virus sample are incubated and then preconcentrated through electrokinesis towards a nanoporous membrane. The sample is then eluted and capture downstream, where the liposomes are lysed releasing fluorescent signal. The authors note that the LOD of 1×10^5 pfu/mL is currently quite high, and suggest several improvements, as well as highlighting the suitability of the device for easy-to-use, rapid and portable systems.

Also, Agrawal et al. used secondary fluorescently labelled antibodies for the fluorescent detection of immunomagnetically captured pathogens, within

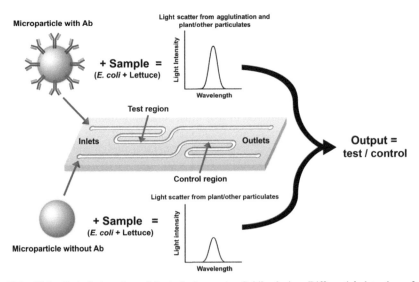

FIG. 10.5 Optical detection of bacteria in a microfluidic device. Differential detection of *E. coli* in iceberg lettuce using a multichannel microfluidic device with antibody-conjugated and nonantibody-conjugated microparticles in a latex particle immunoagglutination assay. *(Fig. 2 from D.J. You, K.J. Geshell, J.Y. Yoon, Direct and sensitive detection of foodborne pathogens within fresh produce samples using a field-deployable handheld device, Biosens. Bioelectron. 28 (2011) 399–406.)*

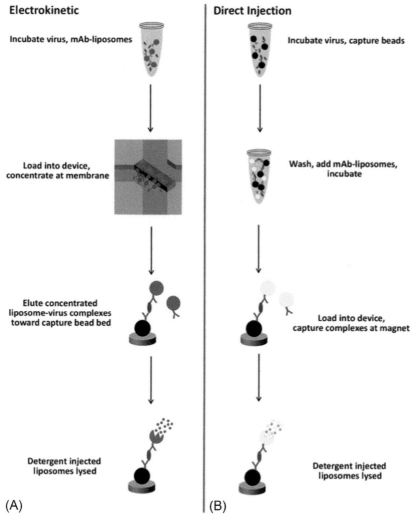

Electrokinetic

Incubate virus, mAb-liposomes

Load into device, concentrate at membrane

Elute concentrated liposome-virus complexes toward capture bead bed

Detergent injected liposomes lysed

(A)

Direct Injection

Incubate virus, capture beads

Wash, add mAb-liposomes, incubate

Load into device, capture complexes at magnet

Detergent injected liposomes lysed

(B)

FIG. 10.6 Virus detection using liposomes. Schematic of assays with and without preconcentration. The assay using electrokinetic preconcentration (A) begins with loading the device with anti-virus pAb-labelled Protein A superparamagnetic beads to create a capture bed and incubating the anti-virus mAb-labelled fluorescent liposomes with virus. The sample is then loaded into the inlet well, concentrated at the nanoporous membrane, and eluted towards the capture bead bed. Following washing, detergent is injected to lyse the liposomes, releasing the fluorescent dye for quantification. The assay without preconcentration (B) begins with incubating an virus sample with the same capture beads as before. The virus–bead complexes are washed and incubated with electrochemical liposomes. This sample is pulled into the microfluidic channel where the detection complexes are captured at a magnet and washed, and the bound liposomes are lysed with detergent. This releases the electroactive species, which undergoes redox cycling at a downstream IDUA. *(Fig. 1 from J.T. Connelly, et al., Micro-total analysis system for virus detection: microfluidic preconcentration coupled to liposome-based detection, Anal. Bioanal. Chem. 401 (2012) 305–323.)*

a circular microfluidic system [74]. Simultaneous detection of *E. coli* and *S. typhimurium* in concentrations ranging from 10^3 to 10^7 cfu/mL was shown. In 2014 a droplet-based microfluidic approach was taken combined with high-throughput droplet imaging to create ScanDrop, a microfluidic method of *E. coli* analysis, which was tested with drinking water and offered the potential for multiplexing along with reduction of reagents required; however, detection time was 8 h [104]. In 2019 a 3D-printed microfluidic device was applied for onsite *E. coli* monitoring, with the ability to add on extra chambers for multiplexed detection. LOD was 10^6 cfu/mL in 6 h [105]. *Cryptosporidium* and *Giardia* were detected by a microfluidic system incorporating multi-angle laser scattering, using an algorithm that enabled 96% accuracy of discrimination based on scattering pattern [106] as well as a diffraction phase microscopy approach, in which the microfluidic system provided a series of trap for protozoan capture and analysis [107]. One of the most promising recent optical microfluidic developments has been the system capable of detecting 10 cells from a 10-mL sample, shown with *E. coli* O157:H7 as this processes a reasonable volume with a low LOD and is packaged into an easy to use portable field-use system that is linked with a smartphone [108] (Fig. 10.7).

10.2.3 In electrical methods

Despite the relative ease of electrode integration on-chip, there appear to be significantly fewer studies showing electrical methods of waterborne pathogen detection with microfluidics. However, many of the systems reported in Chapter 6 are already performed in miniaturised systems; the development of electrical methods to on-chip systems is well-established, and the lacking area is the number of waterborne pathogen applications explored. A few examples are given in this section.

Fig. 10.6 shows an electrical method for virus detection, with electrochemical compounds encapsulated in liposomes, which can be lysed releasing the compounds for detection. Gomez et al. developed a flow-through system to measure the impedance of pathogenic bacteria, testing for metabolic activity as an indicator of viability [109]. They claimed to process low numbers of cells (1–5000) though the sample volume was also very small, being just 30 nL.

Mannoor et al. described a microfluidic impedance system for bacterial detection [110]. In the flow-through set-up the LOD was 10^4 cells/mL as opposed to 1 bacterium/µL in static operation (Fig. 10.8). The authors suggest the difference is due to the influence of flow on the binding kinetics, resulting in reduced binding. Appropriate system design to consider binding kinetics and flow rates is very important [111].

Impedance-based analysis has been a popular area of research in the last couple of years, in both food [112] and water pathogen detection; with

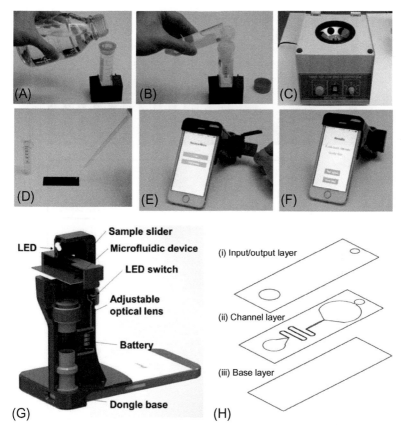

FIG. 10.7 Flowchart of rapid field testing protocol for *E. coli* O157:H7 detection. (A) The 40 μm cell strainer is used for coarse filtration to remove unwanted large particles from 10 mL of containing *E. coli* O157:H7 cells-containing water samples; (B) transfer the water sample into a centrifuge filter (pore size is ~30 nm) for further enrichment of pathogen samples; (C) apply the centrifuge at 2000 g for 7 min, the resulting liquid volume is 50 μL. The antibody-coated microbeads are added into 50 μL of retentate in the centrifuge filter to initiate immunoagglutination reaction; (D) dispense liquid samples containing conjugates onto a microlens-embedded microfluidic chip, where a narrow beam is formed to scan through the conjugates flowing within the microchannel (E, F) Insert a microfluidic device into a smartphone dongle and execute the smartphone application for bacterial counting (G) Schematic of the compact smartphone dongle for rapid waterborne field testing is illustrated, where the microfluidic device is inserted into dongle by using a sample slider. An LED powered by batteries is placed over the microfluidic device for illumination. The adjustable lens in the dongle tube is optically aligned with the CMOS camera on the smartphone; (H) The design of the capillary-driven microfluidic device includes (i) an input/output layer for sample introduction; (ii) a microchannel layer and (iii) a base layer. *(Reproduced with permission from T.-F. Wu, et al., Rapid waterborne pathogen detection with mobile electronics, Sensors 17 (6) (2017) 1348.)*

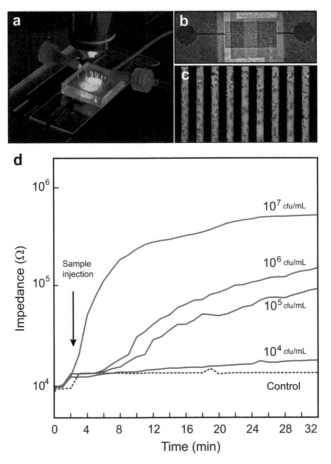

FIG. 10.8 Impedance measurements with a microfluidic device. Real-time binding of bacteria to AMP biosensors. (A) Digital photograph of the microfluidic flow cell. (B) Optical micrograph of the microfluidic channel with an embedded interdigitated microelectrode array chip. (C) Optical image of the embedded microelectrode array after exposure to 107 cfu / mL bacterial cells for 30 min. (D) Real-time monitoring of the interaction of the AMP-functionalised sensor (and an unlabelled control chip) with various concentrations of *E. coli* cells. *(Fig. 5 from M.S. Mannoor, Electrical detection of pathogenic bacteria via immobilized antimicrobial peptides, Proc. Natl. Acad. Sci. U. S. A., 2010.)*

approaches ranging from a complete portable smartphone-linked detection system, focussing on *E. coli* with an LOD of 10 bacteria/mL [113], to species level analysis of protozoan pathogens [114]. The syringe-based filtration technology offered improved LOD compared to a microfluidic positive DEP approach, although the authors report that system optimisation could further lower the 300 cfu/mL LOD [115]. Altinas et al. created an electrochemical biosensor system integrated with microfluidics allowing 50 cfu/mL detection when the assay used was enhanced with nanoparticles [116].

10.2.4   In biosensors

From Chapter 7, it was clear that electrochemical methods of biosensing have been the least applied to waterborne pathogens and the same is true of integration of this type of sensor into microfluidics for waterborne pathogens. A recent review has described integration of microfluidics and biosensors though found few examples for waterborne pathogens [117].

SPR within microfluidics has been shown, though there is little work with waterborne pathogens. Cell-based SPR is less common than small molecule detection and so SPR lab-on-a-chip devices perhaps hold most promise for detection of the outputs of molecular methods. SPR-based biosensors are currently implemented to be applied in field-deployable devices sensing of small molecules, proteins, viruses and whole microbes using a 24-channel SPREETA (Sensata) sensor unit [61].

Microfluidics has been integrated with mass-sensitive cantilever biosensors [118]. Channels have been incorporated onto cantilever structures and used to weigh single cells [119]. For waterborne pathogens, such embedded microfluidic channels on a cantilever have used for the detection of *Cryptosporidium* oocysts [120]. A self-referencing surface-enhanced Raman scattering approach has enabled analysis at an LOD of 10 cfu/mL, with highly specific detection [121]. One interesting recent development exploited the differences in velocity of reference beads and bacteria bound magnetic complexes, using a particle tracking video set-up within a microfluidic system, to enable prediction of *E. coli* concentrations from water samples [122].

10.2.5   In molecular methods

The first microfluidic PCR was shown in a 50-µL silicon chip in 1993 [123]. Since then numerous designs and components for performing molecular methods on-chip have been developed. These are now starting to be incorporated into integrated systems (Fig. 10.9). In addition, many authors have adopted polymeric materials to reduce the turnaround time in device design refinement and to reduce cost. The main advantage of microfluidic PCR is the rapid temperature cycling which can be obtained; in commercial PCR systems this step accounts for over 90% of the operation time. Single-cell detection is possible [124].

For PCR there have been two main approaches to on-chip systems. The first is static PCR where the sample volume is held in a chamber and the temperature cycling is performed by means of some kind of heating system. Initially external heating elements were used though more recent work has shown faster heating with integrated thin metallic film heaters (e.g. 5–30 min compared to 1–3 h) and also with noncontact heating methods. The second is flow-through PCR where the device is designed such that sample travelling in a serpentine channel passes through different temperature zones. The advantage of this approach is that it is faster, delivering results in 90 s–10 min although the disadvantage is

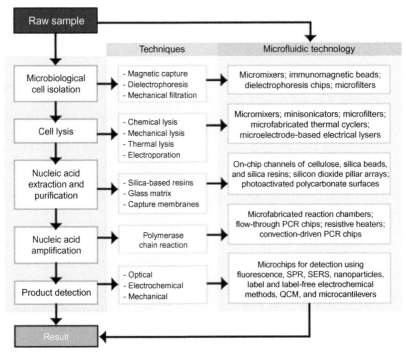

FIG. 10.9 Schematic of stages in molecular detection and the appropriate microfluidic technology. Table 1 from [7].

lower detection sensitivity. One reason for this might be the high surface area in the serpentine channel, compared to static reactor volume, and thus significant adsorption of reagents could occur. There are several commercially available static PCR systems but as yet no flow-through devices.

In addition to these two main approaches other methods such as oscillatory PCR, convective PCR and continuous flow thermal gradient PCR have been developed. A pocket-sized convective PCR system developed by Agraval et al. in 20 was the cheapest prototype molecular method lab-on-a-chip, costing just $10 for the hardware and a few cents per reaction [125]. The continuous flow thermal gradient is the first demonstration of a real-time flow-through system [126]. Another emerging type of PCR is droplet PCR though at present it is very expensive.

Microfluidic systems for isothermal molecular methods have also been developed. NASBA has been shown on chip [127–129]. So far with lab-on-a-chip systems for molecular methods significant progress has been made in the design and operation of amplification and detection stages, with some integration of these two stages. A few systems have presented full-integration with sample processing also performed on-chip but this is an area requiring more development. For more details about the general state-of-the-art regarding microfluidic molecular methods we recommend the 2009 review article by Zhang and Ozdemir [130] and the 2012 review article by Ahmad and Hashsham [131].

10.2.5.1   Waterborne pathogens

While there has been considerable progress developing microfluidic systems that enable on-chip molecular detection, there haven't been as many studies applying these systems to waterborne pathogens [132]. In terms of waterborne viruses, we found that RT-PCR on a microfluidic device with integrated amplification and fluorescence detection, which could be performed within 1 h, was shown for rotavirus in 2011 [133].

Work developing a multiplexed PCR system for bacteria was announced in 2006, with promising early results [134]. Ramalingam et al. presented the simultaneous detection of four waterborne bacteria in a PDMS PCR array, which used capillary flow for sample loading [135]. In 2012, Agrawal et al. showed simultaneous detection of *E. coli* and *S. tyhimurium* on a circular microfluidics design manufactured in PDMS using magnetic nanoparticles to control and concentrate samples [74]. Different systems for bacterial PCR on-chip are still being developed, for example a PDMS system for faecal indicator tracking for three human markers [136].

With regard to protozoa, two publications that relate to molecular sensing of *Cryptosporidium* in miniaturised format describe the performance of NASBA off-chip [137]; only the detection of the mRNAs amplicons was performed on-chip. Esch et al. have developed a fluorescence-based detection assay chip, relying on a sandwich hybridisation of the NASBA product between capture probes and reporter probes [138]. The microfluidic device consists of one channel in a PDMS block bonded to a glass slide with a gold pad at its centre to immobilise the capture probe. The reporter probes were tagged with carboxyfluorescein-filled liposomes giving out better fluorescent intensities than usual fluorophores. This technique gave an LOD of 5 femtomoles of amplicon per test (12.5 μL). The overall time for the full analysis was 1–2 h, including the heat shock and implementation of the NASBA procedure [138].

Multiplexed pathogen detection via molecular methods on microfluidics was reviewed in 2017 [139], describing progress in the stages of detection on-chip, including lysis, extraction and detection concluding that there is relatively little integration of all stages and surprisingly little commercial uptake so far. The authors believed that the lack of all steps being incorporated thus meant few systems could fully be considered point-of-care and that progress to integration would drive commercialisation.

10.2.5.2   Commercial systems

*Crypto*Detect CARD is a platform, shown in Fig. 10.10, with on-chip integrated sample preparation features, developed by Rheonix and reportedly capable of detecting *Cryptosporidium* in raw water samples (Rheonix 2011).

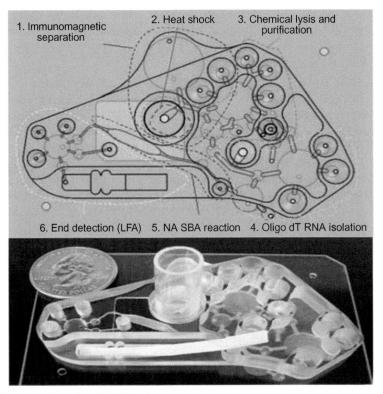

FIG. 10.10   Illustration of the Rheonix system.

The technology involves integrated IMS and washing of the oocysts, heat shock, lysis, extraction, purification and detection of RNA amplicons, using fluorescent liposomes. However, the technology is at an early stage and no LOD or recovery rate was communicated. Furthermore, more sample preparation including sample filtration and concentration would be needed to obtain the 5 mL sample size suitable for this credit-card size chip.

Early Warning Inc. is another company selling an automated platform for on-line monitoring of pathogens including protozoa (EarlyWarningInc 2011). Unlike previous examples, the platform includes a concentrator capable of sampling 10L of water and filtering it to a 10-mL sample. The concentrator unit also features hydrodynamic cavitation to disaggregate clumps. The inclusion of this unit also makes the system relatively large (182 cm × 139 cm × 76 cm) and heavy (around 200 kg), making it difficult for portable field test usage. After concentration and filtration, analytes of interest in the sample are subsequently separated using IMS, lysed and parasite RNA is amplified by NASBA. The detection happens when the hybridisation of target RNA amplicons on specific biomolecular probes generates a current via a guanine oxidation process.

The reported LOD of this device after concentration, filtration and detection on the biosensor chip was 10 oocysts per 10 L. Additionally the total load quantification and viability testing of up to 25 species can be performed on a single chip with a total processing time under 3 h from sampling to results.

Microfluidics has also been applied to the challenge of antimicrobial resistance testing, with miniaturisation and integration of different tools having been attempted to produce handheld or standalone devices for rapid AMR testing using different formats of microfluidics technology such as active microfluidics, droplet microfluidics, paper microfluidics and capillary-driven microfluidics [140]. Lots of systems have focussed on analysis of bacteria or discovery of new antibiotics though others have worked on detection systems, particularly though for patient diagnostics to select treatment [141, 142]. Other systems have been developed specifically for water testing, including tracking beta-lactamase mediated resistance in the environment on a paper microfluidic system [143]. Sandberg et al. developed a multiplexed microfluidic device capable of detection of 39 antimicrobial resistance genes (ARGs), using a PCR approach [144], which was successfully shown with wastewater and drinking water samples. Digital microfluidics, which involves the manipulation of droplets within microfluidic devices (such droplets can act to isolate single pathogens), has been combined with RPA to enable detection of three different ARGs [145], though has not been applied to environmental samples as yet.

10.2.6   Paper-based microfluidics

Paper-based microfluidics is an area of microfluidics which has expanded hugely over the last decade, with numerous examples of different systems reported, which can incorporate operations such as transportation, storage, sorting, mixing and separation. The field grew from the lateral flow assay (LFA), differentiating itself in terms of using patterning for directed flow. Recent reviews of the topic are excellent [3, 146] and give detail on fabrication, operation and applications, all of which will be briefly summarised here before giving information on waterborne pathogen applications, of which there are several examples though much less numerous than healthcare applications of paper microfluidics.

The advantages of paper-based systems are low-cost, ease-of-use, relatively straightforward fabrication along with great suitability for portable, field testing applications and ease of disposal [3, 146], with further details of the advantages of paper given in Fig. 10.11A, though further understanding of properties would accelerate development [147]. Several different fabrication approaches have been developed, though these fall into two main categories, either 2D cutting or shaping of paper or hydrophobic/hydrophilic patterning (Fig. 10.11B). The latter methods use hydrophobic boundaries to define channel edges, of which ink jet printing and wax printing are the most suitable for mass production due to low-cost, simplicity and speed [146].

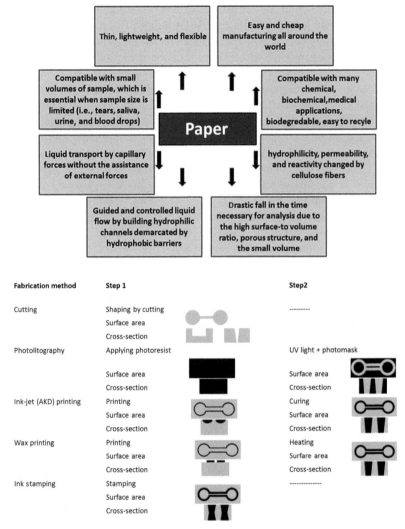

FIG. 10.11 (A) Overview of the advantages of paper. (B) Summary schematic of the two main fabrication approaches for paper microfluidics. *(Reproduced with permission from T. Akyazi, L. Basabe-Desmonts, F. Benito-Lopez, Review on microfluidic paper-based analytical devices towards commercialisation, Anal. Chim. Acta 1001 (2018) 1–17.)*

A detailed description of flow methods and control, along with a description of detection strategies, is given in the review by Akyazi et al. [146]. Briefly, the flow in paper devices is driven by capillary forces, with the advantage of not requiring external pumps but the drawback of challenges in controlling the wicking accurately, particularly for multistep assays [148]. A variety of detection approaches have been adopted in paper microfluidic devices, including colorimetric, electrochemical, fluorescence and chemiluminescence [3] as well as molecular methods of detection [149].

Commercially available devices are on the market, though with the water market being mainly focussed on pH, heavy metals and nitrates, for example work by the group of Henry commercialised by Access Sensor Technologies [150] and the WaterSafe home testing kit, both available from Amazon. The kit does include a faecal indicator monitoring option though longer times are required due to the inclusion of a culturing step. More work has been done in terms of bacterial testing in food than water applications [151], though further progress is required [27]. To lower the LOD extra pretreatment and enrichment is often required [152, 153] and a single step detection device had an LOD of around a million cfu/mL [154].

Virus detection on paper microfluidics has been reported, though the focus has been on Ebola [155], Zika [156], and HIV/Hepatitis C [157] rather than waterborne pathogens, with a couple of exceptions. A variety of approaches have been taken, including RPA amplification [155], LAMP amplification [158] and an electrochemical detection Scheme [157], and multiplexing showed. One interesting recent study on influenza detection incorporated full sample to answer processes on paper [159], though for waterborne pathogens the volumes of water to be analysed will make translation of such technology more challenging. Microfluidics is a growing area in viral diagnostics, with potential to increase detection speed and for field-use [160], and a norovirus paper-based detection system was recently reported, though with the focus on food and clinical applications [161]. A fluorescein-labelled aptamer was associated with carbon nanotubes for use in an FRET-based detection scheme that had a detection time of 10 min and Lo Dof 3–4 ng/mL of virus; this was shown with spiked mussel samples with off-device sample preparation, and as such the system could be adapted for water quality testing. Another norovirus system was able to reach pg/mL LOD using an immunoagglutination scheme and smartphone imaging [162].

A couple of examples of systems for detecting *E. coli* from water samples with paper microfluidics have been reported. One device uses enzymatic activity with a colorimetric substrate, reporting 5–20 cfu/mL LOD, in 30 min, depending on bacterial strain when combined with a immunomagnetic nanoparticle enrichment step, and the addition of 8 h of culturing facilitating 1 cfu/100 mL [163]. Another used antibodies for capture and an antibody-gold nanoparticle label to enable visualisation down to 57 cfu/mL and was applied to water distribution system testing in rural China [164]. DipTest takes a litmus paper approach to *E. coli* testing, concentrating on simplicity and ease of use with a simple dipping procedure to wick sample along the strip to an enzymatic colourimetric detection zone, with the authors reporting an LOD of 200 cfu/mL in 2 h [165]. Other colourimetric technology used catalase with Fentons reagent [166], and another approach was combined with Modified Moore Swabs for sample preparation to manage detection from large volumes of agricultural water [167]. Mie scatter intensity of immunoagglutination was determined via smartphone imaging on a paper device, reporting single cell level detection in 90 s [168].

Protozoan detection on paper microfluidics has been reported by Crannell using an RPA assay, focussing on animal and patient diagnostics for *Cryptosporidium* with an LOD of 1–10 oocysts per reaction [169]. A multiplexed system targeting *Cryptosporidium*, *Giardia*, and *Entamoeba* was also developed [170].

Origami devices have been used to enable molecular methods of detection within paper microfluidics, and operate via folding to process the sample through several steps [171]. Lysis of *E. coli* has been shown [171], and field-use of such systems for malaria testing was recently achieved [172]. The group in Glasgow who developed the malaria system are also working on detection of foodborne pathogens, and with Cranfield University on waterborne pathogens. Many isothermal amplification methods exist, with LAMP being the most popular in food and water testing [173], and the combination of isothermal amplification with paper microfluidics is a rapidly growing area [158]. In some cases, hybrid paper systems are used combining paper segments with, for example PDMS sections. Sample processing is a key concern and area with potential for great future development, which should be targeted as it is a major bottleneck for field implementation of these technologies [173].

Overall, colourimetric type detection schemes seem to be the most popular approach in paper microfluidics for waterborne pathogens, though this has the drawback of typically high detection limits apart from one system reporting single cell level detection. The advantage of this approach is the simplicity and rapidity whereas the use of molecular methods incorporates further steps into the analysis though sensitivity is enhanced and there is greater potential for detailed discrimination, for example faecal source tracking or finding ARGs. The integration of all required steps on-chip is a challenge for this approach, though some devices are now being combined with technologies such as the Whatman FTA card for lysis, extraction and storage of nucleic acids [147]. All types of pathogen have been investigated, though often for other application areas than water, but the field is relatively new and there are plenty of exciting avenues for future development.

10.3. Summary

For a very interesting discussion on the future potential of microfluidics we recommend a 2010 review by Rios et al. [174]. This chapter has presented some basics behind microfluidics and reviewed the applications for waterborne pathogens. Detection to the single-cell level has been shown, though the range of pathogens studied needs to be expanded. For all the detection methods discussed previously in this book there are microfluidic systems for performing these tests, including antimicrobial resistance [140]. Of optical, electrical and biosensor systems, optical detection seems to have received the most attention from the lab-on-a-chip community.

The advantages of performing fluorescent detection on-chip are: the reduced sample volume resulting in a lower background noise signal and therefore improved sensitivity and signal to noise ratio; the small sample volume and control of flow enhancing binding kinetics and increasing sensitivity; and the reduced consumption of reagents [118]. However, there are challenges, especially with the relatively short shelf life of reagents for portable, field devices, and particularly in the production of low-cost, sensitive and portable optics. There has been some progress towards this latter goal with the development of on-chip microscopes and photodiode detectors as well as with alternative optical approaches. Despite this, label-free technologies seem more promising for ease of integration with microfluidics [118].

Molecular methods, discussed further below, exploit many of the advantages of lab-on-a-chip miniaturisation and are likely to be one of the main future directions for microfluidic waterborne pathogen detection. The different detection methods, described in Chapters 5–7 for whole-cell microorganisms can mostly also be applied to molecular detection and so there are many options to perform the on-chip detection.

In general for wider exploitation of microfluidic devices integration of optical and electrical detection components is essential to realise truly portable systems. Either these components need to be included on-chip at low-cost or systems need to be designed where low-cost and portable optical and electrical hardware integrates with a disposable microfluidic chip.

The main microfluidic approaches to sample processing include miniaturisation of antibody-based or filter/sieve-based concentration along with a couple of examples of electrical and hydrodynamic concentration. Dealing with real water samples is challenging, due to potential problems with device clogging as well as obtaining sufficient throughput. However, online monitoring of wastewater using a microfluidic system has been shown, though for phosphate sensing, which is simpler than microorganism analysis [175]. Additionally, recent work has shown the ability to work with 100 L of drinking water. More work is definitely required in this area to enable fully integrated technologies and take advantage of microfluidic phenomena.

Foodborne pathogens seem to have received more attention than waterborne pathogens, though often the pathogens are the same and identical detection procedures are possible. The main difference lies in the sample processing, and it is easier to test wash from food samples as opposed to large volumes of complex environmental water, which might be the explanation for the focus. Additionally, there may be differences in the food and water testing markets, a factor which will be addressed in Chapter 11.

Performing molecular methods on microfluidic chips has many advantages, though has not yet been applied too extensively to the challenge of waterborne pathogen detection. The technology is reaching the stage of integrated systems although more work needs to be done on incorporation of sample processing elements. This challenge is greater for environmental water samples where both the initial sample volumes and the potential number of interferents/inhibitors are high.

Quilliam et al. report that increases in the sensitivity and specificity of detection methods for waterborne pathogens are currently being achieved by combining advances in microfluidics technology and analytical chemistry with molecular and immunological methods [176]. Mairhofer, Roppert and Ertl believe the next generation of pathogen sensing developments will be facilitated by advances in lab-on-a-chip devices [61].

Paper microfluidics are being widely exploited in other application areas, with many exciting examples of portable, easy-to-use systems applied to healthcare applications. While water testing with paper microfluidics has been reported the main focus has been on contaminants rather than pathogens, and the number of reports focussing on waterborne viruses, bacteria and protozoa is low. Hopefully, given the huge potential of this technology more research will be focussed into this area, and particularly in combining paper-based detection with appropriate sample processing approaches.

The global market for microfluidic technology is growing at a great pace and is estimated to be worth US$63 billion by 2027. Given their many advantages, and providing some of the above challenges can be met, it can be expected that in the future microfluidics for waterborne pathogens will have an increasing share in that growing market.

References

[1] D.J. Beebe, G.A. Mensing, G.M. Walker, Physics and applications of microfluidics in biology, Annu. Rev. Biomed. Eng. 4 (2002) 261–286.

[2] J. Merrin, Frontiers in microfluidics, a teaching resource review, Bioengineering 6 (4) (2019) 109.

[3] C. Carrell, et al., Beyond the lateral flow assay: a review of paper-based microfluidics, Microelectron. Eng. 206 (2019) 45–54.

[4] Introduction: the origin, current status, and future of microfluidics, in: Microfluidics: Fundamental, Devices and Applications. pp. 1–18.

[5] K.S. Chow, H. Du, Dielectrophoretic characterization and trapping of different waterborne pathogen in continuous flow manner, Sensors Actuators A 170 (2011) 24–31.

[6] J.Y. Yoon, B. Kim, Lab-on-a-chip pathogen sensors for food safety, Sensors 12 (2012) 10713–10741.

[7] C. Lui, N.C. Cady, C.A. Batt, Nucleic acid-based detection of bacterial pathogens using integrated microfluidic platform systems, Sensors 9 (2009) 3713–3744.

[8] Folk, C., Chen, X., Wudl, F., Ho, C.M., Hydrogel Microvalves With Short Response Time. ho.seas.ucla.edu/publications/conference/2003/232.pdf 2003.

[9] J.M. Ng, et al., Components for integrated poly(dimethylsiloxane) microfluidic systems, Electrophoresis 23 (20) (2002) 3461–3473.

[10] D.C. Duffy, et al., Rapid prototyping of microfluidic systems in poly(dimethylsiloxane), Anal. Chem. 70 (23) (1998) 4974–4984.

[11] S.K. Sia, G.M. Whitesides, Microfluidic devices fabricated in poly(dimethylsiloxane) for biological studies, Electrophoresis 24 (21) (2003) 3563–3576.

[12] K. Haubert, T. Drier, D.J. Beebe, PDMS bonding by means of a portable, low-cost corna system, Lab Chip 6 (2006) 1548–1549.

[13] J. Byung-Ho, et al., Three-dimensional micro-channel fabrication in polydimethylsiloxane (PDMS) elastomer, J. Microelectromech. Syst. 9 (1) (2000) 76–81.

[14] A.M. Christensen, D.A. Chang-Yen, B.K. Gale, Characterisation of interconnects used in PDMS microfluidic systems, J. Micromech. Microeng. 15 (2005) 928–934.

[15] S.R. Quake, A. Scherer, From micro- to nanofabrication with soft materials, Science 290 (2000) 1536–1540.

[16] M.A. Holden, et al., Microfluidic diffusion diluter: bulging of PDMS microchannels under pressure-driven flow, J. Micromech. Microeng. 13 (2003) 412–418.

[17] J.N. Lee, C. Park, G.M. Whitesides, Solvent compatibility of poly(dimethylsiloxane)-based microfluidic devices, Anal. Chem. 75 (23) (2003) 6544–6554.

[18] G. Ocvirk, et al., Electrokinetic control of fluid flow in native poly(dimethylsiloxane) capillary electrophoresis devices, Electrophoresis 21 (1) (2000) 107–115.

[19] M.W. Toepke, D.J. Beebe, PDMS absorption of small molecules and consequences in microfluidic applications, Lab Chip 6 (12) (2006) 1484–1486.

[20] H. Makamba, et al., Surface modification of poly(dimethylsiloxane) microchannels, Electrophoresis 24 (21) (2003) 3607–3619.

[21] M.A. Holden, S.Y. Jung, P.S. Cremer, Patterning enzymes inside microfluidic channels via photoattachment chemistry, Anal. Chem. 76 (7) (2004) 1838–1843.

[22] A. Gliere, C. Delattre, Modeling and fabrication of capillary stop valves for planar microfluidic systems, Sensors Actuators A 130-131 (2006) 601–608.

[23] J.W. Suk, J.-H. Cho, Capillary flow control using hydrophobic patterns, J. Micromech. Microeng. 17 (2007) N11–N15.

[24] N.C. Speller, et al., Green, low-cost, user-friendly, and elastomeric (GLUE) microfluidics, ACS Appl. Polym. Mater. 2 (3) (2020) 1345–1355.

[25] R. Lausecker, et al., Introducing natural thermoplastic shellac to microfluidics: a green fabrication method for point-of-care devices, Biomicrofluidics 10 (4) (2016) 044101.

[26] Y. Xia, J. Si, Z. Li, Fabrication techniques for microfluidic paper-based analytical devices and their applications for biological testing: a review, Biosens. Bioelectron. 77 (2016) 774–789.

[27] M.I.G.S. Almeida, et al., Developments of microfluidic paper-based analytical devices (µPADs) for water analysis: a review, Talanta 177 (2018) 176–190.

[28] T.M. Squires, S.R. Quake, Microfluidics: fluid physics at the nanoliter scale, Rev. Mod. Phys. 77 (3) (2005) 977–1026.

[29] J.P. Brody, et al., Biotechnology at low Reynolds numbers, Biophys. J. 71 (6) (1996) 3430–3441.

[30] P.W. Atkins, Physical Chemistry, fifth ed., Oxford University Press, 1994.

[31] M.A. Unger, et al., Monolithic microfabricated valves and pumps by multilayer soft lithography, Science 288 (5463) (2000) 113–116.

[32] W. Gu, et al., Computerized microfluidic cell culture using elastomeric channels and braille displays, Proc. Natl. Acad. Sci. U. S. A. 101 (45) (2004) 15861–15866.

[33] J. Park, et al., Micro pumping with cardiomyocyte-polymer hybrid, Lab Chip 7 (10) (2007) 1367–1370.

[34] E.W. Jager, E. Smela, O. Inganas, Microfabricating conjugated polymer actuators, Science 290 (5496) (2000) 1540–1545.

[35] D.J. Beebe, et al., Functional hydrogel structures for autonomous flow control inside microfluidic channels, Nature 404 (6778) (2000) 588–590.

[36] B.S. Samel, P. Griss, G. Stemme, A thermally responsive PDMS composite and its microfluidic applications, J. Microelectromech. Syst. 16 (1) (2006).

[37] Q. Yu, J.M. Bauer, J.S. Moore, D.J. Beebe, Responsive biomimetic hydrogel valve for microfluidics, Appl. Phys. Lett. 78 (17) (2001) 2589–2591.

[38] D.T. Eddington, D.J. Beebe, Flow control with hydrogels, Adv. Drug Deliv. Rev. 56 (2) (2004) 199–210.

[39] D.T. Eddington, et al., An organic self-regulating microfluidic system, Lab Chip 1 (2) (2001) 96–99.

[40] N. Li, C.H. Hsu, A. Folch, Parallel mixing of photolithographically defined nanoliter volumes using elastomeric microvalve arrays, Electrophoresis 26 (19) (2005) 3758–3764.

[41] V. Studer, et al., Scaling properties of a low-actuation pressure microfluidic valve, J. Appl. Phys. 95 (1) (2004) 393–398.

[42] R.F. Ismagilov, et al., Pressure-driven laminar flow in tangential microchannels: an elastomeric microfluidic switch, Anal. Chem. 73 (19) (2001) 4682–4687.

[43] W.H. Grover, et al., Development and multiplexed control of latching pneumatic valves using microfluidic logical structures, Lab Chip 6 (5) (2006) 623–631.

[44] R.P.R. Fabio, et al., Multicommutation in flow analysis: concepts, applications and trends, Anal. Chim. Acta 468 (2002) 119–131.

[45] R.E. Dolmetsch, K. Xu, R.S. Lewis, Calcium oscillations increase the efficiency and specificity of gene expression, Nature 392 (6679) (1998) 933–936.

[46] C.H. Hsu, C. Chen, A. Folch, "Microcanals" for micropipette access to single cells in microfluidic environments, Lab Chip 4 (5) (2004) 420–424.

[47] H. Bridle, et al., Automated control of local solution environments in open-volume microfluidics, Anal. Chem. 79 (24) (2007) 9286–9293.

[48] J.M. Ottino, S. Wiggins, Introduction: mixing in microfluidics, Philos. Transact. A Math. Phys. Eng. Sci. 362 (1818) (2004) 923–935.

[49] L.H. Lu, K.S. Ryu, C. Liu, A magnetic microstirrer and array for microfluidic mixing, J. Microelectromech. Syst. 11 (5) (2002) 462–469.

[50] S.M. Shin, I.S. Kang, Y.K. Cho, Mixing enhancement by using electrokinetic instability under time-periodic electric field, J. Micromech. Microeng. 15 (2005) 455–462.

[51] I. Glasgow, N. Aubry, Enhancement of microfluidic mixing using time pulsing, Lab Chip 3 (2) (2003) 114–120.

[52] B.M. Paegel, et al., Microfluidic serial dilution circuit, Anal. Chem. 78 (21) (2006) 7522–7527.

[53] C. Neils, et al., Combinatorial mixing of microfluidic streams, Lab Chip 4 (4) (2004) 342–350.

[54] J. Pihl, et al., Microfluidic gradient-generating device for pharmacological profiling, Anal. Chem. 77 (13) (2005) 3897–3903.

[55] M.R. Bringer, et al., Microfluidic systems for chemical kinetics that rely on chaotic mixing in droplets, Philos. Transact. A Math. Phys. Eng. Sci. 362 (1818) (2004) 1087–1104.

[56] J.C. McDonald, et al., Prototyping of microfluidic devices in poly(dimethylsiloxane) using solid-object printing, Anal. Chem. 74 (7) (2002) 1537–1545.

[57] A.D. Stroock, et al., Chaotic mixer for microchannels, Science 295 (5555) (2002) 647–651.

[58] A.D. Stroock, et al., Patterning flows using grooved surfaces, Anal. Chem. 74 (20) (2002) 5306–5312.

[59] A. Groisman, et al., A microfluidic chemostat for experiments with bacterial and yeast cells, Nat. Methods 2 (9) (2005) 685–689.

[60] K. Kant, et al., Microfluidic devices for sample preparation and rapid detection of foodborne pathogens, Biotechnol. Adv. 36 (4) (2018) 1003–1024.

[61] J. Mairhofer, K. Roppert, P. Ertl, Microfluidic systems for pathogen sensing: a review, Sensors 9 (2009) 4804–4823.

[62] S.A. Jaywant, K.M. Arif, A comprehensive review of microfluidic water quality monitoring sensors, Sensors 19 (21) (2019) 4781.

[63] S. Kumar, et al., Point-of-care strategies for detection of waterborne pathogens, Sensors 19 (20) (2019) 4476.

[64] M. Yew, et al., A review of state-of-the-art microfluidic technologies for environmental applications: detection and remediation, Global Chall. 3 (1) (2019) 1800060.

[65] B.C. Dhar, N.Y. Lee, Lab-on-a-chip technology for environmental monitoring of microorganisms, BioChip J. 12 (3) (2018) 173–183.

[66] P. Yager, et al., Microfluidic diagnostic technologies for global public health, Nature 422 (2006) 412.

[67] P. Chen, X. Feng, B.F. Liu, Microfluidic chips for cell sorting, Front. Biosci. 13 (2008) 2464–2483.

[68] M. Kersaudy-Kerhoas, R. Dhariwal, M.P.Y. Desmulliez, Recent advances in microparticle continuous separation, IET Nanobiotechnol. 2 (1) (2008) 1–13.

[69] M.Z.J. Madou, G. Jia, H. Kido, J. Kim, N. Kim, Lab on a CD, Ann. Rev. Biomed. Eng. 8 (2006) 601–628.

[70] S. Burger, et al., Labslice XL—a centrifugal microfluidic cartridge for the automated biochemical processing of industrial process water, in: 2019 20th International Conference on Solid-State Sensors, Actuators and Microsystems & Eurosensors XXXIII (Transducers & Eurosensors XXXIII), 2019.

[71] U. Dharmarisi, et al., Enrichment and detection of *Escherichia coli* O157:H7 from water samples using an antibody modified microfluidic chip, Anal. Chem. 82 (7) (2010) 2844–2849.

[72] X. Guan, et al., Rapid detection of pathogens using antibody-coated microbeads with bioluminescence in microfluidic chips, Biomed. Microdevices 12 (2012) 683–691.

[73] Q. Ramadan, et al., Flow-through immunomagnetic separation system for waterborne pathogen isolation and detection: application to Giardia and *Cryptosporidium* cell isolation, Anal. Chim. Acta 673 (1) (2010) 101–108.

[74] S. Agrawal, et al., Multiplexed detection of waterborne pathogens in circular microfluidics, Appl. Biochem. Biotechnol. (2012).

[75] L. Diéguez, et al., Disposable microfluidic micromixers for effective capture of Cryptosporidium parvum oocysts from water samples, J. Biol. Eng. 12 (1) (2018) 4.

[76] A.D. Goater, J.P.H. Burt, R. Pethig, A combined travelling wave dielectrophoresis and electrorotation device: applied to the concentration and viability determination of *Cryptosporidium*, J. Phys. D. Appl. Phys. 30 (1997) L65–L69.

[77] C.R. Cabrera, P. Yager, Continuous concentration of bacteria in a microfluidic flow cell using electrokinetic techniques, Electrophoresis 22 (2001) 355–362.

[78] R. Gomez-Sjoberg, D.T. Morisette, R. Bashir, Impedance microbiology-on-a-chip: microfluidic bioprocessor for rapid detection of bacterial metabolism, J. MEMS 14 (2005) 829–838.

[79] B.H. Lapizco-Encinas, et al., An insulator-based (electrodeless) dielectrophoretic concentrator for microbes in water, J. Microbiol. Methods 62 (3) (2005) 317–326.

[80] Y.-K. Cho, et al., Bacteria concentration using a membrane type insulator-based dielectro-phoresis in a plastic chip, Electrophoresis 30 (2009) 3153–3159.

[81] W.A. Braff, Manipulation of bacteria using a three dimensional insulator based dielectrophore-sis, in: Department of Mechanical Engineering, Massachusetts Institute of Technology, 2011.

[82] J. Zhang, et al., Fundamentals and applications of inertial microfluidics: a review, Lab Chip 16 (1) (2016) 10–34.

[83] M. Jimenez, B. Miller, H.L. Bridle, Efficient separation of small microparticles at high flow-rates using spiral channels: application to waterborne pathogens, Chem. Eng. Sci. 157 (2017) 247–254.

[84] M. Jimenez, H. Bridle, Microfluidics for effective concentration and sorting of waterborne protozoan pathogens, J. Microbiol. Methods 126 (2016) 8–11.

[85] K.R. Ganz, et al., Enhancing the detection of giardia duodenalis cysts in foods by inertial microfluidic separation, Appl. Environ. Microbiol. 81 (12) (2015) 3925.

[86] B. Miller, M. Jimenez, H. Bridle, Cascading and parallelising curvilinear inertial focusing systems for high volume, wide size distribution, separation and concentration of particles, Sci. Rep. 6 (1) (2016) 36386.

[87] T. Taguchi, H. Takeyama, T. Matsunaga, Immuno-capture of Cryptosporidium parvum using micro-well array, Biosens. Bioelectron. 20 (2005) 2276–2282.

[88] T. Taguchi, et al., Detection of *Cryptosporidium parvum* oocysts using a microfluidic device equipped with the SUS micromesh and FITC-labeled antibody, Biotechnol. Bioeng. 96 (2) (2007) 272–280.

[89] L. Zhu, et al., Filter-based microfluidic device as a platform for immunofluorescent assay of microbial cells, Lab Chip 4 (4) (2004) 337–341.

[90] J. Liu, et al., Plasma assisted thermal bonding for PMMA microfluidic chips with integrated metal microelectrodes, Sensors Actuators B Chem. 141 (2) (2009) 646–651.

[91] Q. Liu, et al., A novel microfluidic module for rapid detection of airborne and waterborne pathogens, Sensors Actuators B Chem. 258 (2018) 1138–1145.

[92] N.M.M. Pires, T. Dong, Recovery of *Cryptosporidium* and *Giardia* organisms from surface water by counter-flow refining microfiltration, Environ. Technol. 34 (17) (2013) 2541–2551.

[93] L. Chen, J. Choo, Recent advances in surface-enhanced Raman scattering detection technol-ogy for microfluidic chips, Electrophoresis 29 (2008) 1815–1828.

[94] Y.-C. Tung, et al., PDMS-based opto-fluidic micro flow cytometer with two-color, multi-angle fluorescence detection capability using PIN photodiodes, Sens. Actuators B Chem. 98 (2-3) (2004) 356–367.

[95] Y. Li, X.-L. Su, Microfluidics-based optical biosensing method for rapid detection of *Escherichia coli* O157:H7, J. Rapid Methods Autom. Microbiol. 14 (2006) 96–109.

[96] C. Sakamoto, N. Yamaguchi, M. Nasu, Rapid and simple quantification of bacterial cells by using a microfluidic device, Appl. Environ. Microbiol. 71 (2) (2005) 1117–1121.

[97] N. Yamaguchi, et al., Rapid, Semiautomated quantification of bacterial cells in freshwater by using a microfluidic device for on-Chip staining and counting, Appl. Environ. Microbiol. 77 (4) (2011) 1536–1539.

[98] M. Varshney, et al., A microfluidic filter biochip-based chemiluminescence biosensing meth-od for detection of *Escherichia coli* O157:H7, Trans. ASABE 49 (6) (2006) 2061–2068.

[99] X.Y. Karsunke, R. Niessner, M. Seidel, Development of a multichannel flow-through chemi-luminescence microarray chip for parallel calibration and detection of pathogenic bacteria, Anal. Bioanal. Chem. 395 (6) (2009) 1623–1630.

[100] J.-H. Han, B.C. Heinze, J.-Y. Yoon, Single cell level detection of *Escherichia coli* in micro-fluidic device, Biosens. Bioelectron. 23 (8) (2008) 1303–1306.

[101] D.J. You, K.J. Geshell, J.Y. Yoon, Direct and sensitive detection of foodborne pathogens within fresh produce samples using a field-deployable handheld device, Biosens. Bioelectron. 28 (2011) 399–406.

[102] P.R. Marcoux, et al., Micro-confinement of bacteria into w/o emulsion droplets for rapid detection and enumeration, Colloids Surf. A Physicochem. Eng. Asp. 377 (2011) 54–62.

[103] J.T. Connelly, et al., Micro-total analysis system for virus detection: microfluidic pre-concentration coupled to liposome-based detection, Anal. Bioanal. Chem. 401 (2012) 305–323.

[104] A. Golberg, et al., Cloud-enabled microscopy and droplet microfluidic platform for specific detection of *Escherichia coli* in water, PLoS ONE 9 (1) (2014) e86341.

[105] E.C. Sweet, et al., Entirely-3D printed microfluidic platform for on-site detection of drinking waterborne pathogens, in: 2019 IEEE 32nd International Conference on Micro Electro Mechanical Systems (MEMS), 2019.

[106] H. Wei, L. Yang, F. Li, A microfluidic laser scattering sensor for label-free detection of waterborne pathogens, in: International Symposium on Optoelectronic Technology and Application 2016, Vol. 10156, SPIE, 2016.

[107] X. Gu, et al., Microfluidic diffraction phase microscopy for high-throughput, artifact-free quantitative phase imaging and identification of waterborne parasites, Opt. Laser Technol. 120 (2019) 105681.

[108] T.-F. Wu, et al., Rapid waterborne pathogen detection with mobile electronics, Sensors 17 (6) (2017) 1348.

[109] R. GOMEZ, et al., Microfluidic biochip for impedance spectroscopy of biologic species, Biomed. Microdevices 3 (2001) 201–209.

[110] M.S. Mannoor, et al., Electrical detection of pathogenic bacteria via immobilized antimicrobial peptides, Proc. Natl. Acad. Sci. U. S. A. (2010).

[111] T.M. Squires, R.J. Messinger, S.R. Manalis, Making it stick: convection, reaction and diffusion in surface-based biosensors, Nat. Biotechnol. 26 (4) (2008) 417–426.

[112] L. Yao, et al., A microfluidic impedance biosensor based on immunomagnetic separation and urease catalysis for continuous-flow detection of *E. coli* O157:H7, Sensors Actuators B Chem. 259 (2018) 1013–1021.

[113] J. Jiang, et al., Smartphone based portable bacteria pre-concentrating microfluidic sensor and impedance sensing system, Sensors Actuators B Chem. 193 (2014) 653–659.

[114] J.S. McGrath, et al., Analysis of parasitic Protozoa at the single-cell level using microfluidic impedance cytometry, Sci. Rep. 7 (1) (2017) 2601.

[115] M. Kim, et al., A microfluidic device for label-free detection of *Escherichia coli* in drinking water using positive dielectrophoretic focusing, capturing, and impedance measurement, Biosens. Bioelectron. 74 (2015) 1011–1015.

[116] Z. Altintas, et al., A fully automated microfluidic-based electrochemical sensor for real-time bacteria detection, Biosens. Bioelectron. 100 (2018) 541–548.

[117] G. Luka, et al., Microfluidics integrated biosensors: a leading technology towards lab-on-a-chip and sensing applications, Sensors 15 (12) (2015) 30011–30031.

[118] C. Rivet, et al., Microfluidics for medical diagnostics and biosensors, Chem. Eng. Sci. 66 (7) (2011) 1490–1507.

[119] T.P. Burg, et al., Weighing of biomolecules, single cells and single nanoparticles in fluid, Nature 446 (7139) (2007) 1066–1069.

[120] H. Bridle, et al., Static mode microfluidic cantilevers for detection of waterborne pathogens, Sensors Actuators A Phys. 247 (2016) 144–149.

[121] C. Wang, et al., Detection of extremely low concentration waterborne pathogen using a multiplexing self-referencing SERS microfluidic biosensor, J. Biol. Eng. 11 (1) (2017) 9.

[122] A. Malec, et al., Biosensing system for concentration quantification of magnetically labeled E. coli in water samples, Sensors 18 (7) (2018) 2250.

[123] M.A. Northrup, et al., Proc. Transducers, 93, 1993, p. 924.

[124] A.K. Whitea, et al., High-throughput microfluidic single-cell RT-qPCR, Proc. Natl. Acad. Sci. U. S. A. (2010).

[125] N. Agrawal, Y.A. Hassan, V.M. Ugaz, A pocket-sized convective PCR thermocycler, Angew. Chem. Int. Ed. 46 (2007) 4316.

[126] I. Pjescic, Glass-composite prototyping for flow PCR with *in situ* DNA analysis, Biomed. Microdevices 12 (2010) 333.

[127] I.K. Dimov, et al., Integrated microfluidic tmRNA purification and real-time NASBA device for molecular diagnostics, Lab Chip 8 (12) (2008) 2071–2078.

[128] A. Gulliksen, et al., Real-time nucleic acid sequence-based amplification in nanoliter volumes, Anal. Chem. 76 (1) (2004) 9–14.

[129] A. Gulliksen, et al., Parallel nanoliter detection of cancer markers using polymer microchips, Lab Chip 5 (4) (2005) 416–420.

[130] Y. Zhang, P. Ozdemir, Microfluidic DNA amplification—a review, Anal. Chim. Acta 638 (2009) 115–125.

[131] F. Ahmad, S.A. Hashsham, Miniaturized nucleic acid amplification systems for rapid and point-of-care diagnostics: a review, Anal. Chim. Acta 733 (2012) 1–15.

[132] S. Choudhury, Molecular tools for the detection of waterborne pathogens (Chapter 12), in: M.N. Vara Prasad, A. Grobelak (Eds.), Waterborne Pathogens, Butterworth-Heinemann, 2020, pp. 219–235.

[133] Y. Li, C. Zhang, D. Xing, Integrated microfluidic reverse transcription-polymerase chain reaction for rapid detection of food- or waterborne pathogenic rotavirus, Anal. Biochem. 415 (2) (2011) 87–96.

[134] B.H. Weigl, et al., Fully integrated multiplexed lab-on-a-card assay for enteric pathogens, Proc. SPIE 6112 (2006) 11.

[135] N. Ramalingam, et al., Real-time PCR-based microfluidic array chip for simultaneous detection of multiple waterborne pathogens, Sensors Actuators B Chem. 145 (1) (2010) 543–552.

[136] L. Gorgannezhad, et al., Microfluidic array chip for parallel detection of waterborne Bacteria, Micromachines 10 (12) (2019) 883.

[137] S.R. Nugen, et al., PMMA biosensor for nucleic acids with integrated mixer and electrochemical detection, Biosens. Bioelectron. 24 (8) (2009) 2428–2433.

[138] M.B. Esch, A.J. Baeumner, R.A. Durst, Detection of Cryptosporidium parvum using oligonucleotide-tagged liposomes in a competitive assay format, Anal. Chem. 73 (13) (2001) 3162–3167.

[139] I.H.K. Basha, et al., Towards multiplex molecular diagnosis—a review of microfluidic genomics technologies, Micromachines 8 (9) (2017) 266.

[140] S.-U. Hassan, X. Zhang, Microfluidics as an emerging platform for tackling antimicrobial resistance (AMR): a review, Curr. Anal. Chem. 16 (1) (2020) 41–51.

[141] Z. Liu, N. Banaei, K. Ren, Microfluidics for combating antimicrobial resistance, Trends Biotechnol. 35 (12) (2017) 1129–1139.

[142] W.-H. Chang, et al., Vancomycin-resistant gene identification from live bacteria on an integrated microfluidic system by using low temperature lysis and loop-mediated isothermal amplification, Biomicrofluidics 11 (2) (2017) 024101.

[143] K.E. Boehle, et al., Utilizing paper-based devices for antimicrobial-resistant bacteria detection, Angew. Chem. Int. Ed. 56 (24) (2017) 6886–6890.

[144] K.D. Sandberg, S. Ishii, T.M. LaPara, A microfluidic quantitative polymerase chain reaction method for the simultaneous analysis of dozens of antibiotic resistance and heavy metal resistance genes, Environ. Sci. Technol. Lett. 5 (1) (2018) 20–25.

[145] S. Kalsi, et al., A programmable digital microfluidic assay for the simultaneous detection of multiple anti-microbial resistance genes, Micromachines 8 (4) (2017) 111.

[146] T. Akyazi, L. Basabe-Desmonts, F. Benito-Lopez, Review on microfluidic paper-based analytical devices towards commercialisation, Anal. Chim. Acta 1001 (2018) 1–17.

[147] J.R. Choi, et al., Advances and challenges of fully integrated paper-based point-of-care nucleic acid testing, TrAC Trends Anal. Chem. 93 (2017) 37–50.

[148] S.-G. Jeong, et al., Flow control in paper-based microfluidic device for automatic multistep assays: a focused minireview, Korean J. Chem. Eng. 33 (10) (2016) 2761–2770.

[149] J. Ma, et al., Paper microfluidics for cell analysis, Adv. Healthc. Mater. 8 (1) (2019) 1801084.

[150] J.C. Hofstetter, et al., Quantitative colorimetric paper analytical devices based on radial distance measurements for aqueous metal determination, Analyst 143 (13) (2018) 3085–3090.

[151] L.S.A. Busa, et al., Advances in microfluidic paper-based analytical devices for food and water analysis, Micromachines 7 (5) (2016) 86.

[152] M. Srisa-Art, et al., Highly sensitive detection of *Salmonella typhimurium* using a colorimetric paper-based analytical device coupled with Immunomagnetic separation, Anal. Chem. 90 (1) (2018) 1035–1043.

[153] J.A. Adkins, et al., Colorimetric and electrochemical bacteria detection using printed paper- and transparency-based analytic devices, Anal. Chem. 89 (6) (2017) 3613–3621.

[154] J. Park, J.H. Shin, J.-K. Park, Pressed paper-based dipstick for detection of foodborne pathogens with multistep reactions, Anal. Chem. 88 (7) (2016) 3781–3788.

[155] L. Magro, et al., Paper-based RNA detection and multiplexed analysis for Ebola virus diagnostics, Sci. Rep. 7 (1) (2017) 1347.

[156] K. Kaarj, P. Akarapipad, J.-Y. Yoon, Simpler, faster, and sensitive Zika virus assay using smartphone detection of loop-mediated isothermal amplification on paper microfluidic chips, Sci. Rep. 8 (1) (2018) 12438.

[157] C. Zhao, X. Liu, A portable paper-based microfluidic platform for multiplexed electrochemical detection of human immunodeficiency virus and hepatitis C virus antibodies in serum, Biomicrofluidics 10 (2) (2016) 024119.

[158] L. Magro, et al., Paper microfluidics for nucleic acid amplification testing (NAAT) of infectious diseases, Lab Chip 17 (14) (2017) 2347–2371.

[159] N.M. Rodriguez, et al., Paper-based RNA extraction, in situ isothermal amplification, and lateral flow detection for low-cost, rapid diagnosis of influenza a (H1N1) from clinical specimens, Anal. Chem. 87 (15) (2015) 7872–7879.

[160] C. Simpson, et al., Microfluidics: an untapped resource in viral diagnostics and viral cell biology, Curr. Clin. Microbiol. Rep. 5 (4) (2018) 245–251.

[161] X. Weng, S. Neethirajan, Aptamer-based fluorometric determination of norovirus using a paper-based microfluidic device, Microchim. Acta 184 (11) (2017) 4545–4552.

[162] S. Chung, et al., Rapid and reliable norovirus assay at pg/mL level using smartphone-based fluorescence microscope and a microfluidic paper analytic device, in: 2017 ASABE Annual International Meeting, ASABE: St. Joseph, MI, 2017, p. 1.

[163] S.M.Z. Hossain, et al., Multiplexed paper test strip for quantitative bacterial detection, Anal. Bioanal. Chem. 403 (6) (2012) 1567–1576.

[164] S. Ma, et al., Visible paper chip immunoassay for rapid determination of bacteria in water distribution system, Talanta 120 (2014) 135–140.

[165] N.S.K. Gunda, S. Dasgupta, S.K. Mitra, DipTest: a litmus test for *E. coli* detection in water, PLoS ONE 12 (9) (2017) e0183234.

[166] J.-Y. Kim, M.-K. Yeo, A fabricated microfluidic paper-based analytical device (μPAD) for in situ rapid colorimetric detection of microorganisms in environmental water samples, Mol. Cell. Toxicol. 12 (1) (2016) 101–109.

[167] B. Au-Bisha, et al., Colorimetric paper-based detection of *Escherichia coli*, *Salmonella* spp., and *Listeria monocytogenes* from large volumes of agricultural water, JoVE 88 (2014) e51414.

[168] T.S. Park, J. Yoon, Smartphone detection of *Escherichia coli* from field water samples on paper microfluidics, IEEE Sensors J. 15 (3) (2015) 1902–1907.

[169] Z.A. Crannell, et al., Nucleic acid test to diagnose cryptosporidiosis: lab assessment in animal and patient specimens, Anal. Chem. 86 (5) (2014) 2565–2571.

[170] Z. Crannell, et al., Multiplexed recombinase polymerase amplification assay to detect intestinal Protozoa, Anal. Chem. 88 (3) (2016) 1610–1616.

[171] A.V. Govindarajan, et al., A low cost point-of-care viscous sample preparation device for molecular diagnosis in the developing world; an example of microfluidic origami, Lab Chip 12 (1) (2012) 174–181.

[172] J. Reboud, et al., Paper-based microfluidics for DNA diagnostics of malaria in low resource underserved rural communities, Proc. Natl. Acad. Sci. 116 (11) (2019) 4834–4842.

[173] R. Martzy, et al., Challenges and perspectives in the application of isothermal DNA amplification methods for food and water analysis, Anal. Bioanal. Chem. 411 (9) (2019) 1695–1702.

[174] A. Rios, M. Zougagh, M. Avila, Miniaturization through lab-on-a-chip: utopia or reality for routine laboratories? Anal. Chim. Acta 2010 (2010).

[175] J. Cleary, et al., An autonomous microfluidic sensor for phosphate: on-site analysis of treated wastewater, IEEE Sensors J. 8 (5) (2008) 508–515.

[176] R.S. Quilliam, et al., Unearthing human pathogens at the agricultural–environment interface: a review of current methods for the detection of *Escherichia coli* O157 in freshwater ecosystems, Agric. Ecosyst. Environ. 140 (2011) 354–360.

Section D

Applications and evaluation

Chapter 11

Applications of emerging technologies in the drinking water sector

Helen Bridle[a], Lauren Rowe[b], and Graham Sprigg[b]
[a]Institute of Biological Chemistry, Biophysics and Bioengineering, Heriot-Watt University, Edinburgh, Scotland, [b]IMS Consulting, Royal London Buildings, Bristol, England

This chapter explores the application of emerging technologies for waterborne pathogen detection from a market perspective. The water market, particularly in terms of monitoring, is heavily driven by regulation and regulatory compliance. Adoption of new technologies for the detection of pathogens can be challenging within this framework.

This chapter will first describe the application of water safety plans (WSPs) and the water safety frameworks (WSF) in various countries, to set the background in which water monitoring is performed across the globe. There will be a particular focus on the situation in the UK as an in-depth case study.

The chapter will also identify those pathogens which are currently of most concern to the market, providing a perspective from a public health point of view as well as the market view of existing technologies. Finally, the chapter will conclude by discussing future trends and the challenges of bringing new technologies to market in the water monitoring sector.

11.1. Current position of the UK Water industry

While a brief introduction to WSP and WSF was given in Chapter 3, this section will be a more in-depth case study of how they are applied in the UK, as well as giving an introduction to the treatment and management aspects. After all, monitoring of the water for pathogens is, in the end, an aid to the delivery of more effective management of the drinking water supply.

The water and sewerage industry in the UK was privatised in 1989 and a regulatory framework has been in place ever since to ensure fair standards and prices. Public water supply and service, along with the water companies in the UK are regulated by the Water Services Regulation Authority (Ofwat). This is Ofwat for England and Wales; the Water Industry Commission for

Waterborne Pathogens: Detection Methods and Applications. https://doi.org/10.1016/B978-0-444-64319-3.00011-3
367

Scotland and the Utility Regulator for Northern Ireland. Each holds the duty to regulate the financial side, ensuring functions are carried out properly and protecting customer's interests. These authorities make independent decisions but do work closely with Defra, the Consumer Council for Water, the Scottish Environmental Protection Agency (SEPA) and Natural England among others [1]. The Environment Agency (EA), SEPA, the Northern Ireland Environment Agency, and the Drinking Water Inspectorate (DWI) hold responsibilities for environmental regulation and drinking water quality, respectively.

As privatisation over 20 years ago, more than £90 billion has been spent on capital investment into water in the UK; with a further £22 billion invested over the regulatory period from 2010 to 2015 [1]. Water prices are set by regulation authorities in the UK. Costs vary around the country, and between water companies; based on supply, quality, and environmental factors. The cost of water services is required to cover treatment, collection and disposal of the water supply, as well as to fund conservation of the natural environment water supply uses [2]. For England and Wales, Ofwat sets prices for a five-year period.

Looking across the UK, Scotland, and North Ireland both have a single supplier for water and sewerage—Scottish Water and Northern Ireland Water, respectively—that are publically owned. Water in England is supplied by 23 private companies who operate across different areas of the country, these are: Northumbrian Water, United Utilities, Hartlepool (part of Anglian Water), Yorkshire Water, Severn Trent Water, Dee Valley Water, Severn Trent Water, South Staffs Water, Anglian Water, Essex and Suffolk (part of Northumbrian Water), Affinity Water, Cambridge Water, Thames Water, Bristol Water, Wessex Water, South West Water, Sembcorp Bournemouth Water, Cholderton and District Water, Southern Water, Portsmouth Water, South East Water, Sutton and East Surrey Water, and Dwr Cymru, also known as Welsh Water, the private supplier for the whole of Wales.

Water intended for general use in the UK has to be collected from a number of different initial sources including from rivers, surface reservoirs, or underground aquifers. This water then needs to be stored, cleaned, and made consumption-ready before it can be dispersed. Untreated water that is sourced for drinking water in the UK is subject to a number of processes to make it fit for human consumption. These processes involve, but are not limited to, removing any harmful microbes, pathogens, chemicals, taste, or odours. Pathogens of human-origin are considered as the greatest risk to the safety of drinking water. This includes water-borne diseases, such as cholera and typhoid, which are only found in humans but are particularly infectious when present as pathogens in the water supply. 'Pathogens' are considered to include bacteria, such as *Salmonella* and *Campylobacter*; viruses, such as hepatitis A and norovirus; and protozoa, such as *Cryptosporidium* and *Giardia*.

Pathogens are able to enter the water supply in a number of ways. One of the main causes is when surface water becomes contaminated due to the presence of untreated sewage entering rivers or run-off of waste water that contains animal waste. Surface water is also much more likely to become contaminated during storms of periods of heavy rainfall [3]. Contamination of groundwater can also happen, although it tends to be rare due to the subsurface offering self-purification the water and eliminating most pathogens. However, some pathogens do occasionally bypass the soil, for example with leaking latrines or sewers, and are able to enter and contaminate the groundwater system.

The quality of water in the UK has to adhere to the EC Drinking Water Directive (98/83/EC) [4]. The standards defined in this are implemented through specific UK regulation, such as the Water Supply Regulations 2010 and the Private Water Supplies Regulation, 2009 [5–7]. The DWI sets forth an enforcement policy as the independent regulator of drinking water. The policy indicates that the company's objectives and priorities for the UK for the 2010-15 period were:

- Water suppliers deliver safe and clean water.
- The public has confidence in drinking water.
- Legislation for drinking water is fit for the purpose and implemented with the public in interest [8].

The DWI's enforcement policy also indicates that where companies do not comply with the law and their responsibilities the DWI is able to use a variety of civil and criminal options to regulate. In the UK water undergoes a number of different stages of process for treatment. These stages include:

1. Screening and microstrainers: fine solids and fragments are removed as water passes through a series of coarse, fine steel, and plastic mesh.
2. Aeration: using either 'water-fall' or 'air diffusion' aerators to reduce carbon dioxide, dissolved ion, and other organic compounds [9].
3. Chemical coagulation: use of chemicals such as aluminium or iron salts to induce small particulates in the water to agglomerate into larger particulates called 'floc'.
4. Clarification: removal of 'floc' through use of either a horizontal flow of sedimentation tanks or by using millions of air bubbles to float the particles to the top, allowing them to be removed from the surface [4].
5. Filtration: passing water through a series of porous beds.
6. Activated carbon adsorption: use of a granular or powdered form of activated carbon in adsorption vessels to remove any trace taste, colour or organic compounds, and toxins [4].
7. Disinfection: one of the final stages of process—use of strong oxidising chemicals such as chlorine to remove or inactive pathogenic organisms such as bacteria, viruses, or pathogens [4]. Increasingly many water companies are looking to use ultraviolet (UV) light and ozonation to disinfect their water supplies.

A review of the UK drinking water industry in 2011 was released by the DWI in July 2012. The review demonstrates the quality of public water is improving and that compliance to EU standards is now achieved within 99.96% of public supplies. However, private supplies have not achieved as high levels—with 7.2% failing to meet the standards in 2011 [6]. Private water supply users face a high risk of microbial illness was examined in England in 2012 with the conclusion being that young children are most at risk of disease. Structural change within the water industry has occurred over the last few years, for a number of reasons, including an increasingly tough regulatory environment and cost comparisons by Ofwat, as well as wider globalisation and corporate restructuring seen in the UK over the last decade or so. Pressure on water resources and supply of water in the UK is growing all the time with population increase, increased numbers of housing and climate change all contributing. New strategies looking at how to increase sustainability and maintain supply of water are increasingly necessary [10, 11].

11.1.1 Water safety plans

WSPs are the 'most effective means of consistently ensuring the safety of drinking-water supply' ([12], p. 45). The aim of WSPs is to ensure the continued safety and acceptable level of the drinking water supply through using a comprehensive risk assessment and a management approach that encompasses all the steps of supply from catchment to consumer [13]. The development of a WSP will involve numerous stages, including identifying all potential hazards in the water supply from the catchment; assess risk by each hazard or hazardous area; consider control barriers for each risk and review and evaluate both the hazards and control systems over time to improve the plan if necessary [13]. WSPs should be approached carefully and it should be ensured that is not only hazard identification and assessment that is undertaken but plans should also look to envelop all of the operations involved in water supply, including management procedures such as training, communications, and reporting [4].

The UK regulator has been requiring water companies to implement WSPs since the third edition of the WHO Guidelines for Drinking Water Quality was published in 2004. Since then, a regulatory requirement has been in place for all companies to undertake these risk assessments in the UK. Companies were encouraged to undertake them with the knowledge that future investment programmes from the regulator would only support them if they undertook efficient WSPs. The first set of formal risk assessments for the water supply in England and Wales were submitted to the DWI by the 1 October 2008. WSPs in the UK currently follow good practice set by an already mature European water industry but each plan also needs to consider the effect of different locations, as well as the size of the supply; for example, smaller supplies are likely to be simpler to deal with.

In 2008 almost 800 risk assessment reports were submitted to the DWI for a wide variety of treatment works and supply systems across England and Wales and these were then independently checked by the DWI to check for any deficiencies. So far the application of WSPs methodologies in the UK under regulatory frameworks has been considered successful by improving understanding of, and communication, over the issues involved in water safety through effective self-regulation [14]. Scotland's water industry also has the same regulatory framework for producing WSPs but faces a number of different challenges from England and Wales including the variation in types of supply; terrain and relative immaturity of their provider—Scottish Water [15], being still only 10 years old.

11.2.   Application of WSPs and WSF in the rest of the world

The situation in other areas of the world will now be considered, with any major differences between UK and other regions—in terms of policy and implementation—highlighted where appropriate.

The UK is considered to have one of the highest standard drinking water supplies in the world and believed to be at the forefront of the worldwide water industry. However, the UK's water industry does face a number of challenges, including, but not limited to, climate change, growing populations, increased demands for water, and an uncertain economic future. However, the water sector is reasonably well-sheltered from current economic difficulties and capital investment is expected to continue to be around £4 billion per year within the water sector [1]. On a global level, it was determined in 2008 that 42% of the world's population, around 2.6 billion people, still lacked access to improved sanitation; with 1.8 million dying every year from diarrhoeal disease, mainly attributed to unsafe water. Furthermore, the total amount of those affected by ill-health due to water quality is very hard to estimate and is likely to be considerably higher [16]. The WHO sets international guidelines on drinking-water quality that are intended to be used as a basis for regulation of water in countries across the world but the extent to which they are followed and the state of water industries and their associated assets and infrastructure availability varies considerably across the world.

In high-income countries legislation and infrastructure for water quality has helped nearly eliminate pathogens in the water supply. However, despite this continued investment, many of these countries are still prone to outbreak of waterborne diseases, often linked to pathogens. For example, in the USA it has been suggested that waterborne pathogens cause between 12 and 19.5 million cases of illness per year [17]. The costs of reducing the threat of disease from waterborne pathogen are also considerable for developed countries. As an example, an outbreak of Cryptosporidium in the water supply in Sydney, Australia in 1998 cost US$45 million in emergency measures, despite the fact there was no actual link to increased disease indicated during this time [17]. Similarly, a further challenge that remains for high-income countries is the lack of a robust disease surveillance system for these waterborne pathogens [17].

Variations in legislation for water quality and supply across the world is often driven by present need, or otherwise, to rectify following previous outbreaks of waterborne illness. For example, *Cryptosporidium* testing is carried out routinely in both the US and the UK, where regulations ensure it is done. However, in some parts of Europe and elsewhere in the world there is no regulated screening system and outbreaks of *Cryptosporidium* and other waterborne pathogens is a considerable issue. The following section provides a brief outline on the water industry in other parts of the world, where information has been found to be available:

11.2.1 Europe

Water quality standards are derived separately by countries but tend to follow WHO guidelines.

Netherlands: The Netherlands has 11 drinking water suppliers and two bulk suppliers, all represented by the Association of Dutch Water Companies. All these companies, bar one (Waternet) are limited liability companies associated with certain municipalities and provincial bodies [18]. The region government agrees prices but there is no formal framework for setting them. A previously voluntary benchmarking system has become obligatory for the suppliers have been enforced under the Drinking Water Act in 2008 [18]. In general, the operation of the water sector is similar to England and Wales, although a larger proportion of supply is gained from Groundwater by Dutch companies, than here.

Portugal: the water sector has been regulated by the IRAR since 2000, they are responsible for local distribution and retail functions. The average water company size is small compared to England and Wales with the average population served standing at 76,000 [18].

11.2.2 North America

USA: The USA has a large number of water systems that vary considerably in size, ownership, and regulation. Federal policy drives the quality of drinking water but the regulation of systems depends on who owns them and what their location is [18]. The Environmental Protection Agency (EPA) is responsible for enforcing regulations on drinking water in the US and 1974 the Safe Drinking Water Act (most recently amended in 1996) required the EPA to become responsible for determining levels of contaminants (physical, chemical, biological, or radiological) in any drinking water and enforcing regulations on the maximum levels allowed of each contaminant. These regulations also require that if there is no feasible way to measure the maximum concentration of a contaminant, techniques to treat the contaminant will be used regardless [19]. The US now monitors its water quality based on parameters for microbials, volatile organic compounds, synthetic organic chemicals, and inorganics, such as nitrate, lead, and arsenic [20].

US tap water is considered to be relatively extremely clean since Congress passed the Safe Drinking Water Act that allowed the EPA to set limits on contaminants in drinking water. However, there are still dangers. In a 2009 report from the EPA, threats to the US's drinking water are shown as increasing, and although 91 contaminants are monitored under the 1974 Act, there are still a myriad of untested chemicals in existence in water supplies across the country that could have potential health impacts [21]. Water is thoroughly treated in the US similar to that in the UK system—through a process of coagulation, sedimentation, filtering and disinfecting. But this system isn't fool—proof and some microbes are still able to bypass the treatment facilities and remain in the water supply. For example an outbreak of cryptosporidium in 1993 caused illness in 400,000 people and killed at least 104.

Canada: water supply in Canada is through municipally owned infrastructure; subject to finance and quality standards that are set by the government. Each municipality has their own set of standards that must come within government guidelines and each is also responsible for setting water prices, which means there can be a huge range in the cost of bills.

11.2.3 China

China brought into effect new standards for drinking water in July 2012 to update those from the old standards, approved in 1985. The new standards include increased detection for contaminants and better purification methods such as ozonation as well as increasing the number of items tested from 35 to 106, to bring the country in line with international standards [22]. The safety of the water in actual practice is still questioned. Despite the higher standards having been actually made back in 2007 it has taken the government over 5 years to actually prepare for the introduction [22]. Thus, so far, many residents still depend on purifiers and bottled water as the safest option. Putting the new standards into practice is likely to be difficult [23].

11.2.4 Africa

The drinking water situation across Africa is improving; 322 million people across the continent have gained access to an improved drinking water source since 1990 and the number of those using a piped drinking water source onto their premises rose from 147 million up to 271 million between 1990 and 2010. However, the continent is still failing to deliver treated or improved drinking water to all—with 115 million still drawing directly on surface water for their drinking needs in 2010 [24]. The situation in Africa is somewhat different to other parts of the world, regarding the identification and effective removal of pathogens from drinking water. Given the region's limited investment options, money is likely to be spent on improving infrastructure, before improvements to pathogen monitoring and control are considered. It is also important to note that the pathogens regarded as giving cause for concern in Europe, North America and other regions, are well down the priority list in Africa, where much more serious public health problems—often that are a threat to life, not simply well-being—exist.

11.2.5 Australia

The framework for water supply in Australia is reasonably similar to that of England and Wales—with economic regulators setting the price and public bodies regulating the water standards [18]. Owing to the enormous size of Australia, water is regulated within, and by, each state of the country and each has their own regulations and responsibilities for their water sector [18]. Water companies are state-owned, vary by structure and have a considerable size range; some serving only 500,000 customers and some serving up to 1.5 million [18].

Water in Australia is treated in a similar process to that used in the UK prior to deliver to communities. The process, again, includes coagulation, sedimentation, filtration, and disinfection, with chlorine being most widely used for this. Australia has a much wider area to deliver water supplies to than the UK and responsibility lies with each state or territory. Australian Drinking Water Guidelines must be followed and there are number of different agencies involved in water supply systems, including agriculture departments, environment departments, and local authorities; with the drinking water suppliers holding the ultimate responsibility to deliver safe drinking water. There was previously no specific legislation for drinking water—but as of 2003 the various states have been passing legislation to follow the Australian Drinking Water Guidelines efficiently.

Ofwat's international comparisons for drinking water quality demonstrate that the quality of water in England and Wales in relatively very high. England and Wales achieved slightly higher quality compliance in water supplies in 2006-07, than Scotland and Northern Ireland (the only two locations that can be meaningfully compared) although all were considered high quality [18].

11.3. The legislative framework

Strict regulations and legislation regarding pathogen detection and removal has been in place for many years in the UK. This section examines the framework within water companies must operate, how this is enforced and why such regulation is necessary. The UK water industry has gone to great lengths to ensure that pathogens are properly managed. This is partly due to public health concerns but is also a factor of public perception and the quality of drinking water demanded by customers.

Following the privatisation of the water industry in 1989 the Water Industry Act 1991 was passed. The act was amended by the Water Act in 2003. Under this act any undertaker of a water supply must follow all the steps to minimise pollution. Similarly, Ofwat must ensure that any water companies undertake the responsibilities given to them as updated by section 39 of the Water Act in 2003. The legal framework for pathogen detection and removal includes the overarching role of the DWI who have a duty to maintain safety and quality of water through monitoring, as well as providing the government with advice on water supply matters. The Health Protection Agency also contributes in their role to ensure the protection of communities against any infectious diseases or other dangers to public health. The Agency also commission research, analyse data and provide laboratory services to ensure the safety of public drinking water supplies. Finally, under the 1991 Act, local authorities also have a role to play in ensuring the safety of both public and private water supplies within their area [14]. Local authorities must be responsible for staying informed about the quality and sufficiency of supplies in their area.

Regulations in the UK give specific provisions for drinking water, requiring water companies to produce a Water Safety Plan and to follow the recommendations of the WHO Guidelines for drinking water quality. In terms of actual treatment of water–water companies are required to have adequate water treatment systems in place, which they must use to treat and subsequently disinfect all raw water before supplying. The minimum requirement is that the turbidity of water must be < 1 NTU2 prior to disinfection. Methods for disinfection itself are not set out by law but companies are required to advise the DWI of their disinfection methods and procedures [14]. Under UK regulations water companies also have a duty to collect samples at different stages to ensure treatment has been carried out correctly and has been effective. Sample points include at the treatment works and at the consumers' taps. Water companies must supply full details of their monitoring to the DWI, which then has a role to carry out independent checks on these samples to ensure quality is high enough and ensure pathogens are not present [14].

Water under-takers or companies who fail to treat or disinfect their water or that fail to up-keep their treatment facilities sufficiently in the UK will be considered as having committed a criminal offence and enforcement action, such as fining those responsible, is likely to be taken. Section 70 of the Water Industry Act 1991 (as amended) also made it a criminal offence for any water company to supply water that may be deemed as unfit for human consumption under inspection. One recent UK example saw Northumbrian Water Ltd. paying out over £22,000 in fines and costs in September 2012, after being found guilty of supplying customers with discoloured water that had not been sufficiently treated [7].

Some recent developments are likely to add to, or potentially alter, existing regulation and legislation of the UK water industry. The draft Water Bill was published on the 10 July 2012 and outlines that, despite the £98 billion investments in water since privatisation, problems with pollution still exist. The bill sets out the intention for further structural reform of the water industry in the future. At the time of writing, the bill is currently undergoing prelegislative scrutiny but is expected to achieve Royal Assent during the 2013-14 period. When this is achieved the Act will be likely to have an impact on the water industry across the UK. Furthermore, a statement of Obligations released by Defra in October 2012 focused specifically on organisms within the water supply. The statement indicates the Water Supply Regulations require that water must not contain any potential dangerous substance to human health and that undertakers must assess any risk of raw water to ensure their disinfection process is sufficiently robust to remove all pathogenic organisms. Furthermore the statement indicates that the health standards of the 1998 EC Drinking Water Directive have already been effective in reducing exposure to pathogens in water across Europe but will become even stricter in 2013.

11.4. Pathogens of major concern to the market

Pathogens of most concern to the UK water market are *Cryptosporidium* and *Legionella*. There are a number of reasons that these have become the focus of attention for both the regulator and the water companies. In the case of *Cryptosporidium*, it is the fact that the pathogen is difficult to eradicate. In part, the supply of *Cryptosporidium*—free water provides a warning signal to regulators that water supplies have been properly managed, monitored and, where appropriate, treated. Although outbreaks are rare, and seldom life-threatening, they do damage public confidence.

Cryptosporidium is a protozoan parasite that can be transmitted through recreational waters and contaminated drinking water. In humans it can cause a number of side effects including severe diarrhoea and related symptoms [25]. The spread of this pathogen is rare in the UK but it can affect large numbers if it does and therefore needs to be controlled. The unreported rate of cryptosporidium in the UK is believed to be around 60,000 cases per year as of 2010 [17]. Although in 2006 it was reported that only around 8% of cases of disease caused by *Cryptosporidium* were a result of drinking water, it has been identified that there is a correlation between rainfall and cases of cryptosporidiosis—indicating some link between the disease and insufficient treatment of water supplies [26].

Cryptosporidium is resistant to treatment by chlorine and sampling for its presence in drinking water is particularly complicated and lengthy—for these reasons mean there is currently no specific standard set for it under the EU Drinking Water Directive. The directive does state that no organism should be present that would present a threat to public health but does not give any direct mention of, or guidance for, *Cryptosporidium*. As a result of this, companies within the UK water industry tend to adopt a risk-based approach to managing *Cryptosporidium*, assessing whether there are potential sources of the parasite in the catchment area and putting barriers in place to make sure it does not reach the supply in the first place [25]. In the UK, ultraviolet disinfection is a used to remove *Cryptosporidium* oocysts that are resistant to normal chlorine disinfections [25] from our drinking water supplies.

Much of the present policy for monitoring drinking water in the UK comes as a result of disease outbreak caused by waterborne diseases. For example, following a *Cryptosporidium* outbreak in 1999 from a public groundwater, new regulations were introduced. These regulations, The Private Water Supply regulations, now infer that authorities must measure and regularly monitor all supplies to identify potential risk [27]. Cases of cryptosporidiosis (caused by the parasite) are currently still a considerable issue in the UK. In the summer of 2012 the HPA relayed its fears in the increasing number of cases of the illness, as 327 cases were determined between the 11 May and 11 June 2012; compared with only 82 cases for the same period of 2011 [28, 29].

Legionella is a different matter, and has not been previously considered in this book since the transmission route is not primarily through drinking water. However outbreaks over the past 20 years have often been accompanied by high-profile media coverage and, unfortunately, fatalities. The fact that the damage is usually done before a legionella outbreak has been identified, adds to the severity of the problem.

The bacteria are able to multiply and survive at temperatures of between 20°C and 45°C and cause a fatal form of pneumonia called Legionnaire's disease, if the bacteria are inhaled in water droplets [30]. It can also cause two other illnesses—Pontiac fever and Lochgoil head fever. The bacteria are found in natural water systems such as rivers and ponds but can easily get into water systems such as those in large buildings, conditioning systems or communal showers [31]. The latest findings show confirmed cases of Legionnaire's disease in England and Wales were 551 in 2006, 442 in 2007 and 359 in 2008 [32], demonstrating a fall in numbers of the disease over this period. However, the disease is still a considerable issue and risk to health within UK water supplies and, for this reason, water systems on commercial premises need to have a risk assessment for *Legionella* which must be updated at least every two years [33]. Tanks storing drinking water are also required to be cleaned at least once a year within the DWI guidelines, with an inspection also to be carried out on a six month basis [33].

There is a variety of legislation to control exposure to *Legionella* in the UK, including the coverage in the Health and Safety at Work Act 1974, which requires monitoring and inspections, as well as approved codes of practice for water systems in the work place. However, there are no standards for detecting, or removal of, *Legionella* in drinking in the UK, due to the fact risk from drinking water is deemed low. Potable water supplies are not considered the primary cause of the bacteria's spread so water itself is not treated for legionella pathogens; but, instead, focus is kept on maintaining facilities and equipment for water supply and storage and preventing the growth of these bacteria in the first place.

So the drive to improve monitoring and treatment for both these pathogens really comes from three places. First, the regulator; who needs to be confident that public health is protected and that water companies are managing their processes effectively. Second, the water companies; who want to ensure that the confidence of the general public is maintained. (In the case of *Legionella*, it is often businesses that have no part to play in water supply that are also liable to monitor and control this pathogen.) Third, the general public needs to be assured that what comes out of the tap won't harm them.

11.5. Public health policy implications for detection and treatment

The treatment of water is closely regulated in the UK. Both the treatment of wastewater as well as the supply of potable water for human consumption are subject to regulation and control. The public is at varying risk from pathogens in water across the UK and the role of location and supply types is important in determining risk. It is generally considered that the greatest microbial risks from water are through the ingestion of water that has been contaminated with human or animal faeces.

Predominantly, contamination is caused when wastewater comes in contact with fresh or coastal water contaminating it with faecal microorganisms that include various pathogens [34] Furthermore, there is a potential risk of microorganisms and pathogens being present in water used for recreation on beaches that could cause a considerable risk to public health. This may happen that run-off water from nearby agricultural functions or urban areas can bring pathogens into the beach area. Water companies are responsible for minimising the chance of untreated sewage entering the sea but sometimes it is hard to avoid, such as in instances of leaking septic tanks or overflowing sewers in the beach vicinity, and thus there is always an inherent risk to the public using these areas [29].

Use of water in agriculture also provides a potential entry point for bacteria and protozoa. Water is predominantly used in agriculture for crop irrigation, produce washing and livestock drinking. The sources of the water used are predominantly rivers and streams but also boreholes, ponds, lakes, and only a small amount (about 2%) from mains supplies. A portion of this water is likely to contain pathogens that could go on to infect livestock or contaminate crops. It is therefore necessary for the quality of water used for irrigation to be controlled and assessed prior to use to prevent pathogens from water passing to humans in this manner [35].

In the UK a high percentage of outbreaks of disease caused by pathogens in water are associated with private supplies. In particular, *Cryptosporidium*, *Giardia*, and *Campylobacter* have been associated with private supplies. Reasons for the contamination may include water being derived from surface water or springs that are within the agricultural catchment and therefore prone to inundation of contaminated agricultural run-off [36]. In addition, water used as a source for small supplies may be subject to a higher probability of containing pathogenic microorganisms making it vitally important that a multi-barrier approach is used for treatment to ensure all different pathogenic agents are removed [36]. The number of outbreaks resulting from public supplies is considerably lower than from private supplies—believed to be due to the fact that they tend to employ a multi-barrier treatment system more often that private supplies.

Public health plays a key role in, and considerably determines, targets for water quality and the definition of what constitutes 'safe' in drinking water. Targets for public health are essentially translated into the standards for water suppliers set through legislation [17].

11.6.   Detection and treatment from market perspective

A range of water monitoring, pathogen detection and treatment protocols and systems have been developed over the past 20 years, both to satisfy the requirements of water supply and water treatment companies and to ensure that the regulator is satisfied that potable water is of a sufficiently high standard to be supplied to customers. Various approaches are reviewed here.

11.6.1   Detection

Water needs to be monitored before human consumption to ensure there is no presence of pathogens including bacteria, viruses, and protozoa that could potentially cause disease and illness if digested. The detection of pathogens is important and can be done with a variety of methods. There are a variety of current systems or methods available for detection in the water supply sector within the UK [37]. The sector mainly relies upon the use of faecal indicator monitoring as a safety check of the finished drinking water. As noted in earlier chapters this method has been well-used for many years and has contributed to advancing drinking water quality. However, there is sometimes a lack of correlation between pathogen presence and faecal indicator markers.

Cryptosporidium is the only pathogen for which direct detection is performed. The current system used for the detection of *Cryptosporidium* in the UK involves the use of a filtration system. Currently, those recommended under UK guidance for the drinking water section in the UK are either the IDEXX or Pall filtration systems. However, the majority of the market share is claimed by IDEXX. While the method is relatively expensive and time-consuming the long-term market dominance of IDEXX, along with the conservative nature of the water industry makes it challenging for new companies to enter the market. Other difficulties include the relatively small market size and the potentially high cost of entry for a new supplier.

Established methods for detecting pathogens in water include:

1. *Polymerase chain reaction (PCR)*: A method developed in the 1980s. It involves using nucleic acid amplification technology and has widespread use in bacterial detection. It entails isolating, amplifying, and quantifying a short DNA sequence in the targeted bacteria's genetic material [37]. The unique

DNA of the pathogen is detected so scientists can construct oligos to detect the pathogens. These methods are advantageous in that they are very sensitive; however they are not always capable of reliably detecting or quantifying viable organisms because they may also detect the nucleic acid of noninfectious microbes [38].

2. *Culture and colony counting methods*: The oldest bacterial detection technique and still the widespread, standard method. It is considered to be the most accurate and reliable method but is time-consuming due to the amount of time required to wait for results (often 2-3 days for initial results, then a further 7 days for confirmations). The method involves culturing and plating then using different selection methods are applied to detect particular bacteria, such as particular substrates that confer a particular colour in growing colonies and therefore can be detected with optical methods [37].

3. *Immunology-based methods*: These methods provide powerful analytical tools. For example 'immunomagnetic separation' can be used to capture and extract a pathogen by placing an antibody coated in magnetic beads into the bacterial suspension holding it [37]. Drawbacks of this method include that it is time-consuming and the methods used may not be as sensitive in detecting pathogens as PCR methods [39].

Existing technology for monitoring and detection pathogens in the UK water industry does have shortcomings in that the majority of the work is time-consuming and labour-intensive as well as not always been entirely 100% reliable and costly. For example, identifying *Cryptosporidium* in water supplies takes UK water companies anywhere between 5 and 6 h for a high risk, priority sample right up to 3 days for a low risk one. Furthermore the cost of one *Cryptosporidium* sample is between around £70 to £110 for UK companies; and with some companies taking out almost up to 500 samples per week this is a very substantial cost.

Although traditional approaches can be deemed sufficiently sensitive, their slow pace and the amount of effort taken to undertake them means there is need to look for potential alternate options. One new approach being considered and studied is the use of 'biosensors'. This involves the use of biological or biologically derived material to detect pathogens [37]. Biosensor technology has the ability to combine a variety of molecules into a single chip and is able to monitor water in real-time [40].

'DNA microarrays', are another emerging technology that uses specialised biosensors that utilise special oligonucleotides to screen nucleotide mixtures. However, this method is likely to face challenges in that the relative abundance of pathogens is likely to vary making isolation more difficult. Furthermore, many authors argue there needs to be a move towards 'molecular' methods for detecting pathogens. There is currently an emerging market for more sophisticated gene technologies for pathogen detection—however, these potential changes would require an increase of skilled staff as well as a need for the system to be both flexible and reproducible to allow it to be used in different countries and locations [41]. Despite this the advantages are numerous; molecular methods have enhanced sensitivity and an increased ability to distinguish genotypes [40]. These techniques provide effective analytical tools to detect pathogens—including new emergent strains—and can be used to evaluate the microbiological quality of water.

11.7. Market adoption of emerging technologies

The market for water testing is growing rapidly at present and is predicted to reach US$1.6bn by 2022 with a CAGR of 7% from 2017 to 2022, according to a report by Markets and Markets. Companies with a significant presence in the microbiological water quality testing market include Bio-Rad Laboratories, Inc. (U.S.), Döhler GmbH (Germany), Agilent Technologies, Inc.(U.S.), Thermo Fisher Scientific, Inc.(U.S.), Perkinelmer, Inc.(U.S.), Shimadzu Corporation (Japan), and Sigma-Aldrich Corporation (U.S.). Other major players in the market include Accepta Ltd. (U.K.), Lamotte Company (U.S.), Danaher Corporation (U.S.), Millipore Sigma (U.S.), 3M Company (U.S.), and Idexx Laboratories Inc. (U.S.).

The market segments comprise instruments, test kits and reagent manufacturers, and while instruments accounted for the largest market share in 2016, test kits are predicted to grow more quickly from now on, driven by offering the advantage of rapid results. In 2016, North America was the largest market but in the future strongest growth is expected in Asia-Pacific regions due to increasing growth in these economies. Along with India, China and Japan, Brazil, and Argentina are also seen as primary targets of the water testing industry.

The market is driven by factors such as the development of new detection technologies as well as the requirement to upgrade equipment along with increased urban waster increasing microbial contamination in water and a stringent regulatory environment. Challenges for new companies and detection approaches to enter the market include high capital investment (for many systems), and a reluctance to adopt new techniques.

One of the main problems facing developers of new technologies for the monitoring and treatment of water is the high level of inertia that has to be overcome in order to enter a market successfully. There are several reasons for this, and these will vary from country to country. However, in regions where water supplies are generally treated to a high standard and safe to drink, there are two main barriers to the adoption of new detection technologies.

The first is scalability. What works well as a laboratory test is often difficult to scale up to the size required for commercial use. To be adopted widely, a pathogen detection technology needs to enable samples to be tested simply, quickly, with a high level of repeatability and a low level of error. Before a customer is prepared to adopt any new monitoring or measurement procedures, they need to be convinced that results are reliable and testing is repeatable.

While many laboratory-scale detection systems could potentially be developed to meet these criteria, the second barrier to entry often poses the greatest problem, which is the willingness—or not—of water companies to adopt a new screening technology. Existing methods are generally not new or revolutionary, they are not necessarily the best solution on the market, but they are accepted by regulators and therefore have been adopted, usually on an industry-wide basis. The water industry is a conservative one and pathogen detection is not an area where any individual organisation is likely to go it alone and adopt a new method, unless there are clear and obvious advantages over existing techniques. It is also important for manufacturers to appreciate that adoption of a process or technology in one country does not automatically guarantee that other countries will adopt it.

The market is not driven solely by cost. While monitoring can be an expensive business for a water utility, the costs of dealing with an outbreak as well as the reputational damage caused mean that the water companies primarily want reliable and fast methods. Monitoring is also undertaken as it is regulatory requirement, and current methods are well-established in this regard. While the regulator in the UK is positive towards developments in monitoring technology, the water companies themselves are often conservative. The view that existing levels of monitoring are sufficient presents a challenge for new technologies offering improved limits of detection. If these extend beyond requirements and will merely highlight potential problems, water companies might not be interested until the regulator enforces adoption of improved methods.

In terms of opportunities for the development of new testing systems and treatment procedures, one of the major drivers is likely to be unit cost of testing. Another driver is an increasing interest indirect detection of pathogens, that is moving away from the general faecal indicator monitoring to look specifically at the presence of particular pathogens, and perhaps even particular species. Another factor is the availability of samples in volume. The need to concentrate samples to ensure a sufficiently high number of pathogens are present for effective testing can be time-consuming, expensive in terms of both manpower and materials and is not always practical.

Once a technique has been tested and proven, the problems for the technology innovator are still not over. Convincing the appropriate national regulator/water industry that the new test or treatment procedure is at least as good as that used currently can often be a lengthy process. Validation studies are likely to be required and will need to demonstrate good performance of the new method over a time period of at least a few months. In addition, validation with different tap waters is likely to be an important part of this process, demonstrating the method is reliable despite a range of input conditions. Conducting such studies is likely to prove challenging for many companies looking to enter the water market. The development of water innovation parks could be one key way to enable small businesses to trial new technologies, and assist in overcoming this barrier.

11.7.1   Future trends

11.7.1.1   Potential developments of pathogen detection and removal systems

Chlorination has been most commonly used to disinfect drinking water in the past due to its low cost, the fact it is harmless to humans in the concentrations used and the fact it also helps to protect the networks of pipes it travels through [42]. However, for smaller or more rural locations in the UK the use of UV disinfection is becoming more widespread. UV disinfection is becoming increasingly more important within water treatment and has the opportunity to be expanded at a global scale. A white paper published by Siemens in 2008 demonstrates the market for UV is a substantial growth area in the water treatment market—worth around $500 million globally. The market for UV is also predicted to grow by 10% for industrial, commercial, and residential waters, and by up to 15% or 20% for municipal and wastewater markets [43]. Although the US currently leads the market for use of UV in water disinfection, regulations, and recommendations are now increasingly being made for its use within the UK water industry—particularly now, with the release of guidance for its use in public water supplies released by the DWI in 2010.

UV disinfection has the advantage in that it is effective for treatment of waterborne pathogens, such as *Cryptosporidium*, that are resistant to chlorine and, when applied in conventional does, has no known by-products formed. This gives UV the advantage over chemical disinfectants, such as chlorine and ozone [44]. However, disadvantage of using UV for disinfection is that organic matter present in water can sometimes affect the transmission of the UV, and therefore reduce its efficiency [42]. This is a matter that still needs to be addressed if UV use is to become more widespread.

Kiesel et al. [45] highlight a new development by PARC, a Xerox company that pioneers technology platforms, for assessing water quality. They are developing a new, robust microfluidic platform to provide rapid identification and quantification of pathogens in water. The technique is expected to enable 'on-the-flow' detecting of pathogens with a high level of signal-to-noise discrimination [45]. The company has assembled and tested a working prototype for their 'microfluidic-based flow cytometer', which was assembled at a total cost of less than $350 [45]. Alternatively, several authors have suggested developments to integrate systems and detect for multiple pathogens in drinking water over the past few years. It is believed that if this technology could be coupled with PCR amplification techniques the sensitivity of signal detectors could be increase by up to 106-fold.

The Aqua Research Collaboration (ARC) is also in the process of developing new systems for improving water quality. Their vision is to develop 'membrane technology' focusing on hybrid technologies to improve water quality through rejecting of micropollutants and enhancing the biological stability of drinking water [46]. Membrane technology plays a role in removing pathogens, as well as particles and organic matter to help improve water quality. There is currently a huge market for methods using high-pressure membranes—with Global Water Intelligence indicating that there is a $450 million market in desalination worldwide, and that is expected to double by 2016. ARC is keen to develop these technologies further as the use of membranes, such as for ultrafiltration, is one of the fastest-growing sectors within water treatment [46]. Likewise, they are looking to improve water quality through enhancing membrane technology by improving cost-effectiveness of methods, reducing operational problems associated with waste streams and cutting the energy requirements of the technologies. They hope to achieve this through a new hybrid combination of different treatment processes alongside considerable research into improving the performance of the membranes themselves [46].

Detection systems are also being developed, as is highlighted in Section 3 of this book, though with these systems being at various stages of development and frequently with further work required to translate the research from lab to industry. Water companies themselves are also driving innovation, for example Scottish Water, Severn Trent, Northumbrian Water, and United utilities exploring new ways to utilise flow cytometry. UNICEF are also operating a large programme, the Rapid Water Testing project, at present (2017-24) focussing on the challenges of rapid detection of microbial contamination in drinking water supplies with a focus on applicability to use in UNICEF programmes. A target product profile has been developed incorporating a range of diagnostic technology requirements, including power and safety considerations, performance characteristics (speed, sensitivity, specificity), and operating parameters (life span, ease of use, dimensions, environmental impact). The aim is to accelerate technology development, adoption, and create sustainable markets.

11.7.2 Barriers to adoption

One of the main barriers to, and considerations for, adopting new technologies for water treatment, is the need to follow government guidance and regulations for the water industry to become more sustainable. There has been a plethora of new regulations and legislation for the water sector over the last few years; from the Flood and Water Management Act (2010) to the EA's strategy 'Water for people and the environment' (2009). These new initiatives, along with the Water Framework Directive at a European level, all highlight the need for the water sector to innovate with a consideration of the need to be sustainable, take a long term perspective in managing risks and also work towards mitigating climate change impacts [47]. Under the light of these regulations and recommendations, the water industry now needs to innovate and ensure that new technologies for water treatment adhere to new guidance on sustainability. Increased innovation in solutions for water treatment technology is needed to help reduce carbon output, make systems sustainable and also reduce costs where necessary [47].

Low-carbon solutions for water treatment need to be developed to balance the demands of the water sector, as well as the need to reduce carbon under international regulations and legislation for climate change. The water industry emitted over five million tonnes of greenhouses during 2007/2008 throughout all its functions for dealing with the treatment and supply of clean water, as well as wastewater and sewage. The Climate Change Act 2008 set out an 80% reduction target for carbon emissions in the UK by 2050—something which the EA and the government strongly believe the water industry needs to help contribute to [48]. In light of this recent legislation, water companies in England and Wales now have to provide details of potential carbon use and greenhouse gas emissions when looking for investment in future activities [48] and it has, thus, become very important for water industries to consider this when developing new technologies for water treatment. However, these changes mean UV treatment may prove to be a useful technology, as it typically only uses < 20 kWh/Ml of energy, compared to between 250 and 500 kWh/Ml typically used in high lift pumping methods [44]. The fact it has considerably lower energy usage than many other treatment methods could mean it is an important method to harness under in the future under stricter technology regulations.

CIWEM highlights that both the regulators and the industry within the water sector need to work together for sustainability and begin to move away from the industry's reliance on traditional, and often carbon-intensive, technological fixes [44]. It is evident that public health and health-based standards need to take a fundamental role in the regulation of potable water quality in the future. Collaboration needs to be made between those developing treatment methods within the water industry and the policy-makers for health [17]. Developing a reliable and high-quality knowledge base of environmental waterborne pathogens should be a first crucial step; followed by strategic investment and funding to help apply the knowledge to formulate the best new methods for water treatment [17].

The UK currently holds a vision for the future and for 'taking responsibility of water'. The vision aims to ensure that, by 2030, the UK is a key contributor to global water security and is using sustainable, integrated solutions that can be applied to the global water market [49]. A 'UK Water research and innovation partnership' was established in mid-2011 with an aim to capture, share, and enhance the capacity and the skills across all different groups concerned with the water industry—from the private sector to academia and the NGO community. They want to then harness these skills within the UK to help deal with the international issues of water quality and water security [49].

11.8. Conclusions

The water sector is viewed by many technology and equipment suppliers as a stable and profitable market. Emerging methods for pathogen detection and treatment are being developed and generally welcomed by the industry.

The large number of pathogens that can contaminate drinking water supplies provides challenges for water companies, particularly in terms of reliable treatment processes. However, the market is reasonably mature and a number of reliable, repeatable, and reasonably cost-effective procedures are currently available, and widely used, to ensure the safe treatment of potable water.

Opportunities do exist for new detection and treatment methods. However, there can be significant barriers to entry. For developers of new approaches, it is essential to conduct experiments simulating real conditions, that is to show that new methods are able to deal with large volumes of tap water samples. Access to water innovation parks or links with industry are likely to prove vital for the developer of new technologies. At regional levels, the regulator and governmental organisations can stimulate innovation in the area of pathogen detection by providing funding for water innovation parks and validation studies. The ongoing work by UNICEF to validate new technologies for their use in the field could be another route for emerging technologies to enter the market. Assistance in route to market is likely to play an important role in the next few years as Section 3 of this book showed a lot of detection technologies and particularly combined approaches are starting to offer impressive levels of performance.

References

[1] P. Hipwell, The UK Water Industry—A Brief Overview, 2010, Available from: http://www.thewaterplace.co.uk/briefoverview.htm. Last accessed 26 March 2013.

[2] Water UK, Water Prices, 2013, Available from: http://www.water.org.uk/home/resources-and-links/uk-water-industry/waterprices. Last accessed 21 March 2013.

[3] Ministry of Health, Pathogens and Pathways, and Small Drinking-Water Supplies, 2007, Available from: http://www.google.co.uk/url?sa=t&rct=j&q=&esrc=s&source=web&cd=1&ved=0CC8QFjAA&url=http%3A%2F%2Fwww.waternz.org.nz%2FFolder%3FAction%3DView%2520File%26Folder_id%3D96%26File%3D101207_moh_pathogens_and_. Last accessed 26 March 13.

[4] FWR, What is Water Treatment? 2010, Available from: http://www.euwfd.com/html/water_treatment_and_supply.html. Last accessed 26 March 2013.

[5] DWI, Drinking Water 2011, 2012, Available from: http://dwi.defra.gov.uk/about/annualreport/2011/index.htm. Last accessed 22 March 2013.

[6] DWI, Drinking Water 2011, 2012, Available from: http://dwi.defra.gov.uk/about/annualreport/2011/london&se.pdf. Last accessed 26 March 2013.

[7] DWI, Northumbrian Water LTD Plead Guilty to Supplying Water Unfit for Human Consumption, 2012, Available from: http://dwi.defra.gov.uk/press-media/press-releases/20120911-nne-horsley.pdf. Last accessed 22 March 2013.

[8] DWI, Securing Safe, Clean Drinking Water for All, 2010, Available from: http://dwi.defra.gov.uk/about/our-strategic-plan/dwi-enforcement.pdf. Last accessed 22 March 2013.

[9] GE Power and Water, Water and Process Technologies, 2012, Available from: http://www.gewater.com/handbook/ext_treatment/ch_4_aeration.jsp. Last accessed 26 March 2013.

[10] EA, Statement of Obligations, 2012, Available from: http://www.defra.gov.uk/publications/files/pb13829-statement-obligations.pdf. Last accessed 22 March 2013.

[11] EA, Water Resources Strategy—Water for People and the Environment, 2012, Available from: http://www.environment-agency.gov.uk/research/library/publications/40731.aspx. Last accessed 26 March 2013.

[12] WHO, Guidelines for Drinking-Water Quality, fourth ed., WHO Press, Switzerland, 2011, p. 45.

[13] J. Bartram, L. Corrales, A. Davison, D. Deere, D. Drury, B. Gordon, et al., Water Safety Plan Manual: Step-By-Step Risk Management for Drinking-Water Suppliers, World Health Organization (WHO); International Water Association (IWA), Geneva/London, 2009. accessed 22 March 2013.

[14] DWI, Drinking Water Safety, 2009, Available from: http://dwi.defra.gov.uk/stakeholders/information-letters/2009/09_2009Annex.pdf. Last accessed 26 Match 2013.

[15] J. Fawell, J. Littlejohn, J. Watkins, 2 Development of Drinking Water Safety Plans in Scotland ENV3/04/03, Drinking Water Quality Division, Edinburgh, 2005.

[16] UN Water, Water Quality, 2011, Available from: http://www.unwater.org/downloads/water-quality_policybrief.pdf.

[17] J. Bridge, D. Oliver, D. Chadwick, H. Charles, J. Godfray, A. Heathwaite, et al., Engaging with the water sector for public health benefits: waterborne pathogens and diseases in developed countries, Bull. World Health Organ. 88 (11) (2010) 873–875.

[18] Ofwat, International Comparison of Water and Sewerage Service 2008 Report., 2008, Available from: http://www.ofwat.gov.uk/publications/internationalcomparators/rpt_int_08.pdf. Last accessed 22 March 2013.

[19] EPA, 2012 Edition of the Drinking Water Standards and Health Advisories, 2012, Available from: http://water.epa.gov/action/advisories/drinking/upload/dwstandards2012.pdf. Last accessed 22 March 2013.

[20] Centre for Sustainable Systems, U.S. Water Supply and Distribution, 2012, Available from: http://css.snre.umich.edu/css_doc/CSS05-17.pdf. Last accessed 22 March 2013.

[21] R. McLendon, How Polluted is U.S. Drinking Water? 2011, Available from: http://www.mnn.com/earth-matters/translating-uncle-sam/stories/how-polluted-is-us-drinking-water. Last accessed 26 March 2013.

[22] China Daily, 2012. Available from: http://usa.chinadaily.com.cn/. Last accessed 26 March 2013.

[23] CCTV, China's New Drinking Water Standards in Effect, 2012, Available from: http://english.cntv.cn/program/newshour/20120701/105621.shtml. Last accessed 22 March 2013.

[24] AMCOW, Status Report on the Application of Integrated Approaches to Water Resources Management in Africa, 2012, Available from: http://www.amcow-online.org/images/docs/2012%20africa%20status%20report%20on%20iwrm.pdf. Last accessed 22 March 2013.

[25] Water UK, Cryptosporidium, 2011, Available from: http://www.water.org.uk/home/policy/positions/cryptosporidium. Last accessed 22 March 2013.

[26] DWI, Cryptosporidiosis: a report on the surveillance and epidemiology of *Cryptosporidium* infection in England and Wales, 2006, Available from: http://dwi.defra.gov.uk/research/completedresearch/reports/DWI70_2_201.pdf. Last accessed 22 March 2013.

[27] UK Groundwater Forum, Industrial and Urban Pollution of Groundwater, NERC, London, 2013, pp. 1–10.

[28] HPA, Update 8 June: Increase in Cases of Cryptosporidiosis, 2012, Available from: http://www.hpa.org.uk/NewsCentre/NationalPressReleases/2012PressReleases/120608Cryptoupdate/. Last accessed 22 March 2013.

[29] HPA, Bathing Water and Beach Risks, 2012, Available from: http://www.hpa.org.uk/Topics/InfectiousDiseases/InfectionsAZ/BathingAndBeaches. Last accessed 26 March 2013.

[30] HSE, Managing Legionella in Hot and Cold Water Systems, 2012, Available from: http://www.hse.gov.uk/healthservices/legionella.htm. Last accessed 26 March 2013.

[31] R. Hicks, Legionnaires' Disease, 2009, Available from: http://www.bbc.co.uk/health/physical_health/conditions/legionnaires1.shtml. Last accessed 22 March 2013.

[32] EU-OSHA, Legionella and Legionnaires' Disease: A Policy Overview, 2011, Available from: https://osha.europa.eu/en/publications/literature_reviews/legionella-policy-overview.pdf. Last accessed 22 March 2013.

[33] AquaLegion, Legionella—Facts and FAQ's, 2012, Available from: http://www.aqualegion.com/legionella-information-bank/frequently-asked-questions-on-legionella-and-waterhygiene/. Last accessed 22 March 2013.

[34] J. Cabral, Water Microbiology. Bacterial Pathogens and Water, 2010, Available from: http://www.ncbi.nlm.nih.gov/pmc/articles/PMC2996186/. Last accessed 22 March 2013.

[35] S. Groves, N. Davies, M. Aitken, A Review of the Use of Water in UK Agriculture and the Potential Risks to Food Safety, 2002, Available from: http://www.foodbase.org.uk/results.php?f_report_id=194. Last accessed 26 March 2013.

[36] The Scottish Executive, 2006. Available from: http://home.scotland.gov.uk/home. Last accessed 26 March 2013.

[37] O. Lazcka, J.F. Del Campo, X.F. Munoz, Pathogen detection: a perspective of traditional methods and biosensors, Biosens. Bioelectron. 22 (2007) 1205–1217.

[38] M.D. Sobsey, D.A. Battigelli, G.A. Shin, S. Newland, RT-PCR amplification detects inactivated viruses in water and wastewater, Water Sci. Technol. 38 (12) (1998) 91–94.

[39] T. Langerholc, P.A. Maragkoudakis, J. Wollgast, L. Gradisnik, A. Cencic, Novel and established intestinal cell line models—an indispensable tool in food science and nutrition, Trends Food Sci. Technol. 22 (2011) S11–S20.

[40] D. Bouzid, K. Steverding, M. Tyler, Detection and Surveillance of Waterborne Protozoan Parasites, 2008, Available from: http://www.uea.ac.uk/~wm077/My%20papers/bouzidetal2008.pdf. Last accessed 22 March 2013.

[41] A. Patel, S.H. Kang, P.A. Lennon, Y.F. Li, P.N. Rao, L. Abruzzo, et al., Validation of a Targeted DNA Microarray for the Clinical Evaluation of Recurrent Abnormalities in Chronic Lymphocytic Leukemia, 2008, Available from: http://www.ncbi.nlm.nih.gov/pubmed/18161787. Last accessed 22 March 2013.

[42] CIWEM, Disinfection of Water Supplies, 2012, Available from: http://www.ciwem.org.uk/policyand-international/policy-position-statements/disinfection-of-water-supplies.aspx. Last accessed 22 March 2013.

[43] Siemens, Trends and Developments in the UV Water Treatment Industry White Paper, 2008, Available from: http://www.water.siemens.com/SiteCollectionDocuments/Product_Lines/Industrial_Process_Water/Brochures/Trends-and-Developments-in-the-UV-Water-Treatment-Industry.pdf. Last accessed 02 February 2013.

[44] CIWEM, Ultraviolet (UV) Disinfection of Drinking Water Supplies, 2008, Available from: http://www.ciwem.org/policy-and-international/policy-position-statements/ultraviolet-(uv)-disinfection-of-drinking-water-supplies.aspx. Last accessed 22 March 2013.

[45] P. Kiesel, J. Martini, M. Huck, M. Bern, N. Johnson, Opto-fluidic detection platform for pathogen detection in water, in: CLEO: 2011-laser applications to photonic applications, OSA technical digest (CD), Optical Society of America, 2011. paper CWL1.

[46] Aqua Research Collaboration (ARC), Membrane Technology, 2011, Available from: http://www.arc-online.eu/pillar-1-collaborative-research/membrane-technology/. Last accessed 22 March 2013.

[47] CIWEM, Regulation for a Sustainable Water Industry, 2010, Available from: http://www.ci-wem.org/media/158640/RSWI%20FINAL.pdf. Last accessed 22 March 2013.

[48] EA (Environment Agency), Limiting Climate Change: Water Industry Carbon Reduction, 2010, Available from: http://www.environment-agency.gov.uk/static/documents/Research/(16)_Carbon_water_mitigation_FINAL.pdf. Last accessed 22 March 2013.

[49] BIS, Taking Responsibility for Water, 2011, Available from: http://www.lwec.org.uk/sites/default/files/Taking%20Responsibility%20for%20Water%20Full%20doc.pdf. Last accessed 22 March 2013.

Chapter 12

Conclusions

Helen Bridle

Institute of Biological Chemistry, Biophysics and Bioengineering, Heriot-Watt University, Edinburgh, Scotland

12.1. Summary

Although it was announced in spring 2012 that the Millennium development goal, of halving the proportion of people without access to an improved water source, had been achieved, nearly a billion people still lack access to safe drinking water. Such water, which poses no significant risk to health during a lifetime of consumption, is considered an essential human right. Contaminated drinking water causes huge negative economic and health repercussions. The World Health Organisation (WHO) reports that microbial contamination of water is the primary concern in both developing and developed countries.

Waterborne pathogens include viruses, bacteria and parasites, several of which are highly infectious, robust and long-lived in the environment, as well as being resistant to standard methods of water treatment. Viruses are the smallest of these pathogens, typically around 20–300 nm in diameter, which makes them difficult both to remove and detect. In addition, viruses are highly infectious and often long-lived in the aqueous environment, with norovirus, for example, having been shown to remain infectious after over 2 months in groundwater. Furthermore, many viruses are resistant to disinfection, particularly norovirus, which has demonstrated resistance to chlorination, and adenovirus, which has remained viable even after UV treatment.

Bacteria have sizes on the order of a few micrometres, are often less infectious, with some notable exceptions (e.g. *Escherichia coli* O157:H7 and *Shigella*) and are more susceptible to chlorine disinfection.

Parasites are the largest of the waterborne pathogens and comprise protozoa and helminths. With regard to helminth infections, this has decreased significantly over recent years as the causative agents (e.g. the host within the water environment) are easily removed by filtration. Protozoa, however, remain a problem due to a low infectious dose, longevity in the environment, and resistance to water treatment methods.

Existing methods of monitoring for waterborne pathogens mainly rely upon detection of faecal indicators. This strategy assumes that faecal indicators,

Waterborne Pathogens: Detection Methods and Applications. https://doi.org/10.1016/B978-0-444-64319-3.00012-5

showing that waters have been faecally contaminated, will therefore also indicate the presence of any pathogens. The advantages of this approach are that the tests are cheap and easy to perform. However, studies have indicated a poor correlation between the detection of indicators and the presence of microbial pathogens and there are also concerns that this approach does not allow a valid identification of the pathogen.

The alternative to indicator monitoring is undertaking of tests to directly identify particular pathogens. This is extremely challenging due to the low concentrations of pathogens in large water volumes; the requirement to detect a single organism in a 100 mL water sample has been compared to the problem of finding a single coffee bean in 40,000 Olympic-sized swimming pools. Furthermore, it is often necessary to determine the species and viability/infectivity of a microorganism to determine the pathogenic potential. Indeed, for some pathogens, identification beyond the species level is required, for example *Vibrio* species.

Various methods of direct detection exist, including cell culture-based methods, immunological methods, microscopy and nucleic acid amplification approaches. Each has advantages and disadvantages and the most suitable choice of monitoring approach is likely to depend upon why monitoring is being undertaken. Monitoring plays several key roles in the design and implementation of water safety plans and can be applied for surveillance or for operational or investigative means. However, whatever the application, improvements in methods for water quality monitoring are required and have been the subject of much research in recent years.

Section 2 of the book has covered developments in sample processing, a key stage in any monitoring protocol. A variety of sampling and analytical procedures for the recovery of viruses, bacteria and parasites from water exist, each with their advantages and disadvantages. Many factors making up the physicochemical quality of the water, including but not limited to the pH, conductivity, turbidity, presence of particulate matter and organic acids can all affect the efficiency of recovery of microorganisms. There is no single universally superior method; the choice of the appropriate approach depends upon the type of microorganism, the type of monitoring to be undertaken, and the eventual detection method, as well as factors such as efficiency, constancy of performance, robustness, cost, and complexity. For example, in operational monitoring, rapid and simple-to-use methods are preferable, in investigative monitoring, the key focus is most likely to be efficient removal of interferents to enable species-level determination whereas in surveillance monitoring, recovery rates are essential to identify pathogens at extremely low concentrations.

Techniques for the recovery and concentration of viruses include ultrafiltration, adsorption-elution using filters or membranes, NanoCeram filters, glass wool or glass powder, two-phase separation with polymers, flocculation, the use of monolithic chromatographic columns, ultrafilter centrifugation, immunomagnetic separation and polyethylene glycol (PEG) precipitation. The recovery rates generally do not exceed 60%. Bacteria are commonly processed using ultrafiltration, centrifugation and membrane filtration although flocculation, immunomagnetic separation, and density gradient centrifugation are also used and initial studies have explored the potential of NanoCeram and glass wool. The use of membrane filtration is popular as the membranes can be immediately employed in traditional culture-based detection techniques. In general, recovery rates for bacteria appear to be higher than for viruses, though there is a high degree of variability observed depending upon factors such as water quality, operator skill, secondary concentration efficiency, etc. Parasites can be concentrated using ultrafiltration, flocculation, centrifugation and membrane filtration; they can also be isolated using immunomagnetic separation. Recovery rates are often comparable to those achieved for bacteria.

Section 3 of the book has presented a range of developing and emerging technologies for waterborne pathogen detection, giving an overview of the latest research advances. Chapter 5 discussed optical detection technologies. Fluorescent based techniques are well-established, with some, though not always successful, developments toward automation and commercialization of existing fluorescent approaches, for example TECTA and the Shaw water systems. Although automated approaches enable rapid, on-site detection with a reduction in technician time, disadvantages such as the need for expensive reagents and the lack of information on species and viability remain. Spectroscopic techniques, including infrared and Raman spectroscopy avoiding the use of any labels, have an immediate advantage in terms of simplified sample preparation and reduced reagent cost. 'Fingerprint' spectra from single cells have been recorded, thus offering the opportunity of species and viability testing to the single organism level. However, more evidence is required to determine spectral variability to improve understanding of the sensitivity and robustness of this type of technique. In addition, a spectral library is needed characterising the results of numerous pathogen screens, thus enabling identification of an unknown sample.

Chapter 6 introduced electrochemical and electrical methods of detection of waterborne pathogens. After an introduction to the various technologies, the chapter discussed the advantages and gains to be expected from miniaturisation and integration of these types of systems. Electrical approaches have been applied to viruses, bacteria, and protozoa, though most literature reports are not directly concerned with waterborne pathogens. Impedance and dielectrophoresis for protozoan detection appears to be the most common waterborne pathogen application. These methods offer the advantage of viability discrimination, and have recently been applied at a single cell level. The main concern relating to this type of technology is whether reproducibility can be assured to enable scale-up from laboratory demonstrations to real-world applications.

Chapter 7 presented a range of biosensor technologies which have been applied to the detection of waterborne pathogens. All main types of biosensor transduction approaches, e.g. optical, electrical, and mass-sensitive, have been studied. Bacterial biosensors have been well-researched area, and limits of detection have come down in recent years with some single cell level examples reported and several systems reporting detection sensitivity to about 10-100 cfu/mL. In addition, a photonic crystal virus biosensor and the piezoelectric excited millimetre sized cantilevers for bacteria and protozoa are close to demonstrating single microorganism level sensitivity. Particularly, *E. coli* is often employed as the test organism for new biosensor approaches. Optical and electrical detection schemes have proved most popular for bacteria, and with optical dominating for viruses. Biosensors for viruses are a more recent area of research, with fewer examples found. However, protozoan biosensors are fairly well-represented in the research literature, with mass-sensitive approaches dominating, possibly due to their greater size and mass, relative to viruses and bacteria. Advantages of biosensors are speed and the potential for portability. Systems integrated with aspects of sample processing have been developed which is a trend that should continue as while the limits of detection are improving there needs to be a connection to the ability to process larger volumes of water. Other challenges to be overcome include robustness and long-term operation, biofouling and the ability to deliver species and viability based information.

Chapter 8 has given an overview of the types of molecular detection methods that have been applied to waterborne pathogen monitoring. The chapter also highlights the key demands placed on sample processing due to the increased uptake of molecular methods; since environmental contaminants can interfere with the amplification and detection of molecular products, successful detection is reliant upon efficient removal of these interferents. Molecular methods offer the advantage of efficient speciation and identification of pathogens, though quantification/sensitivity and viability determination can be more challenging. Targets that have developed around rRNA genes have had the most success at determining the infectivity of the pathogen since RNA degrades faster than DNA after the death of a cell. The lack of RNA detection, however, may still occur in viable but not culturable cells and initiate incorrect risk assessment decisions. While many methods have been designed over the past decade, in the case of DNA-based methods, no consensus on a universal primer set for the detection of common waterborne pathogens has been established nor have the target genes been agreed on.

Molecular methods have been developed for all the key waterborne viruses, though at present none of these are practical for routine, large-scale monitoring. Low copies of adenovirus have been detected from a variety of water types, indicating the potential of the approach. One particular challenge is that most of the target viruses are RNA viruses and it is therefore especially essential to select primers or probes which target highly conserved regions. In terms of bacteria, many of the primers target the 16SrRNA gene which is present in all organisms and contains hypervariable regions that can be targeted down to the genus or species level, and this approach has been used with polymerase chain reaction (PCR) for *E. coli*, *Salmonella*, *Campylobacter*, and *Legionella*. The best sensitivity demonstrated so far is $4\,cfu\,mL^{-1}$, with many studies reporting far higher limits of detection. Detection of protozoa with molecular methods has also been developed in PCR and microarray formats. Detection limits of 1-5 (oo)cysts have been reported although this sensitivity required two rounds of amplification.

Chapter 9 discussed how nanotechnology can be applied to enhance the sample processing and detection schemes presented in previous chapters, as well as enabling new approaches. Nanotechnology is predicted to have a huge impact on pathogen diagnostics in a wide variety of fields over the next decade and has been the subject of much recent work, though the majority of the studies are not specifically related to waterborne pathogen detection. Bacteria, and particularly *E. coli* have been well studied with fewer reports of detection of viruses and protozoa. Gold nanoparticles enable new colorimetric detection schemes, which are especially advantageous for cheap diagnostics, e.g. for developing country applications. Other new methods of sample processing and detection have also been invented, for example depletion flocculation or the magnetic relaxation switch approach. However, there is a variability in the success of utilising nano-materials which needs to be further understood. For several studies which report signal enhancements using nanotechnology, this is often just one or two orders of magnitude, whereas others have shown a large benefit particularly in recent biosensors applications. Given the recent growth in the research area though, further developments and improvements can be expected.

Chapter 10 has considered the potential of miniaturisation and the role that lab-on-a-chip devices could play in waterborne pathogen monitoring. For all the detection methods discussed above, there are microfluidic systems for performing these tests, though not always for waterborne pathogen applications. Performing fluorescent detection on-chip offers many advantages, including: the reduced sample volume resulting in a lower background noise signal and, therefore, improved sensitivity and signal to noise ratio; the small sample volume and control of flow enhancing binding kinetics and increasing sensitivity; and the reduced consumption of reagents. Performing molecular methods on chip is an excellent way to exploit many of the advantages of lab-on-a-chip miniaturisation and this is likely to be one of the main future directions for microfluidic waterborne pathogen detection. As yet, however, the potential of microfluidics is not widely exploited in the waterborne pathogen monitoring arena, and indeed more examples of foodborne pathogens can be found in the literature. Given that many of these pathogens are similar, indicates that there is scope for wider use of microfluidics. One potential challenge is the sample processing of large volumes of complex environmental water, due to potential problems with device clogging as well as obtaining sufficient throughput. However, despite this concern, examples have been found of microfluidic approaches to waterborne sample processing. And more recently paper-based microfluidics have been expanding for the type of application, which has particular advantages for remote monitoring in low-income regions due to their low-cost and ease of use.

Chapter 11 explored the application of emerging technologies forwaterborne pathogen detection from a market perspective. The watermarket, particularly in terms of monitoring, is heavily driven by regulation and regulatory compliance. Adoption of new technologies for detection of pathogens can be challenging within this framework.

12.2. The future of waterborne pathogen monitoring

Drawing conclusions from the above summary, it is clear that a wide range of sample processing and detection techniques are the subject of ongoing development for all types of microorganism. At present, the performance of these approaches requires further improvement before wider adoption by the water industry is possible.

In terms of sample processing, improvements are required to increase recovery rates, to ensure the efficient removal of interferents, and to develop methods which are low cost, robust, and easy to operate. New materials is one key area of potential future development in sample processing, including materials/chemicals for flocculation, filters (e.g. NanoCeram or glass wool), elution protocols, and chemical additives or surface modifications to prevent adsorption. Further research is required to optimise the use of these materials, to fully understand their interactions with pathogens, and perhaps to work toward cheaper, greener materials.

In general, the issue of sample processing has not received as much research attention as detection technologies. However, this is a critical stage in monitoring protocols, often determining the overall success, and it is, therefore, essential to continue research in this area, particularly in characterising performance of different methods, development of new materials and processes, and also in automation of processes, bringing the advantages of reduced reliance on operator skill, reduced variability, and reduced risk of cross-contamination of samples. Another important factor to develop is that of analytical controls, which enable validation and quality assurance of the sample concentration and enrichment processes.

Those developing new methods of detection should ensure that consideration is given to sample processing and the challenge of environmental water samples, rather than solely developing laboratory methods effective for clean water samples. Furthermore, in general, limits of detection remain too high and improvements in sensitivity are required. Optical, electrical, biosensor and molecular methods have all been demonstrated for all types of microorganisms, although bacteria have been most well-studied. Some optical, electrical and molecular approaches enable detailed characterisation of pathogens, to the subspecies level as well as viability determination, and these approaches will be very useful in certain types of monitoring. Alternatively, other methods such as biosensors can be rapid and portable and therefore have great potential in other applications.

Optical approaches, and in particular the spectroscopic methods, offer the promise of single organism fingerprinting to obtain detailed information on pathogen characteristics. However, improved understanding of the relation between pathogen species, viability, etc. and the spectra obtained is required, as well as packaging into low-cost, easy-to-operate systems. Furthermore, for all optical methods, instrument design and miniaturisation are essential to obtain more widespread acceptance in the water quality monitoring arena.

Electrical methods of detection have been applied to a wide range of viruses, bacteria, and protozoa although little work has concentrated directly on waterborne pathogens. Given the potential for detailed characterisation of single microorganisms, expanding the range of waterborne pathogens studied is a crucial research task. Reproducibility and cost are other key issues to be addressed.

Some biosensor technologies are now reaching toward single organism detection, which are highly promising results. However, while biosensors offer speed and sensitivity there are still many challenges to be overcome, including issues of sample processing, biofouling and robustness. Future research to enable species and viability discrimination would also be extremely beneficial for the real world application of biosensing technology to waterborne pathogen detection. Other biosensing techniques for viruses and protozoa, perhaps particularly, electrochemical methods for viruses and both optical and electrochemical methods for protozoa, should also be explored.

Molecular methods have been developed for viruses, bacteria, and protozoa offering speciation, and potentially, an indication of viability. An important upcoming task is agreement on the primers and probes most suitable for use with different waterborne pathogens, and eventually validation of these choices in moving toward regulatory approved methods. Lowering detection limits is also essential, especially in obtaining information from environmental samples with low pathogen concentrations. For all of the above techniques, nanotechnology and miniaturisation are likely to play an important role in the future development of methods.

There are many exciting opportunities in the field of nanomaterials and nanotechnology, and more research should concentrate on applications of these to waterborne pathogens. The same is true for microfluidics and miniaturised systems. System integration is important and several useful examples have been demonstrated combining flow systems and/or nanotechnology with existing detection schemes which has improved performance. Some approaches are now integrating aspects of sample processing with detection systems which is a trend that should continue, especially to assist in the translation of technologies developed for food and medical detection into water applications. In general waterborne pathogens have received less focus than other areas and there is plenty of potential for learning from alternative application areas, particularly food samples where work has been demonstrated in complex samples.

There is also growing awareness of the issue of antimicrobial resistance (AMR) spread through the water environment and some examples of detection schemes for this problem have been investigated. However, there is an urgent need for improved sensor systems for AMR and this is a key area of future development to be able to understand and mitigate the aquatic component of AMR.

Even excellent new technologies for waterborne pathogen monitoring might struggle to reach the market. There are several barriers to entry. One is evidently the cost of testing. Another factor, particularly problematic in the highly regulated and conservative water industry, is the validation of new methods. Alternative approaches to sample processing and detection need to demonstrate scale-up from the laboratory, be repeatable and reliable across a range of water sample types, and be easy to operate.

Access to water innovation parks or links with industry are likely to prove vital for the developer of new technologies. At regional levels, the regulators and governmental organisations could perhaps also help to stimulate innovation in the area of pathogen detection by providing funding for water innovation parks and for validation studies.

Looking ahead, it is challenging to predict how future waterborne pathogen monitoring will take place. A vision of networked sensors reporting on microbial aspects of a water supply for operational decision making still seems a long way from reality; it seems more likely that there will first be the adoption of new laboratory tests/commercial systems in water industry analytical facilities. However, it is clear that more direct detection of pathogens will be required and the methods described in this book will be needed to deliver on this requirement. Furthermore, more detailed pathogen analysis is already sought after and techniques enabling simple speciation and viability determination will prove popular. Many methods are showing promise in delivering on sensitive, rapid direct detection or detailed characterisation. Few approaches, as yet, offer both advantages and there is unlikely to be any one size fits all solution to either sample processing or detection.

Index